LABORATORY INFORMATION MANAGEMENT SYSTEMS

- Concepts, Integration and Implementation

Edited by

R. D. McDowall

Smith, Kline and French Research Ltd.

SIGMA PRESS
Wilmslow, United Kingdom

HALSTED PRESS
a division of JOHN WILEY & SONS, Inc.
605 Third Avenue, New York, N.Y. 10158
New York • Chichester • Brisbane • Toronto • Singapore

First published in 1987 by **Sigma Press**
98a Water Lane, Wilmslow, SK9 5BB, England.

British Library Cataloguing in Publication Data
McDowall, R.D.
Laboratory information and mangement systems
1. Management information systems
2. Laboratories -- Management -- Data Processing
I. Title
502'.8 Q180.57

Library of Congress Cataloguing in Publication Data (applied for)

ISBN 1-85058-083-9 (Sigma Press)
ISBN 0-470-20947-X (Halsted Press)

Distributed by

John Wiley & Sons Ltd., Baffins Lane, Chichester, West Sussex, England.

Halsted Press, a division of John Wiley & Sons, 605 Third Avenue, New York, NY 10158, USA.

Acknowlegments

The following are trademarks of the respective companies:

IBM	International Business Machines
MS-DOS	Microsoft Corporation
DEC and Vax	Digital Equipment Corporation
Lotus 1,2,3	Lotus Development Corporation
dBase III	Ashton-Tate
Wang	Wang Corporation
Gem	Digital Research

Printed in Malta by Interprint Ltd

Typeset by Sigma Press

Preface

Laboratories are under pressure to make the most of their resources and to improve efficiency. For whatever reason, laboratory automation, usually involving computers or microprocessors, is the main solution to this problem.

Laboratory Information Management Systems (LIMS) are a new form of automation which use databases and information management techniques applied to the laboratory. There is great interest in LIMS as a means of meeting the challenges of greater efficiency in the laboratory, containing costs and reducing administration whilst complying with the requirements of Good Laboratory Practice (GLP).

As scientists become aware of laboratory automation and the benefits it can bestow, there are problems. Traditionally, analytical instruments have been purchased off-the-shelf and used with little or no modification. However, with the emergence of complex laboratory automation, prospective purchasers are faced with specifying the tasks to be performed. Often, this places the scientist in a difficult position as he is not often certain of what the equipment should do. A LIMS is a case in point. Here a multi-disciplinary approach is essential to interface computer hardware and software programs with the instrumentation and the data produced from them in a drive to improve the efficiency of the laboratory.

Little has been written about LIMS to date, but interest in the subject is growing. This book has been written to provide a practical guide to the selection and installation of this type of laboratory automation, the various paths a prospective user could take and the framework needed for success. It is intended for the users of such systems and has been written by scientists with practical experience of LIMS.

This book is divided into five sections. The first is intended to give an introduction to LIMS, together with the benefits the technology can offer. The management considerations and the decisions to be taken if a laboratory wishes to install and integrate such a system are contrasted with the effort and support required to ensure that the installation and operation of a LIMS is successful.

A prospective user then needs a background knowledge of computing in order to understand the technology. The second section provides this; from a discussion of computer hardware and the user interface to the users' requirements of a database that is at the heart of a LIMS. This section is completed with a discussion of interfacing, communication and data capture which are vital areas as the full benefits of a LIMS are only obtained when data are transferred to a system electronically.

The third section is concerned with the implementation of a LIMS. It begins with a chapter on specifying the requirements and functions of the proposed system and ends with a chapter on evaluation. These are tasks that require the upmost care as a good specification and evaluation of the alternatives available will form the basis of a successful system. Between these two chapters are contributions from scientists from a wide range of disciplines, who have devised and built their own systems; a chapter detailing the services available from a software house and five chapters from companies which sell commercial systems. The purpose is to give the reader a wide view of the alternatives that are possible.

The fourth section is concerned with validation and the needs of computing in a regulated environment. Validation is vitally important; computer systems do not work the first time without problems. Firstly acceptance testing followed by validation is intended to demonstrate that the system will actually do what the users require and produce documentation to show this. Two complementary chapters are concerned with the interaction of LIMS (and by implication other computer systems) with the requirements of working in industries that are regulated by government agencies.

The final section consists of a single chapter which summarizes the present status of LIMS and the directions developments may take in the future.

Finally, my thanks are due to Alan Braithwaite, Ray Dessy, Karen Feldman, Steve Feldman, Adrian Hampshire, Roger James, Ken Leiper and John Vann who gave useful comments, advice and help during the preparation of this book.

R.D.McDowall.

List of Contributors

Peter Berthrong
Product Line Manager, Beckman Instruments Inc,
160 Hopper Ave, Waldwick, New Jersey 07463 USA

Mike Bertram
Director, Application Development Facility, Department of Information Sciences,
Smith Kline & French Laboratories, M.S. L331, 709 Swedeland Road, Swedeland
Pennsylvania 19479 USA

Alan Braithwaite
Senior Lecturer in Analytical Chemistry, Department of Physical Sciences,
Trent Polytechnic, Clifton Lane, Nottingham NG11 8NS England

Betsy Brown,
Scientific Systems Manager,
McNeill Pharmaceutical, Spring House, Pennsylvania 19477 USA

Harmon Brown
Director of Technical Research
Nelson Analytical Inc, 10061 Bubb Road, Cupertino, California 95014 USA

Ross Burgess
Principal Consultant,
Digitus Ltd, 16/17 Clerkenwell Close, London EC1R 0AA England

Steve Crisp
European Data Systems Manager, Hewlett Packard Gmbh, Waldbronn Werk,
Hewlett Packard Strasse, Postfach 1280, D 7517 Waldbronn 2 West Germany

Dai Darkin
Senior Chemist, Department of Physical Organic Chemistry,
Smith Kline & French Research Ltd, The Frythe, Welwyn, Herts AL6 9AR England

Adrian Hampshire
Director of Computing, Longman Cartermill Ltd, Technology Centre,
St. Andrews, Fife KY16 9EA Scotland

Mike Head
Sales Director, VG Laboratory Systems,
Tribune Avenue, Altrincham, Cheshire WA14 5TP England

Bob Hillhouse
Managing Director, VG Laboratory Systems
Tribune Avenue, Altrincham, Cheshire WA14 5TP England

Peter Johnson
Chairman and Managing Director, Smith Kline & French Research Ltd,
The Frythe, Welwyn, Herts AL6 9AR England

Gareth Jones
Scientific Director, Wessex Water Authority, Passage Street,
Bristol BS2 0JQ England

John Knox
Project Manager, Fraser Williams (South West) Ltd,
9 Church Street, Twyford, Reading, Berks RG10 9DN England

Gloria Lawlor
Technical Writer, Nelson Analytical Inc, 10061 Bubb Road,
Cupertino, California 95014 USA

Lou Lesser
Field Support Manager, Perkin-Elmer Corporation, Main Street,
Norwalk, Connecticut 06859 USA

Bob McDowall
Senior Analyst, Department Drug Analysis, Smith Kline & French Research Ltd,
The Frythe, Welwyn, Herts AL6 9AR England

Alan McLelland
Principal Biochemist, Department of Biochemistry, Glasgow Royal Infirmary,
Alexandra Brook, Glasgow G4 0SF Scotland

Graham Martin
New Ventures Manager, Imperial Chemical Industries,
Chemicals and Polymers Group, Billingham, Cleveland TS23 1LB England

Don Mattes
Manager, Application Development - Lab Automation,
Department of Information Sciences, Smith Kline & French Laboratories, M.S. L331,
709 Swedeland Road, Swedeland, Pennsylvania 19479 USA

Mike Milano
US Product Manager, Perkin-Elmer Corporation, Main Street,
Norwalk, Connecticut 06859 USA

Johann Mollner
Labsam Project Engineer, Hewlett Packard Gmbh, Waldbronn Werk,
Hewlett Packard Strasse, Postfach 1280, D 7517 Waldbronn 2 West Germany

Graham Murkitt
Analyst, Department Drug Analysis, Smith Kline & French Research Ltd,
The Frythe, Welwyn, Herts AL6 9AR England

John Pearce
Senior Analyst, Department of Biochemical Pharmacology,
Glaxo Group Research Ltd, Priory Street, Ware, Herts SG12 0DJ England

Chris Triner
Principal Scientist, Wessex Water Authority, Passage Street,
Bristol BS2 0JQ England

Alan Williams
UK Product Manager, Perkin-Elmer Ltd, Post Office Lane,
Beaconsfield, Bucks HP9 1QA England

Contents

Section 1: Concepts

Section 2: Essentials of Computing

Chapter 10 Interfacing and Communication to Analytical Instrumentation

Chapter 11 Chromatographic Data Acquistion

Section 3: Implementation and Choices

Chapter 12 The Request for Proposal and The Functional Specification

Section 4: Validation

Chapter 23 Validation of LIMS

Chapter 24 LIMS and Good Laboratory Practice

Chapter 1

Introduction to Laboratory Information Management Systems

R.D.McDowall

1.1 The Analytical Laboratory

The main function of any analytical laboratory is to accept samples, carry out the required analyses and report the results (1). To this end, the primary goals of any analytical laboratory are:

1. The productive use of human and equipment resources.

2. Fast turnaround of samples.

3. The production of high quality analytical results.

There may be some secondary aims which are different for individual laboratories, such as compliance with regulatory agencies. However, the performance of any laboratory is usually judged by the quality of the results and the speed at which they are produced, but these criteria only reflect the efficient use of resources within that laboratory. How does the manager obtain the information to show that the laboratory is functioning effectively?

Today's laboratories are processing more samples and obtaining more information and data from them. Often, staff numbers are constrained and cannot expand to match the growing workloads so the throughput per person must rise as a result. This places a great strain on staff as often the way in which the data are obtained must conform with the Good Laboratory Practice (GLP) guidelines. This administration can be slow, tedious, and labour intensive and results need to be checked many times, which leads to delays. Thus, there is a need to minimize administration, maximize efficiency and free highly trained scientists to concentrate on their work.

A survey by Jones (2) of some analytical laboratories in the United Kingdom found that 73% of the laboratory managers who responded thought that the efficiency of their laboratories could be improved. One of the reasons given for this inefficiency was that staff were dissatisfied with repetitive work. This survey also highlighted the variety of methods employed in these laboratories to report results, including verbal and written reports. These reporting methods could be very inefficient. In the case of verbal reporting, the results may be liable to misinterpretation, there being no written record to refer to.

To relieve the problems described above whilst maintaining a quick response, high quality work and regulatory compliance, the analyst inevitably turns to laboratory automation.

1.2 Laboratory Automation

There are many reasons for this: the major one is the improvement of performance. This may be achieved through improved specificity or sensitivity of a procedure, faster analysis, greater throughput of samples, more effective data handling, better communications, or simply to save labour costs. (3). There are other motivating factors, including planning for the future, a need to comply with legislative requirements or to keep up with competitive companies.

Automation enables the laboratory to keep pace with the workload and ensures the achievement of good, consistent precision and accuracy of analytical procedures at a realistic cost. Automation also frees the scientist to monitor the analysis and develop further automation rather than carry out mundane tasks such as checking for transcription errors.

Laboratory automation is not a goal in itself, but is the means of achieving an objective and is a process for solving some laboratory problems (4). The development of automation in analytical chemistry can be viewed as a series of steps in which the solution of today's bottleneck creates another problem for the future to solve (5). The process is illustrated in Table 1.1.

Automation of the laboratory is an evolutionary process. It is rare to be presented with either sufficient finances or the opportunity to build an automated laboratory from first principles (3). Normally, the constraints on automation within the laboratory begin with the purchase of the first item of analytical instrumentation. Therefore, every item of equipment purchased should be bought with automation, and communication with other equipment and computers, in mind.

Laboratory automation involves two areas: instrument automation and laboratory management automation (4), each of which will now be discussed.

1.2.1 Instrument Automation

The aims of this type of automation are increased sample throughput, more efficient data capture, and faster data processing. This can take many forms: autosamplers have been available for many years for a variety of instrumental techniques, both flexible robotic automation (6-8) and dedicated laboratory automation (9,10) instruments are becoming more accepted. Here the aim is to allow the preparation and the processing of samples for instrumental analysis quickly, consistently and efficiently.

Table 1.1
The development of automation in analytical chemistry

Date	Barrier to Automation	Solution	New Problem
pre 1960	Data generation	Electronic control	Miles of chart paper
1960s	Data acquisition	Digitizers	Data manipulation
1970s	Data reduction	Minicomputers	Many computers Data transfer
1980s	Data management	Workstations LIMS	Management of technical data
1990s	Data interpretation	Artificial intelligence Expert systems	?

Quoted from Betteridge (5).

A trend in analytical instrumentation over the past ten years has been the widespread introduction of computers and microprocessors (11). This advance has had the effect of increasing the speed of data production as well as the amount per analysis. Indeed, the development of techniques such as Fourier Transform Infra Red (FTIR) has proceeded because of the availability of inexpensive computers. As most analytical instruments are now available with a dedicated microcomputer, automation is becoming concerned with control of equipment and communications with other computers.

However, the production of large amounts of data as a result of instrument automation has caused problems in the analysis, interpretation and administration of these observations. The analytical instruments generating data exist as 'automation islands' (12) within the laboratory with little or no interconnection between them. Data management is now essential to overcome these problems and effectively bridge the gap between the islands.

1.2.2 Laboratory Management Automation

A Laboratory Information Management System, more generally known by the abbreviation LIMS, is the means of achieving management automation.

It has been developed to manage the data and information arising from the increase in the workload and the computer aided analytical equipment that has occurred over the past decade (see Table 1.1). This type of automation has been developed by applying relational databases (used previously to enhance productivity in a variety of business disciplines) to laboratory problems. This produces a very effective unit for the management, manipulation and dissemination of information and data both inside and outside the laboratory.

In a survey of British laboratory managers (2), one finding was that 71% of managers had not yet made an assessment of the costs of installing a LIMS in their laboratories. This finding, the article concluded, was not surprising considering that little material has been published on LIMS and the relatively short time that LIMS have been commercially available. Additionally, the present systems are the first generation available and many scientists may not be willing to commit themselves until they have seen the problems and benefits from their more adventurous colleagues.

1.2.3 Planning Laboratory Automation

Laboratory automation must be a carefully thought out process. It is imperative that an automation strategy is developed that integrates the whole laboratory; this should be implemented in a professional manner with checkpoints to review progress. Planning is the key to success, as a laboratory automation project can generate more problems than it solves. This is particularly true with the introduction of a laboratory management computer system. Figure 1.1 shows the life-cycle of a LIMS project, whether the system is developed in-house or purchased.

As can be seen, there are many stages to go through before the system is fully operational, thus acquisition of a LIMS must not be undertaken lightly. The requirements must be fully defined, the responsibilities clearly stated and good communications should be set up between all relevant parties. All aspects of the life cycle will be discussed in the following chapters of this book.

1.3 Laboratory Information Management Systems

1.3.1 What is a LIMS?

In essence, a LIMS is a database tailored to the analytical laboratory so that it can handle the data generated by the analysis of samples. It is intended to integrate sample information (eg. source, batch number, sampling information) with the results from instruments which will drastically reduce administrative tasks and speed the production of the final report. Gibbon (13) has defined a LIMS in general terms, as a computerized system designed to provide on-line information about the analytical laboratory and the samples assayed within it. This information should include acknowledgement of receipt, the current location of samples, how the analysis is progressing (status of samples) and the results to be reported to the submitter. A LIMS is intended to be flexible in that it can be customized or tailored to an individual laboratory's needs.

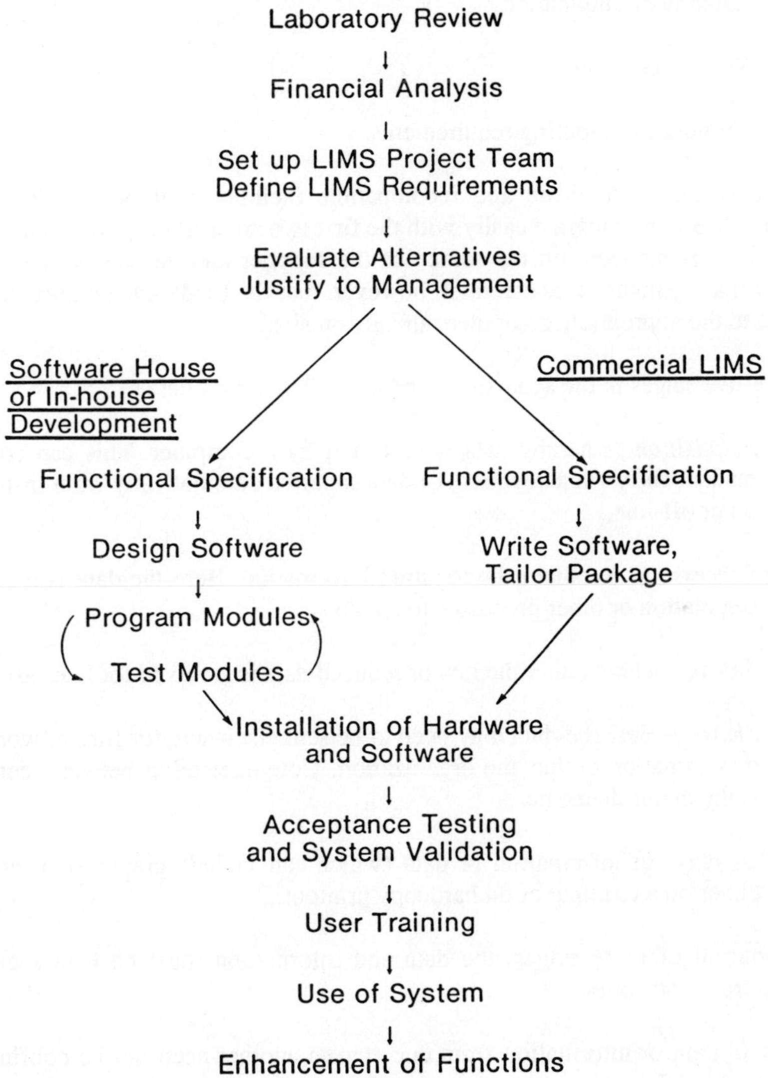

Figure 1.1 Life-cycle of a LIMS project

In many respects the name, Laboratory Information Management System, is a misnomer as these first generation instruments do not manage information, they manage data. Their information handling power is relatively weak in relation to data handling. The situation will improve over the next few years.

1.3.2 LIMS and Laboratory Automation

A LIMS should integrate three areas:

1. Laboratory automation.

2. Word processing.

3. Corporate computing requirements.

otherwise separate, duplicate and incompatible facilities will be developed (14). Whilst a LIMS may integrate easily with the first two areas, it may come into conflict with the third, dependent on the source of the supplier and the views of the site or computer management. It is essential, however, that the LIMS can communicate and pass data to the appropriate computers already on site.

There are five stages in the acquisition and use of laboratory data (15):

1. *Acquisition* whereby data is collected by a computer. This can consist of manual entry via a keyboard or data captured automatically from instruments on or off-line.

2. *Processing* to obtain the required information. Here the data is reduced by calculation or other processes to results.

3. *Storage* where either the raw or reduced data is archived for later use.

4. *Sharing* Here the data may need to be sent elsewhere for further working or dissemination within the organization. Communication between computers is the major theme here.

5. *Display* of information or data (which can include graphs or histograms) either on a terminal or on hardcopy printout.

Throughout all of these stages, the data and information must be kept secure and protected from corruption.

The flow of data or information from one step to another need not be confined to a single computer. It can, for instance, take place amongst several types of computers, intelligent instruments or workstations. However, a LIMS is intended to integrate most, if not all, of the above functions and become the hub of a fully automated integrated laboratory.

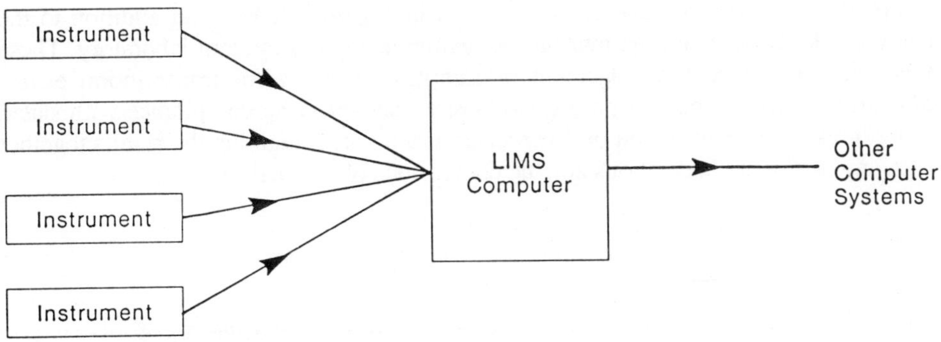

Figure 1.2 Integration of a LIMS within an analytical laboratory.

This is shown in Figure 1.2 where the analytical instruments are connected to a LIMS computer to pass results and data. Note, that the flow of data is one way. This is deliberate as the majority of instruments are not capable of receiving data from a LIMS unless they are very new. This situation will change during the next few years as more intelligence is built into analytical equipment. Within the LIMS the data will be matched to the corresponding sample information within the database. This enables speedier production of reports that can be transmitted to other computers within the organization for further analysis.

1.3.3 Laboratory Information Flows

To discover further how a LIMS can be used, the analytical laboratory environment needs to be considered. Although every laboratory tends to be different, in that its function may be quality control, development or research in many widely differing industries, the nature of analysis is such that the basic sample flow follows the pattern depicted in Figure 1.3. Here, the aim of the analysis is to take a sample and produce a set of results. To achieve this, the sample may undergo a preparation stage, followed by various analytical procedures prior to a result being generated. As shown in the figure, re-analysis may be required if the result does not come within the expected range.

Other tasks associated with this flow are sample labelling, generation of worksheets, collection and approval of the experimental data from which the validated result will be generated. This scheme is representative of the analytical information level within the laboratory.

The situation shown in Figure 1.3 is rather simplistic as laboratories do not assay single samples in isolation. In reality, it is more complex, usually many samples are being examined simultaneously, they may be at different stages of analysis at any given time, and many samples undergo more than one test. A better scheme of the information flow in the laboratory is shown in Figure 1.4. Here, in addition to the analytical level flow, are the managerial requirements of a modern laboratory. These functions include acknowledgement of samples, checking for transcription errors, answering queries about the progress of assays, and ensuring compliance with stated procedures. The culmination of this process is a report containing the results together with any pertinent observations and the conclusions of the analyst.

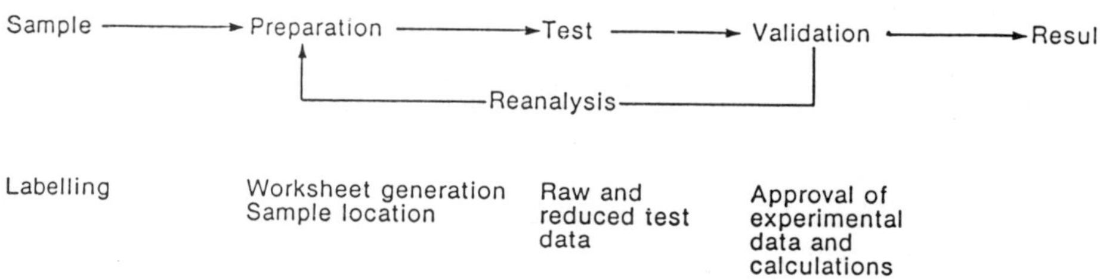

Figure 1.3 Information flow associated with a sample at the analytical level

At a higher level of management information there is the need to know details of the workload, ie. the number of assays of a particular type, in order to use resources effectively within the laboratory. Scott (1) describes the nature of this information with examples of the type of planning possible, including trends in sample load and the services required by each project to give information regarding the efficient deployment of staff and laboratory resources. Activity management (eg. work schedules, deadlines, project management and updating progress from several different projects) are also areas that can benefit from the power of a LIMS computer (16). Therefore, superimposed over the analytical information there is a complex managerial level of information. The challenge is how to obtain this information easily and use it to the full.

There is much data associated with a sample that has the potential to be utilized for greater efficiency in management of the laboratory. If this can be easily extracted, the managers could have better vision of how the laboratory is performing, and hence how better to use the resources at their disposal.

A LIMS, therefore, should be a computer system that effectively integrates the analytical and management levels of laboratory information. It should provide, where appropriate, the computational power for the analyst at the bench and sample management information for the manager from the database. This view is an extension of the definition of a LIMS given by Gibbon (13) above, and will become the more accepted type of package available, especially with the more powerful processor power and software that will be available in the future to encompass both requirements. That is not to say that the computer will be required to collect and store all laboratory data. The prime purpose of a LIMS is to collect and disseminate information. The difference between data and information will now be discussed.

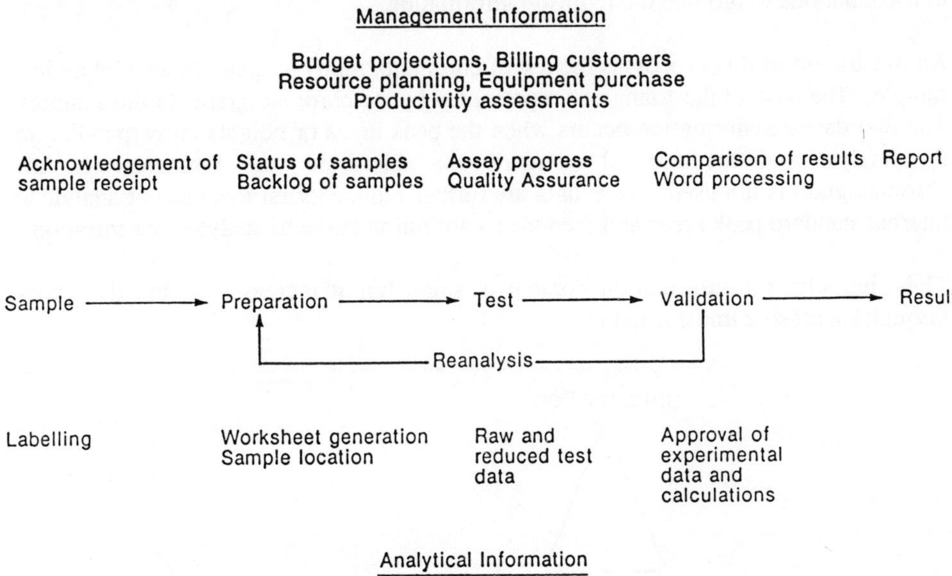

Figure 1.4 Information flows in an analytical laboratory

1.4 Data, Information and Knowledge

1.4.1 Information Versus Data

The analytical process is a means to an end; it is involved in providing information as an aid to making a decision. For example, the questions posed could be: does a particular batch of raw material meet its specification, or is a new drug formulation bioequivalent with the standard one, or what is the significance of the concentration of a compound in a water sample? The analyst, through the experimental techniques and procedures at his disposal, gains the data, which are processed into information, to answer the questions and help make the decisions.

Modern instrumentation for chemical analysis generates and collects vast amounts of data. This is facilitated by computer control of both experimental variables and high-speed collection of multiple channels of data. This in turn allows complex measurement principles to be implemented. Unfortunately, the data explosion that has accompanied the evolution of modern chemical instrumentation has not provided a corresponding information explosion. Little attention, to date, has been paid to the development of information extraction methods, usually this has to be done by another computer or, worse, by hand (17).

Information content, as opposed to the data collected from modern analytical instruments, is usually very small. This is shown in Figure 1.5. Here, the base of the triangle represents the original data which is reduced through successive transformations to provide the requisite information.

An illustration of this process is the use of chromatography to quantify an analyte in a sample. The base of the triangle is represented by the chromatograms of the samples. The first data transformation occurs when the peak areas or heights corresponding to the analyte and the internal standard are abstracted; the remainder of the chromatogram is not used. These data are further reduced, first to a ratio of analyte to internal standard peak areas and then via a calibration curve to analyte concentration.

Thus, initially, the information content is small but it increases as the data goes through successive transformations.

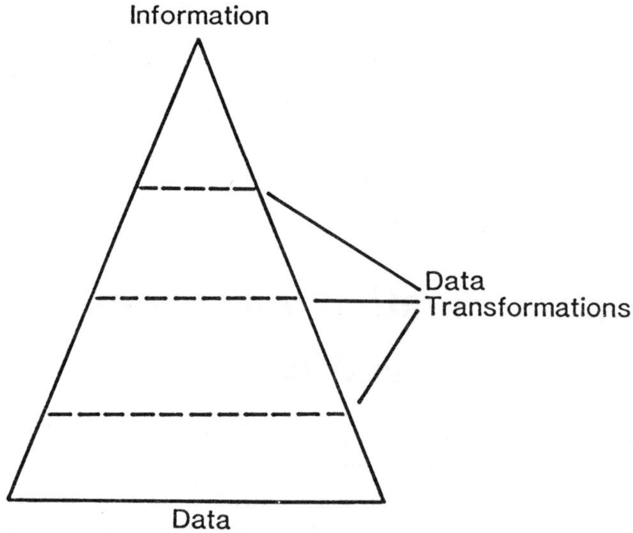

Figure 1.5 Data versus information

It is through the control and management of the processes of data reduction and information management that a LIMS achieves its role. In order for the database to work efficiently, much thought should be given to if and when data should be transferred to it. The amount of laboratory data that should be stored in the database is an issue which is not easily decided (13). A simple wet chemistry laboratory is a relatively straightforward matter as all data can be transferred to the LIMS. However, a modern instrumental laboratory where there are many varied techniques, eg. NMR,

HPLC or GLC is a different situation. If all laboratory data is transferred then the database becomes very large and will become slow to access. Thus, it is essential to decide when in the analytical process data should be passed to the LIMS for inclusion in the database. Should all data be passed to the LIMS or should only reduced data be passed to the LIMS with the primary reduction taking place at a workstation outside the LIMS but networked to it? The latter is probably the best method as it gives the analyst the power where needed at the bench and uses the database to the full.

1.4.2 Knowledge

Once sufficient information has been obtained it can be turned into knowledge or fact; at present this process can only be achieved by humans. Eventually, with the development of artificial intelligence, which is expected to become an integral part of systems design during the next decade, the possibility of knowledge acquisition becomes exciting. Hence, a LIMS that could derive knowledge from its operation would be of great benefit to its users. The areas that would derive the greatest benefit from this would be the organization of the laboratory and its resources.

1.5 LIMS Functions

The functions of a LIMS are achieved by a series of application programs that interact, access and update the entries in the database, as shown in Figure 1.6.

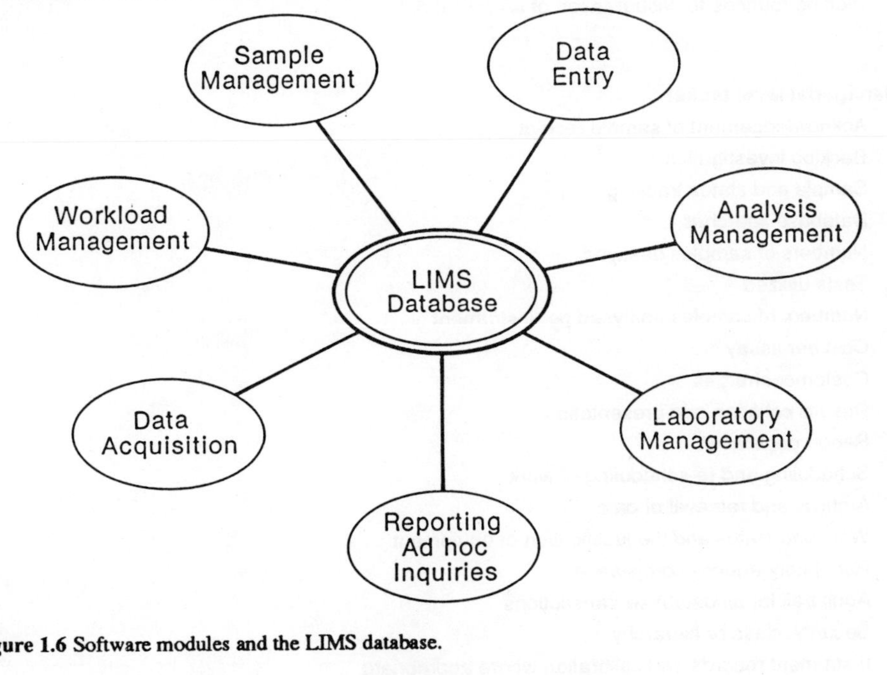

Figure 1.6 Software modules and the LIMS database.

Table 1.2 summarises some of the various functions that could be available in a LIMS at both the analytical and managerial levels. These tasks can be broken down into several broad areas which are discussed briefly below.

Table 1.2

A List of LIMS functions at the analytical and managerial levels

Analytical level tasks:

Sample number generation
Bar code label generation
Sample log-in
Verification of data format entered into the computer
Worksheet generation
Construction and checking of calibrated curves
Direct data acquisition from chromatographs
Data collection from analytical instruments
Entry of instrumental readings
Manual results entry
Interpretation of calibrated curves and quality control samples
Interpretation and acceptance of sample data
Routine automatic calculations
Plotting routines for visualisation of analytical data

Managerial level tasks:

Acknowledgement of sample receipt
Backlog investigation
Sample and status tracking
Database searches
Numbers of samples assayed
Tests utilized
Numbers of samples analysed per instrument
Cost per assay
Customer charges
Results collation and presentation
Report generation
Scheduling and re-scheduling of work
Archival and retrieval of data
Workload status and the justification of equipment
Regulatory agency compliance
Audit trail for all database transactions
Security: class or hierarchy
Instrument records and calibration where appropriate

1.5.1 Sample Management

This program module is responsible for tracking the sample from entry into the laboratory to the completion and approval of all the tests assigned. During this process an acknowledgement of sample receipt form can be automatically generated and the tests appropriate to the sample assigned. A unique LIMS number can be generated at log-in to the computer and labels, often incorporating bar code identification, can be printed (18). This will reduce, if not remove problems with sample identification later in the analytical process. At least two LIMS have the ability to record the locations of the samples via bar codes so that they can be tracked if desired (see Chapters 14 and 20). If the sample is sub-divided, the software can track each portion until analysis is completed.

1.5.2 Sample Status

Associated with the sample management module is the status of the sample which can be either a numerical value or abbreviation associated with the record within the database. It is automatically updated by the applications software after the sample has completed a particular phase of analysis. This is then used to monitor the progress of the samples through the laboratory. Reporting modules can reference this status field to generate backlog reports or status summaries about any sample.

1.5.3 Data Entry and Acquisition

Results and observations can be entered manually through the keyboard or automatically from instruments on-line. The software programs are responsible for the entry of these results into the analytical records of the appropriate sample. If a LIMS has chromatographic data acquisition software, then programs can also be used to transfer the results to the database. Data reduction, from raw data to the final result, may also be undertaken by the database software; it may be done automatically, or with human intervention, depending on the complexity of the operation.

1.5.4 Sample Analysis

Samples often undergo multiple tests and each test might require one or more repeats to be performed. The purpose of this module is to ensure that the sample is not released until each of the required tests has been accepted and validated. Often, re-analysis may be required (see Figure 1.3) and the additional results will be collated by this software. If standard methods are used they can be stored on-line and only modified by authorised personnel.

1.5.5 Workload Management

Scheduling work enables the most efficient use to be made of valuable laboratory resources. This can take varied forms such as the generation of laboratory worksheets for staff to use whilst preparing samples for analysis. Automated planning and scheduling of work is a useful feature, particularly if there is a problem with the stability of a substance or if a series of experiments have been planned. The system can automatically inform the scientist that the next experiment is due or samples are overdue for analysis. Setting and altering the priority of samples is also possible, so that urgent samples are placed at the top of the work list whilst less important work can wait. Such a system should also take into account the age of the sample in calculating its priority.

1.5.6 Reporting

Report programs collate all the test results for the appropriate sample and can compare these with ranges anticipated or with allowable values preset within the database. If the sample fails the overall specification any problem areas can be highlighted, provided that these have been defined in the database. If required, the results can also be passed to word processing systems either resident on the LIMS computer or external to it. This will reduce the need to check transcription errors and is faster and more efficient than re-typing.

The database has the facility for the user to integrate it using either predefined searches for common enquiries or by an ad hoc enquiry for unique information.

1.5.7 Laboratory Management

The following basic management functions are available in a LIMS: financial and budgetary information, such as the cost per analysis, automatic billing to customers and the time taken per analysis. Reports of the numbers of samples assayed by particular tests can be used to assess if the resources of the laboratory are being used efficiently. However, the tasks offered are little more than those using arithmetic manipulation, full laboratory management capability must await future LIMS development.

1.6 References

1. Scott, R.P.W., 'The Analytical Laboratory Computer as a Management Aid', *Trends in Analytical Chemistry*, 1 (1982) 219 - 221

2. Jones, K., 'Laboratory Information Management Systems - a Survey', *Journal of Automatic Chemistry*, 7 (1985) 136 - 140

3. Squirrell, D.C.M., 'The Why, When and Wherefore of Automatic Analysis', *Analytical Proceedings*, 22 (1985) 79-80

4. Liscouski, J.G., 'Planning an Approach to Laboratory Automation' in *Computers in the Laboratory, Current Practice and Future Trends*, ed Liscouski J.G., American

Chemical Society, Symposium Series 265. American Chemical Society, Washington D.C. (1984) pp 1-9

5. Betteridge, D.J., 'Towards Intelligent Automation', *Analytical Proceedings*, 24 (1987) 106-108.

6. Pearce, J.C., Allen, M.P. and McDowall, R.D., 'Robotics in Drug Analysis', *Bioactive Analytes, CNS Drugs, Peptides and Enantiomers* (Methodological Surveys in Biochemistry and Analysis Vol 16). Eds Reid, E., Scales, B. and Wilson, I.D. Plenum Press, New York, (1986) pp 293 - 296

7. Leiper, K.J., 'Robotics - a Helping Hand?', *Trends in Analytical Chemistry*, 4 (1985) 40-43

8. Owens, G.D. and DePalma, R.A., 'Comparison of Approaches to Robotics in the Laboratory', *Trends in Analytical Chemistry*, 4 (1985) 32-39

9. Verillion, F. and Glandian, R., 'A New Robotic Unit for HPLC Sample Preparation and Injection', *International Biotechnology Laboratory*, 4 (1986) 3 pp26-34

10. McDowall, R.D. and Pearce, J.C., 'Comparison of Manual and Automated Methods of Liquid-Solid Sample Preparation for the Determination of Drugs in Plasma by HPLC/UV', *Proceedings of the International Conference on Analytical Developments in the Pharmaceutical, Biomedical and Forensic Sciences* - Verona June 1986, eds. Piemonte, G., Frigerio, A., Plenum Press, New York, in press

11. McLafferty, F.W., 'Trends in Analytical Instrumentation', *Science*, 226 (1984) 251-253

12. Purvis, J.R.III, 'Communications Standards Facilitate Laboratory Automation', *Pittsburgh Conference on Analytical Chemistry and Applied Spectroscopy*, (1986) Abstract 892

13. Gibbon, G.A., 'Trends in Laboratory Information Management Systems', *Trends in Analytical Chemistry*, 3 (1984) 36-38

14. Dessy, R.E., 'Managing the Electronic Laboratory Part 1', *Analytical Chemistry*, 56 (1984) 725A-731A

15. Liscouski, J.G., 'Distributed Laboratory Data Collection and Management', *American Laboratory*, 15 (1983) September 127-132

16. Blumenthal, M. and Hirsch, G.B., 'An Effective Solution for Laboratory Management', *Intelligent Instruments and Computers*, 3 (1985) (3) 32-34

17. Perone, S., 'Coping with the Information Explosion Provided by Modern Chemical Instrumentation' in *Computers in the Laboratory Current Practice and Future Trends*, ed Liscouski, J.G., American Chemical Society, Symposium Series 265, American Chemical Society, Washington D.C., 1984 pp 99-107

18. McDowall, R.D., Pearce, J.C. and Murkitt, G.S., 'Laboratory Information Management Systems: Part 1 Concepts', *Journal of Pharmaceutical and Biomedical Analysis*, in press.

Acknowledgment

The author wishes to thank Ken Leiper for permission to use Figure 1.5

Chapter 2

Management Considerations in the Purchase of a Laboratory Information Management System

P. Johnson

2.1 Introduction

The proposal to purchase a LIMS may emanate from higher management, but in most cases will arise from the perceived needs of the laboratory staff themselves, often through a local 'champion'. The fundamental reasons may vary from the trivial to the serious, from 'every other laboratory has one and therefore we must also', through 'no other laboratory has one but we should be the pathfinders', to 'a LIMS is the only way to achieve a meaningful increase in productivity'. Clearly, it is only the last argument which is likely to have any impact on senior management.

2.2 Justification of the System

Arguments related to increases in productivity are not easily presented. The out-dated methods which a LIMS seeks to replace may in themselves restrict any quantitative analysis (especially over a number of past years) which will be required to produce a trend analysis of the workload. It will also be necessary for laboratory staff to demonstrate not only the increase in workload, but also how such increases have been handled, whether by staff recruitment or productivity improvements, or both.

Management will require an analysis of the full impact on operations of the LIMS installation. Fundamental questions include effects on staff and their motivation: more staff, the same number of staff or fewer staff (with all that may imply)? Even if it is true, it is rarely acceptable to argue that the proposed purchase will effect a staff reduction through removal of the need to ask for extra staff in the future. It is also inevitable in the short term, and even permanently, that an increase in specialist staff to run the system will be required in both the laboratory and computing groups.

Laboratory 'champions' should also be able to present in a reasonably non-technical or at least jargon-free manner the perceived advantages of their operation of the LIMS installation. Will they be able to do more work, or do their work better with less room for error, or a combination of these? What can they predict that management will be able to see at the end of the first full year of operation in contrast to the projected picture in the absence of LIMS? Can the combined advantages of staff levels, productivity and scientific standards be translated into 'real money' and hence meaningful return on investment calculations? Furthermore, what will be the impact on other laboratory operations of the capital money invested in LIMS which will therefore not be available for other purchases, and what will be the effect of the company's or institution's depreciation practices on the day-to-day revenue budget for laboratory operations? It is the clear responsibility of senior management to balance the budget against the requirements of different, and often competing, laboratory operations.

The responsibility for the impact of the LIMS installation on the local or departmental environment lies with the laboratory head or departmental manager. Nevertheless, he or she should be able to outline for higher management, concise but informative statements addressing the impact in a number of areas. These will vary with the institution concerned but might include: consequences for laboratory or computer space; general day-to-day operations and sample tracking; database management and its impact on searching archival and comparative information; the important impact on the speed and standard of report generation; the impact on compliance with various statutory regulations, such as Good Laboratory Practice; future impact in terms of the addition to the system of further laboratory instruments; compatibility with other systems existing or proposed in the institution; relationship to central and distributed computing equipment and support expertise.

2.3 The LIMS Project Team

Senior management will also require assurance that all relevant staff within the institution have been consulted, and that a team effort of laboratory and computer staff has been applied to the detailed specification which will be presented to prospective suppliers of the required system. This may be best achieved by the formation of a LIMS Project Team, with a formally appointed Project Leader, which will work to a defined Project Plan and will produce action minutes from regular meetings. The Team will need to be fully informed of local management practices in relation to capital equipment justification, computer policy, software purchase and key target dates in relation to any approval committees and the financial accounting period.

The responsibility for choice of individual manufacturer should also be with laboratory experts who have to work with the system on a day-to-day basis. Nevertheless, senior management could reasonably be expected to need reassurance on a number of points which would include: the past record of the manufacturer in relation to other equipment in the institution, including truthful and factual presentation of its products and good maintenance record; impressions of other customers (if they exist) with regard to their LIMS installation; arrangements for both prompt call-out maintenance and future upgrading (and compatibility) of the equipment; delivery dates; competitive price quotations and available discounts; and the impact of discounts on the future attitude of the supplier. It is also important to establish, especially if the institution is one of the first customers for the particular equipment, whether the manufacturer wishes to involve the institution in its advertising or marketing exercises, including visits to the installation by potential customers. In the author's experience it is wise to resist such arrangements or, if advantages to both parties can be identified, to agree very specific ground rules.

Finally, senior management should ask for information on the wider implications of the LIMS in relation to resource management, to avoid the labour intensive manual methods which were used to justify the system in the first place. Management at the departmental level, and by implication at the more senior level, should be able to obtain quantitative information on the number of assays per project, per laboratory, per analyst, per instrument, per unit time. Where projects are competing for instruments or manpower, interactive analysis of the impact of delays and/or substitution of other project work, is essential. The way in which the LIMS can be used to generate information which can be incorporated into, or be compatible with, other project planning or critical path analysis systems should therefore be clear. Important subsidiary uses could include staff availability for resource allocation, major project target dates, interim stage and report tracking dates, staff training records and equipment maintenance records.

2.4 Summary

In summary, senior management agreement to the considerable financial outlay on a LIMS will be dependent on the ability of the laboratory staff to demonstrate, and where possible quantify, a return on investment arising from benefits in: increased efficiency and productivity; reduced or limited increase in staff levels in relation to increased workload; increased scientific standards and the production of more reliable data; improved production of reports; and provision of management analysis functions.

Chapter 3

Decision Criteria for LIMS

Alan Braithwaite

3.1 Introduction

The introduction of microcomputers and intelligent instruments into analytical laboratories, together with automation of analytical procedures, has produced a revolution in the work of the analytical laboratory (1, 2). The net effect has been to change the pattern of data collection and processing, storage and evaluation of results and formatting of the analytical report. The next era in computerization of laboratory procedures, LIMS, has now arrived. Here, the aim is to integrate the automation of instrumentation and data handling with the distribution of information within the overall structure of an organization (3). It is therefore most important to carefully consider all aspects of LIMS and see how such a system could be advantageously integrated into the structure of an organization. Various LIMS configurations may be implemented, ranging from a dedicated multifunction laboratory data collection and reporting system, to an integrated company wide system.

Throughout this chapter the term 'laboratory' is used to describe the collection of analytical instrumentation and associated activities that together provide the service and method development requirements of an organization. The 'laboratory' may be a collection of smaller laboratories situated in adjacent rooms, scattered in various locations on a large site, the 'dispersed laboratory'; or be a collection of automated instruments and measuring devices situated at remote locations for monitoring purposes or for on-line process analysis, the 'sampleless' laboratory.

A LIMS can be pictured as a three-tiered system of integrated modules (Figure 3.1) each having a specific range of tasks to perform yet functioning under the overall control of the core LIMS software. The lower tier consists of the instrumentation and dedicated microcomputer hardware and software required to generate the analytical data. The intermediate tier is concerned with management of the laboratory and report generation, whilst the upper tier involves integration of management and communications activities with a company-wide computer network.

The decision criteria to be considered range from the evaluation of analytical methodology and the types of instrumentation required, via management requirements to integrated computer networks. Ultimately this will result in an integrated multi computer based system that incorporates all the requirements necessary to collate sample information, control instrumentation, collect and process data, co-ordinate procedures for sample analysis and communicate with a company wide computer network for information transfer and data access.

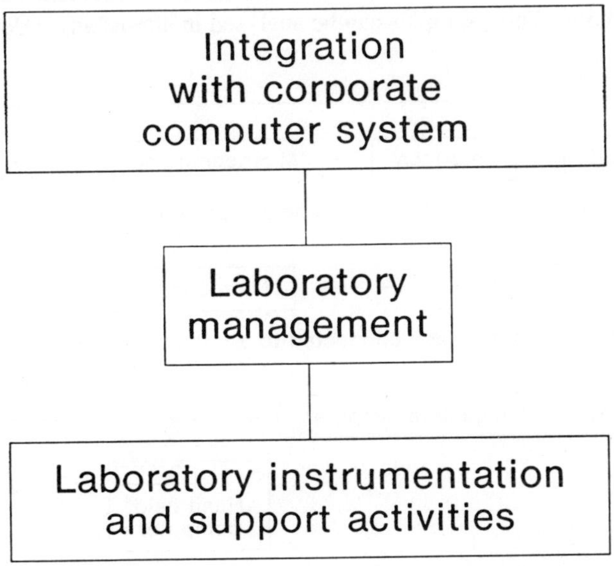

Figure 3.1 Tiers of a LIMS

3.2 Instrumentation and LIMS

Instrumentation, sensors and measuring devices generate the raw analytical data that is subsequently processed by the LIMS. It is essential that this data is converted to a digital format, either within the instrument, or in a suitable interface for local data processing by a micro or minicomputer before transfer to the LIMS (4 - 7). Instrumentation is connected to a microcomputer via a bidirectional interface, the function of which is to convert the analogue and status signals of the instrument to a digital format as required by the microcomputer input port. It also transfers any signals issued by the microcomputer to control instrument functions and variable settings (8). The microcomputer may be connected via a separate interface to an analogue instrument or connected directly to an instrument already equipped with digital output, Figure 3.2 (references 9, 10).

Many present day instruments use microprocessors for direct digital control of instrument functions. Such instruments have digital output facilities and could be controlled by a LIMS or laboratory computer via RS-232C communications. Thus control and instrument variables for a specific analysis could be down loaded from a LIMS or laboratory computer prior to sample analysis. The necessary procedural and methods software would be stored on a local database. This type of instrumentation is flexible in use and is ideal for a LIMS as it can be programmed as required for a specific analysis. Such instruments are able to take full advantage of this flexibility since the parameters can be automatically set up for individual samples and the analysis carried out. Thus, samples can be analysed in almost any order.

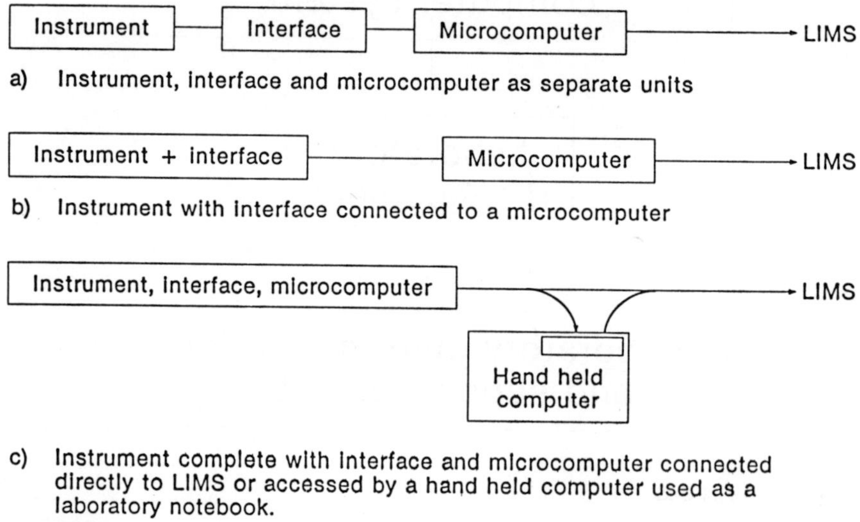

Figure 3.2 Instrument interfacing to LIMS

A further requirement of the instrument computer system is the dedicated software required. This has three functions:

1. Control of the instrumentation and data collection.

2. Data handling and processing specific to the analytical technique and sample being analysed.

3. Communications and information transfer to the LIMS network (1).

The relevant software may be resident in the instrument, often referred to as an intelligent instrument, stored in a micro or minicomputer dedicated to the instrumental technique, or stored in a LIMS computer. There are a wide range of possible computer configurations, but the most flexible LIMS systems will employ satellite computers for this purpose using a local area network for communications served by a LIMS computer. A more detailed discussion of this aspect of LIMS is contained in Section 3.4.

3.3 Planning Considerations for a LIMS Laboratory

Generally, a LIMS is designed to integrate the activities of a laboratory complex with management requirements and corporate communications and computing facilities. However, in order to ensure that the proposed LIMS will be able to achieve the specified aims it is necessary to consider the overall activities of the laboratory. These include the techniques and instrumentation employed and the organization and the role of the laboratory complex within the corporate plan. Introducing a LIMS will therefore require careful study and judicial planning since considerable changes in the overall organization and activities of the laboratory are generally necessary (11, 12).

One of the first decisions to be considered when automating the activities of a laboratory concerns the siting of the laboratories and the structure and fabric of the buildings. Automation has a number of prerequisites some of which are not compatible with the typical, conventional laboratory. For example, the installation of mini and microcomputers and the associated equipment requires clean and, if possible, air conditioned facilities with numerous power points. As computers and digital electronic equipment are sensitive to mains borne interference and switching spikes, a regulated filtered mains electricity supply is required. Generally, considerably more power points are needed with the new equipment. This requirement, together with the need to have easy access to the rear of instruments, and a change in the area of bench space required to accommodate and operate the equipment, frequently means reorganization and/or refurbishment of the laboratories is advisable.

Therefore, it is most important to consider the layout of the buildings and the internal configuration of fixtures and fittings almost as carefully as the chemistry. A further point that requires investigation concerns the compatibility of any new equipment and instrumentation with existing analytical procedures, including sample handling and preparation. Thus, any changes in equipment from balances, instrumentation and sample handling procedures to computer systems and communication protocols require careful research to take full advantage of technological developments and avoid problems of incompatibility.

The most practical, and in the long term, the most cost effective approach would be to redesign, where necessary, the analytical procedures, install new instrumentation as appropriate and implement a laboratory computer network. Adapting a traditional laboratory often places too many constraints on the design considerations of the LIMS and therefore may not prove the best solution. Again, it may be more cost effective in the long term to redesign and refurbish the laboratory, including new apparatus, sample handling systems and instrumentation. A primary requirement therefore is to define the main service functions and projected role of the laboratory. The aims and objectives of the LIMS as far as the practical aspects of the sample analysis and local data handling are concerned can then be drawn up. Some of the questions to be researched are discussed below.

3.3.1 Sample Handling and Analytical Procedures

The first point concerns the sample handling requirements and the instrumentation necessary to implement the projected aims of the laboratory within the corporate plan. Many traditional laboratories have been designed for a specific technique, for example, gas chromatography or IR and UV spectrophotometry, or perhaps to service a single production line or type of product. With automation of sample handling now available and data handling and report generation automated through the use of data stations and microcomputers, the restrictions mentioned above on the scope of the activities of a laboratory are removed (13). It is now quite feasible to simultaneously prepare a sample for analysis by a variety of methods or to plan the analysis of samples by a sequence of techniques. Multicomponent analysis using techniques that require the minimum of sample preparation are ideal for the automated laboratory and are receiving increasing interest, for example, near infra red spectrophotometry and multi-sensor ion specific electrodes.

The main question therefore is whether the facility should be a multi or mono technique laboratory, a laboratory for a single product line or type, or for all types of samples. Once the activities of the laboratory have been defined it may be worthwhile to consider a redesign of experimental procedures for some of the projected samples. This would produce not only an increase in efficiency but would also take advantage of the newer instrumentation, and provide more analytical information on the sample components.

3.3.2 Intelligent Instrumentation

The main problem with instrumentation is transferring the required data to the LIMS. This can be quite an involved process if the full capabilities of a LIMS are to be realised. It is therefore worth considering equipping the laboratory with the new generation of intelligent instruments to simplfy communication. These instruments are readily available for most analytical techniques and are able to communicate with a wide range of computer systems from personal computers to minicomputers. Therefore, such instruments can readily be integrated into a LIMS network with the minimum of interfacing problems.

3.3.3 Personnel and Management

When the overall activities and organization of the laboratory have been decided the personnel and management requirements can be defined, not forgetting secretarial assistance: an important support service which is often overlooked. The management aspects of a LIMS may be considered under two headings:

1. Laboratory organization.

2. Analytical services and corporate requirements.

The first aspect is concerned with developing effective and efficient management of the analytical procedures and supporting activities carried out in the laboratory. These include use and maintenance of equipment, preparation and transmission of reports, ordering and storing of reagents and consumables, and maintaining good laboratory practice.

The second management area concerns the interface between the laboratory and other departments within the organization. This involves communications with personnel and finance departments, manufacturing plants and senior management.

It may prove rather difficult to incorporate some aspects of the senior management area into the LIMS. However, it is essential that all the data and information relating to the performance and requirements of the laboratory are available in a form which makes the corporate management aspects an easier task. Secretarial assistance is usually necessary, particularly to support the corporate management requirements. The level of this support should be less than traditionally required as much of the laboratory information is handled by the LIMS and satellite computers using direct data entry methods (3). Corporate aspects of laboratory management include the maintenance of financial and personnel records and information. It is feasible to include these as modules in the LIMS, however, access to this information would be restricted and security of data could be a problem. Also, it may prove difficult to ensure that all the legal and company requirements could be met.

3.3.4 Reports

Production of reports is a major requirement of any laboratory and also a factor to be considered in defining the configuration of the computing requirements discussed below. Identification of the activities of the laboratory, as described in Section 3.3.1 will have highlighted the reporting requirements. These will generally fall into one or more of the following groups (14):

1. Local reports, to be collated and stored in the laboratory.

2. Reports to be transmitted to those requesting information on the analytical results.

3. Reduced data reports for archiving.

4. Preparation of bar charts, graphs and trends analysis.

Preparation of the reports requires careful planning to ensure all the necessary information is transmitted in the correct format. The LIMS software therefore needs to include report generation software together with supporting software for editing and wordprocessing. Data analysis and spreadsheet software for graphical and bar chart representation are also useful and add flexibility. The ability to transfer files between software modules is essential, and a further advantage is for the LIMS to be able to accept and download files from personal computers and workstations (IBM PC format) in the form of ASCII files, or from the common spreadsheets and databases such as Lotus 1,2,3, or dBase III.

3.3.5 Databases

Data is stored for laboratory, corporate and regulatory agency requirements. Whilst a microcomputer database may be satisfactory for local data, a flexible relational database running on a minicomputer, is more satisfactory for user access. It is important to define all the requirements of this database; to design the system to achieve these aims and for retrieval of information in a specified format. This will ensure that stored data can be rapidly searched and the desired information extracted in a meaningful and useful report. It is essential to ensure that the database is secure and the procedures used are compatible with corporate policies. Also, access to company and national library and literature databases and communications via national and international networks should be discussed at this early stage. If access is required, this places more emphasis on efficient communications between computer systems.

3.3.6 Computing Resources

A LIMS network will probably incorporate a central LIMS minicomputer with a range of micro or minicomputers acting as intelligent interfaces or work stations. There are a wide range of configurations that may be considered and a number of alternative instrumentation systems that could be employed. It is therefore essential to consider the aims and activities of the laboratory as defined under the above criteria and decide upon the overall configuration of the LIMS. The computing resources required, particularly at laboratory level, to implement the LIMS can then be identified. These include computers for laboratory management, local data collection and storage of instrument results, data storage and filing requirements at laboratory level and use of personal work stations. The latter can include a personal computer, hand held computer or terminal. Communications between computers is an important feature of LIMS and therefore the introduction of local area networks (LANs), mainframe interfaces and telecommunication facilities required should be identified at the same time (2, 3, 12).

The overall response and speed of the LIMS is an important consideration if the defined objectives are to be achieved. Frequently, computer systems are installed which prove too limiting within a year or so, that is, the processor is not capable of handling the usage and data rates required to provide an efficient service. A frequently used maxim is to define the system based on current expectations of performance required and then to install a system with twice the capabilities. This is a somewhat unscientific approach. A better one is to specify a system that has a flexible architecture. Systems based on a minicomputer (eg. VAX 11/785) serving a network of mini or micro computers, (eg. micro VAX and VAXmate) not only maintain compatible operating systems and protocols, but also have an architecture where more central computing power in the form of a VAX cluster can be added for parallel processing, in addition to extending the number of computers on the network.

3.3.7 Education

Finally, a LIMS will inevitably introduce a wide range of new equipment and new ways of carrying out the work of the laboratory. This includes different analytical procedures and techniques and probably will require staff to learn new skills, particularly as computer users. Thus, education of all users is an important requirement that can easily be overlooked in the planning stages. Documentation and user manuals which contain the essential information in a readable, clearly presented format are also important. At least two members of staff should have expert knowledge of the system and be involved with a coordinated education policy, not just to familiarize staff with the LIMS, but also to explain their role within the LIMS and the expertize and support available. These aspects are expanded in Chapters 5 and 6.

3.4 The Integrated LIMS

3.4.1 Requirements for Integration of a LIMS

Integration of a proposed LIMS with the existing computers is essential if it is to be fully effective and have access to the relevant databases and communications networks. Both analysts and computer experts need to carefully consider the practical levels of integration from a rather limited insular system to a fully integrated one that can be accessed throughout the organization. It is useful to have a working knowledge of the configuration, organization, operating requirements and jargon of typical computer systems (15, 16). This will prove useful for specifying the LIMS and also in understanding large computer installations. A further benefit is that chemists and computer personnel can work together in a constructive manner so that informed decisions can be made on integrating the LIMS into a corporate computer system.

Schematic diagrams of complex integrated computer installations are shown in Figures 3.3 and 3.4, and present an idea of the complexity of such systems. There are, in effect, three levels to the integrated system (Figure 3.3). The central computing facility contains a cluster of mainframe and mini computers connected together by very high speed links, served by high speed disk units for mass storage, databases and archive data storage (Figure 3.4).

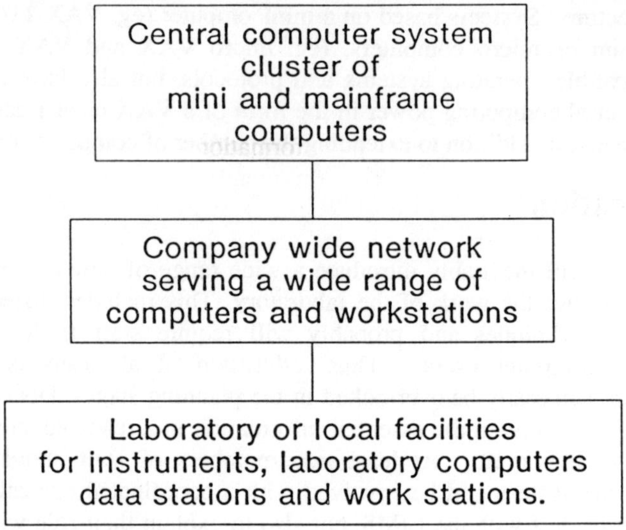

Figure 3.3 Summary of integrated computer systems and networks

The central cluster is connected to the second level of computer facilities via one or more network controllers whose main function is to connect the users with the software packages and files required and with the appropriate computer in the cluster to perform these tasks. The distributed access points may be RS-232C serial lines, a network or a number of multiplexers connected to the central cluster by high speed multiplexed lines. The latter are often used for communications with remote sites. Flexibility is the keyword at this level and a wide range of computers, work stations and terminals can be easily connected into the system. The main requirement is for each device on the network to have a standard RS-232C port and the appropriate communications software.

Transfer of files and data between computers or terminal points is straightforward and is handled by the main cluster software. Thus, mailing of information and reports, searching databases, central archiving of data and using specific software packages are functions that are transparent to the user. The connections onto the network are referred to as nodes or terminal points and are only limited by either the capacity of the network controller or the amount of data or traffic on the network. A gateway to national and international networks is a further requirement and enables the user to access and search databases of business and scientific information, for example, Chemical Abstracts, libraries of spectral data, and environmental databases. National and international networks are also used by multinational companies to communicate with their locations abroad and this will often involve satellite links (3).

The third level is shown in Figure 3.4 as the local or laboratory level. Laboratory computers and instrumentation may be connected directly onto the company wide network or connected via a local area network (LAN) and network interface. This interface may be an intelligent node or a laboratory computer configured to act as a file server to the laboratory LAN and as a controller of a gateway onto the company wide network. The latter approach is more effective from a management point of view and does not tie up communications lines unnecessarily.

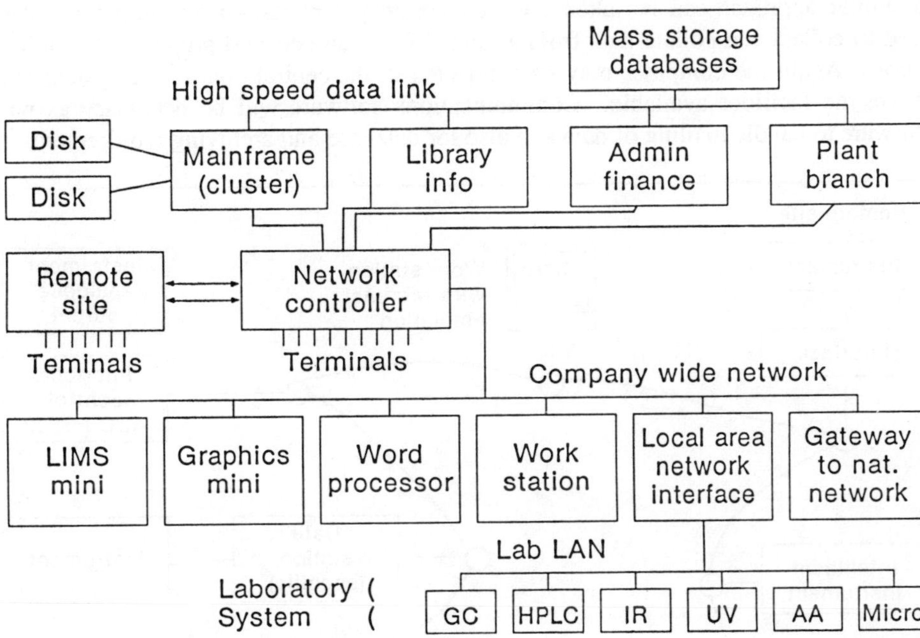

Figure 3.4 Diagram of an integrated computer installation

3.4.2 Levels of Integration for a LIMS

The possible levels of integration of a LIMS within the laboratory and with the corporate computer system can now be considered.

Laboratory level approach to a LIMS A rather basic system which could use existing instruments and data stations can be regarded as an extension of a facility which is already in use. Such systems are based on small computers and are usually dedicated to a specific technique although they may be able to serve more than one instrument. With suitable programming they are able to produce laboratory reports for distribution in addition to providing local data storage. The data station may be connected via a RS-232C serial line into a company wide computer network and, with suitable software, use the communications facilities and access software packages, wordprocessors and databases.

If a microcomputer based terminal to the central computer system is also sited in the laboratory, then separate access to the data files created by the data station can be obtained for editing purposes, data manipulation and trend analysis and searching of databases. Careful planning and writing of the appropriate software can therefore achieve some aspects of a LIMS, albeit on a rather limited scale.

A similar approach can be taken with a laboratory data station or computer that is used to collect and collate data from a range of instruments and prepare the required reports. Again the computer may be connected to the central computing system and utilize the facilities available. Communications software will be necessary as will software to handle mailing of data and also for database and archiving requirements.

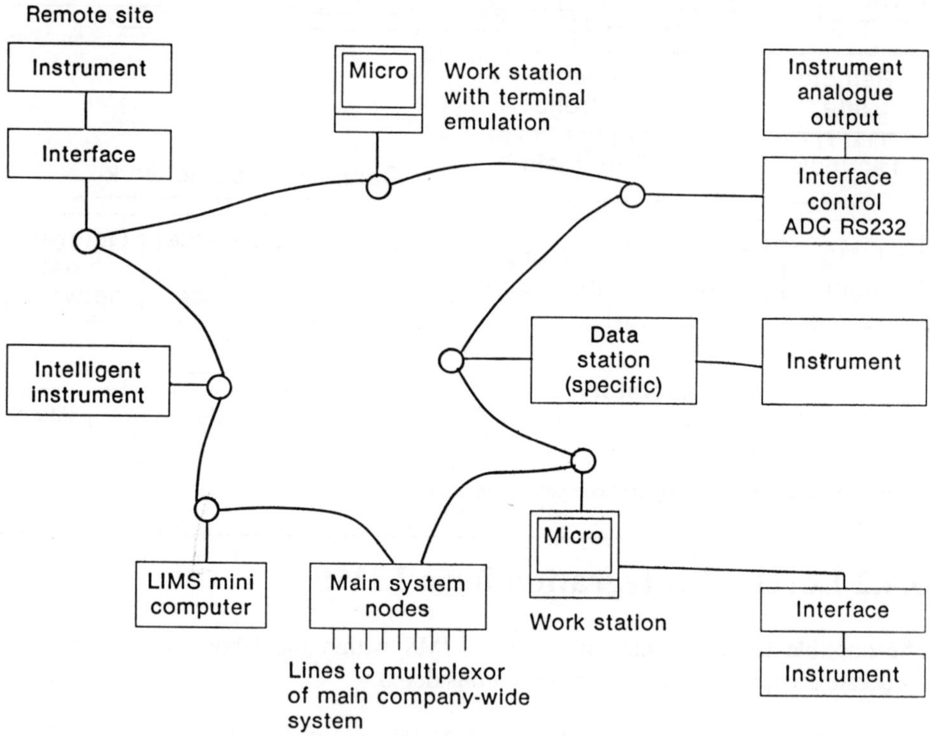

Figure 3.5 Laboratory local area network

Both the above approaches are extensions of existing practices and therefore do not have quite the flexibility to take advantage of a radical rethinking of laboratory activities that a new LIMS would require. However, it is quite feasible to build on the two approaches above and use the modified data station or laboratory computer as a module connected onto a laboratory LAN or into a LIMS computer, as shown in Figures 3.5 and 3.6.

Local area network based system A local area network (LAN) is a convenient method of linking together a variety of instruments, computers, work stations and printers as shown in Figure 3.5. This is a practical, flexible approach whereby existing instruments equipped with a suitable interface can be utilized. Further instruments can be added at a later stage and older instruments replaced as necessary. Microcomputers, wordprocessors and workstations can also be included on the LAN. A microcomputer equipped with a hard disk unit is required to act as file server, that is, to manage the LAN and act as the main storage device for software, users' files, local reports and general information.

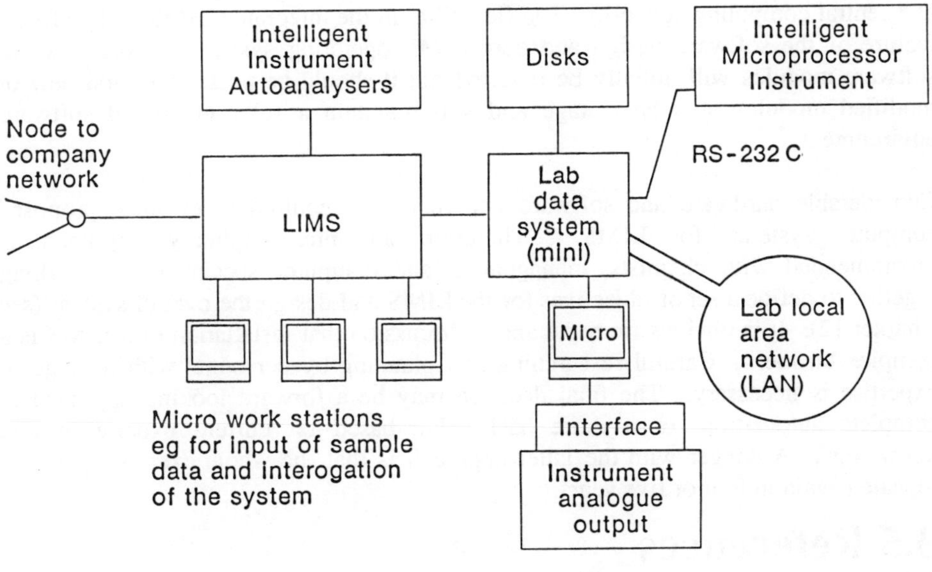

Figure 3.6 Configuration of a LIMS and laboratory computer system

Production and distribution of reports, local data storage and handling of information concerned with managing the analytical services of the laboratory can be accommodated on a LAN based system, a 'first stage' LIMS. However, the LAN will be linked into the central computer network with full access to all the attendant benefits of software, database and archiving facilities. A minicomputer that can also run a full set of LIMS software modules may be included at the outset or added at a later stage in the development. The LAN approach therefore confers a wide degree of flexibility on a LIMS and enables additional modules and instrumentation to be included as the need arises and as experience of the LIMS increases.

The LAN is a particularly flexible approach to a LIMS allowing the system to be built up in stages. However, a cautionary warning is appropriate: a 'piecemeal' approach is not recommended. It is most important that the criteria discussed in Section 3.3 are evaluated and a full specification and plan of the LIMS requirements is drawn up before any steps to implement any stage of a LIMS or pseudo LIMS is carried out.

Complete integrated LIMS A fully integrated LIMS is centred on a minicomputer running a set of software modules designed to carry out the objectives defined in the master plan. The LIMS computer will have the capability of carrying out all the hardware and software functions mentioned in the previous two sections and shown in Figure 3.5. The hardware flexibility required for linking to a range of instruments and computers should not be a problem. Suitable minicomputers will have a set of input/output ports to link directly with intelligent instruments, data stations, other minicomputers, microcomputers and work stations, local area networks and central computing networks. The flexibility in the integrated LIMS is therefore a feature of the software design and core LIMS operating system. A basic set of software modules will initially be required but it should be possible to add new or modified modules at a later stage and still maintain a fully integrated software environment.

Considerable hardware and software experience is required to evaluate proposed computer systems for LIMS. Therefore an interdisciplinary approach is recommended with chemists, management and computer specialists all working together to define a set of objectives for the LIMS and design the overall system (see Chapter 12). It is obvious from the above discussion that installation of a LIMS is a complex business. Careful researching and planning by personnel with a range of expertise is necessary. The final decision may be a forward looking approach to complete automation, or a more basic plan based on current experience and knowledge. A danger with the latter approach is that the whole exercise could be repeated again in four or five years.

3.5 References

1. McLafferty, F.W., 'Trends in Analytical Instrumentation', *Science*, 226 (1984) 251-253.

2. Liscouski, J.G., 'Planning an Approach to Laboratory Automation', *Computers in the Laboratory, Current Practice and Future Trends*, American Chemical Society Symposium, No. 265, American Chemical Society, Washington D.C., (1984) pp1-9.

3. Dessy, R.E., 'Laboratory Information Management Systems Part 1', *Analytical Chemistry*, 55 (1983) 70A-80A.

4. Dessy, R.E., 'Robots in the Laboratory Part 1', *Analytical Chemistry*, 55 (1983) 1100A-1114A.

5. Braithwaite, A., 'Developments in Microcomputer Systems and Laboratory Instruments', *UV Group Bulletin*, 10 (1982) 26-40.

6. Malcolme-Lawes, D.J., *Microcomputers and Instrumentation*, Plenum Press, New York,(1984) Chapters 1 and 6.

7. Braithwaite, A., 'Microcomputer Hardware and Software for Collecting and Processing Data', *European Spectroscopy News*, 51 (1983) 24-27.

8. Dessy, R.E., 'The PC Connection Part 2', *Analytical Chemistry*, 58 (1986) 793A-804A.

9. Dessy, R.E., 'The PC Connection Part 1', *Analytical Chemistry*, 58 (1986) 678A-689A.

10. Borman, S., 'Laboratory Computer Networks', *Analytical Chemistry*, 56 (1984) 408A-413A.

11. Braithwaite, A., 'Decision Criteria for LIMS', *Laboratory Practice*, 35 (1986) 11-18.

12. Dessy, R.E., 'Laboratory Information Management Systems Part 2', *Analytical Chemistry*, 55 (1983) 277A-303A.

13. Lochmuller, K.R., Lung, K.R.and Cushman, J., 'Laboratory Robotics: The Advent of 'Soft' Instrumentation', *Journal of Chromatographic Science*, 23 (1985) 429-436.

14. Liscouski, J.G., 'Distributed Laboratory Data Collection and Management', *American Laboratory*, 15 (1983) September 127-132.

15. Bishop, P., *Computer Science*, Nelson, London, (1982).

16. *Understanding Computer Science*, Texas Instruments Learning Centre, PO Box, 225012, MS54, Dallas, Texas.

Chapter 4

Benefits and Justification of a LIMS

R.D.McDowall

4.1 Introduction

The analytical laboratory is often on the critical path in many organizations, as the assay report is needed before decisions can be taken. Therefore a chapter on the benefits of Laboratory Information Management Systems should be viewed in the overall light of speeding the production of that report, and ensuring that the data contained therein is correct and precise. This chapter is intended to give a general insight into the advantages of a LIMS. In part, the benefits presented here may be used by readers to justify a system in their own laboratories. However, these will not be comprehensive and will invariably be extended as the technology develops. Included in this chapter is an overview of the methods that can be used to justify a system.

4.2 Benefits of a LIMS

The revolution in analytical instrumentation promises high productivity providing that the analysts involved are correctly educated and trained (1). Once this is achieved the benefits of a LIMS can be realized.

There is little written in the scientific literature by users about the benefits of LIMS, reflecting that these are early days in the development of this type of laboratory automation. Five articles by users have been published on the applications of LIMS (2-6). These discuss the installation of LIMS within three laboratories with an overview of the systems used and a little insight to the benefits obtained. Benefits can be difficult to assess and quantify as each laboratory has its individual requirements. However, some generalizations can be made.

Major benefits are realized by laboratories that have successfully installed LIMS. These are:

- Data storage.
- Data manipulation.
- Data integrity.
- Productivity benefits, that are simply the sum of the three benefits above (7).

Some benefits applicable to each area are listed in Table 4.1 and are now examined in turn.

Table 4.1

Some benefits of Laboratory Information Management Systems

Data Storage
- Storage of data collected automatically
- Storage of manually entered data
- Comparison of results with acceptable ranges
- Electronic filing of methods and tests
- Sample status
- Sample location
- Fast retrieval of results
- Rapid report generation

Data Integrity
- Automatic sample numbering
- Verification of data entry
- Automatic test assignment for routine samples
- Bar codes for sample identification
- Automatic calculation of results
- Security of data (audit trail, passwords, database access)
- Reduction of errors and need for transcription error checking
- Compliance with regulatory guidelines

Data Manipulation
- On-line data capture
- Data reduction
- Graphical presentation of data
- Automatic calculation of results

Productivity
- Automatic generation of sample receipt forms
- Automatic label printing
- Worklists for laboratory staff
- Automatic billing of customers
- Integration with word processing and other computers for efficient data transfer
- Backlog reports
- Reduction of administration and paperwork
- Speed and reliability of reporting results
- Efficient use of laboratory resources
- Faster project or product development (R&D)
- Quicker product release (Q.A.)
- Time saving

4.2.1 Data Storage

Data storage is the fundamental basis from which all other benefits of LIMS are derived. At the heart of the system is a database which is equivalent, in part, to an electronic or automated laboratory notebook. The data therein are easily accessible to many more users, often simultaneously, than any material held in a laboratory notebook. It is the logical relationship between the stored data that makes it relatively easy to search the database for results or information that would be impossible or very tedious to find with a manual system. The data contained therein are arranged in a logical order for easy access. Software modules around the database enable automatic update of records and extraction of data quickly and efficiently, often without the need for human intervention.

These modules facilitate searching of analytical data held on-line. The automatic update of the records within the database means that sample status can be searched to determine progress of the analysis in answer to customer queries (6). In a LIMS where the location of samples has been implemented it should be possible to find the sample quickly and efficiently minimising delays due to physically searching storage areas.

If the data retention policy is such that old data is kept within the database, then searching it to find trends in the workload or simply to keep the data as a reference for comparison with new samples, or to use as a means of accepting or rejecting material is possible. Moreover, if previous results are stored on-line, they can be used to compile year-end figures for all projects and thus gain a better view of the laboratory's workload (3).

Data storage encompasses many other facets: data collected automatically by the LIMS is cross-referenced in the database which enables it to be easily retrieved. This applies to newly acquired data as well as that already held on-line. In connection with this function, the methods used to analyse this data are also stored within the database.

Collation and reporting of results are areas where a LIMS excels. When using validated computer programs the required information can be extracted from the database and rapidly included in reports. If required, limits can be set on the results for particular assays and the system programmed to highlight any results outside them, thus bringing them to the attention of the analysts for verification or retest. This is especially important in a Quality Control (QC) laboratory where the system can be used to compare the results obtained experimentally with the range of those stated in the specification and to produce a certificate of analysis (Figure 4.1). A report can be printed automatically and if these results are out of limits they can be highlighted for the supervisors to check and decide if it is a significant result.

SMITH KLINE & FRENCH LABORATORIES LIMITED

Welwyn Garden City Hertfordshire England AL7 1EY . Telephone Welwyn Garden (0707)325111

CERTIFICATE OF ANALYSIS

For : DEMONSTRATION ONLY
Material: LACTOSE
Batch : 327136

Indent No:
Anal Ref : 86R 1461/2

We have tested a sample of LACTOSE marked as above.
Our results are as follows :

Description	A white odourless crystalline powder.
Chemical Identity	Positive
Chemical identity(b)	Positive
Colour of Solution	Colourless
Clarity of Solution	Clear
Escherichia coli	Absent in 10g
Acidity/Alkalinity	< 1.50(cm3)
Sulphated Ash	0.05%m/m
Specific Rotation	52.26Degrs
Arsenic	< 1.0 ppm
Lead	< 0.5 ppm
Ethanol Soluble Matter	0.24%m/m
Total Count(TSA at 30C)	5. CFU per gr
Total Count(SDA at 25C)	5. CFU per gr

This sample complies with SK&F Specification 163

Signed .... C.Chem M.R.S.C.

A W CAMERON ,Quality Control Manager
FRI., 27 FEB., 1987

Figure 4.1 A certificate of analysis (courtesy A.D.Henderson)

4.2.2 Data Integrity

The integrity of data is fundamental to any scientific discipline, but this need is often reinforced by regulatory agency guidelines. Compliance with these requirements is crucial to the acceptance of studies supporting the registration of a new drug or agrochemical; this can mean an increasing number of samples for analysis. Here LIMS can be of great value. All changes to the database are monitored by an audit trail which ensures that any modification is recorded. This provides a measure of security to the database as it provides a means to show if a record has been changed and by whom. Furthermore, the use of validated programs ensures adherence to any procedures laid down in the relevant Standard Operating Procedure (SOP). Although the benefits in this area are difficult to quantify, their importance should not be understated. More detailed discussions of the interaction between a LIMS and Good Laboratory Practice (GLP) will be found in Chapters 24 and 25.

The LIMS can be set up to generate sequential samples automatically. In cases where this is advantageous, bar coded labels can be printed and attached to the samples and any aliquots derived from it. The bar codes can be read at a later stage to unequivocally identify the sample prior to entry of results into the LIMS. This is a very positive benefit as it enables the scientist to have greater confidence in the ensuing results, as mistakes in sample identification can be virtually eliminated. This philosophy has been taken further in the implementation of some LIMS where some keyboard entries have been eliminated in preference to using bar codes for certain tasks.

Verification of data entry by using references within the database is a very powerful, yet simple method of ensuring data integrity. This facility should be used at every opportunity as the data or entries can be rigorously checked and this builds confidence in the system and the results produced. For instance, if data is expected in the format x.yy the system will check that the format of the result is what is expected. If a comma instead of a full stop (eg. 5,93) were entered in the field, this will be rejected. Note that it cannot check that the result is correct. A similar way of validating entries can be used for alphanumeric fields utilizing the correct logical arrangements of the data sets. This process is shown in Figure 4.2 where the analyst has requested a test which is not appropriate to the sample. When the return key is pressed, this is automatically rejected by the LIMS, inviting the user to re-enter the field. When this is done successfully, Figure 4.3, the system allows the user to proceed to the next field.

This process can be taken a stage further by automatic test assignment to a sample. For instance, if a sample from a given source always undergoes the same tests, these can be automatically assigned to the sample upon its entry into the computer system. Alternatively, if the users do not wish to be constrained by this way of working it is possible to define tests specific to that sample source which cannot be allocated to samples from other sources. Then at log-in, only those tests deemed essential are assigned to the sample. The second method of working is more flexible and may be more economical with laboratory resources, but incurs more keyboard use and training of staff, compared to the first which allocated all tests regardless. It is the judgement of the individual laboratory management and users which decides the method of working they prefer.

The integrity benefits of a LIMS are at their greatest with regard to the reduction of both errors and transcription error checking. Sample information is stored within the database and updated by the software modules; results can be transferred to the database from analytical instruments and incorporated in the records for the appropriate sample. This automatic transfer means that little transcription checking needs to be done if the programs have been validated. This means that a final report concerning any sample will be a true representation of the data and any calculations performed within the database. Thus, transcription error checking can be greatly reduced but not entirely eliminated. However, the checking that is now done is at a higher level than previously, being concerned with the review of the results rather than detailed checking of the figures and the calculations involved to obtain the result as was the case previously.

Sample Entry

Project : AX101
Source : Pilot Plant
Process : Hydrogenation
Submitter : ADC
Tests : HPPC

Note test HPPC does not exist – please re-enter

Figure 4.2 Verification of data entry

Sample Entry

Project : AX101
Source : Pilot Plant
Process : Hydrogenation
Submitter : ADC
Tests : HPLC

Figure 4.3 Verification of data entry

Where results have been typed into the computer, then there is no alternative to checking all results manually before issuing the report in order to avoid errors entered during typing. It can be seen that the best performance of a LIMS occurs when analytical instruments are directly linked to it for automatic transfer of results. This in turn fully integrates the sample information with the corresponding results for greater data integrity, and greatly simplifies the administration and ensures the quality of the results generated from the laboratory.

Automatic calculations are easy to implement in a LIMS. They can be used to convert data easily, efficiently and reliably, avoiding the need to check for error, and building more confidence in the system and the results it produces. As most scientists appreciate, the checking of analytical data is a very tedious task where mistakes are easily made or missed. The benefits of data integrity can remove the need for a large amount of transcription error checking that is necessary in a modern analytical laboratory.

As an illustration, a LIMS can help the analyst investigate various analytical parameters such as sampling error, bias, precision and control samples. For instance, data from quality control samples can be plotted on control charts and the control limits (± 2 and 3 standard deviations) are calculated by the program from historical results stored on-line. This will warn the analyst of any results that are outside acceptable limits (8).

4.2.3 Data Manipulation

Data manipulation allows the user to transform or reduce data into information efficiently and without error. This can take two main forms: automatic calculation and graphical manipulation.

Automatic calculation of data and results has been shown to be a feature of data integrity benefits. Once confidence has been established, their full potential in data manipulation benefits can be realized. It is possible to use algorithms in a number of ways: to convert the salt weight of an analyte into its equivalent base weight for calculation of a standard solution concentration, dilution factors can be used for routine calculation of results and conversion factors to change units of the results upon issue of the final report. The point is that data can be transformed or manipulated easily and speedily, without human intervention and without error by the system. Therefore, a LIMS should be designed to take advantage of this benefit whenever possible. An instance of the use of automatic calculations is the on-line capture of chromatographic data which can be reduced, interpreted, and passed to a LIMS database automatically. If required, the data can also be presented graphically in a number of ways to show the reproducibility of chromatography, integration and interpretation.

To illustrate the use of graphical data presentation, the acceptance and validation of large numbers of plasma samples from clinical and pre-clinical studies in drug development can cause long delays before the emergence of the final report. In the author's laboratory this process was speeded up by plotting the calibrated curve followed by the results of each set of subject samples on a screen. Moreover, to aid the process, colour was used to display the data and make interpretation easier.

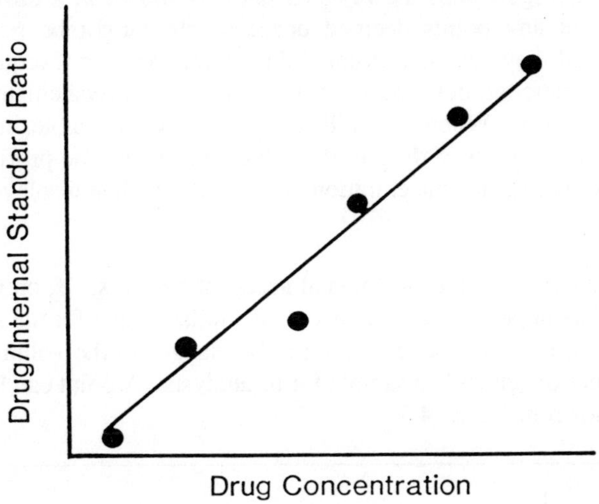

Figure 4.4 Display of a calibrated curve.

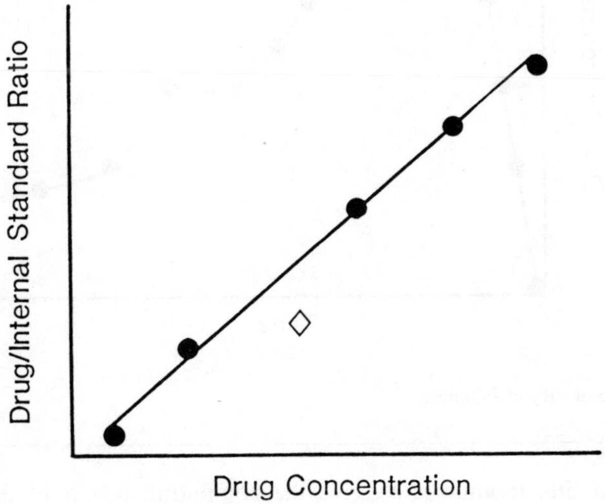

Figure 4.5 Interpretation of a calibrated curve.

The software displays the calibrated curve in the form of a graph of the drug/internal standard ratio versus drug concentration (Figure 4.4) with the slope, intercept, correlation coefficient of the line calculated from a linear regression. A prompt asks the analyst if all the points are acceptable, if not a table showing the individual points is displayed and the operator can reject any standard. If this is done, when the graph is next displayed (Figure 4.5) the rejected point is shown in a different colour and symbol. Note that any points deemed unacceptable cannot be removed from the display and indeed they can be reinstated during this process if required. With each new display the intercept, correlation coefficient, etc. are recalculated and shown on the side of the screen. When the calibrated curve is acceptable, it is then used to calculate the concentration of drug in the subject samples. The profile of the drug in the subject's samples (drug concentration versus time) is then displayed on the screen, (Figure 4.6).

Here the analyst can view the profile and accept the results. If, on visual inspection one or more points appears wrong then a table similar to that for the standard samples appears. Here the operator can enter, not only changes to the volume of the sample, but can also reject or schedule a sample for re-analysis. A point can be removed from the profile as shown in Figure 4.7.

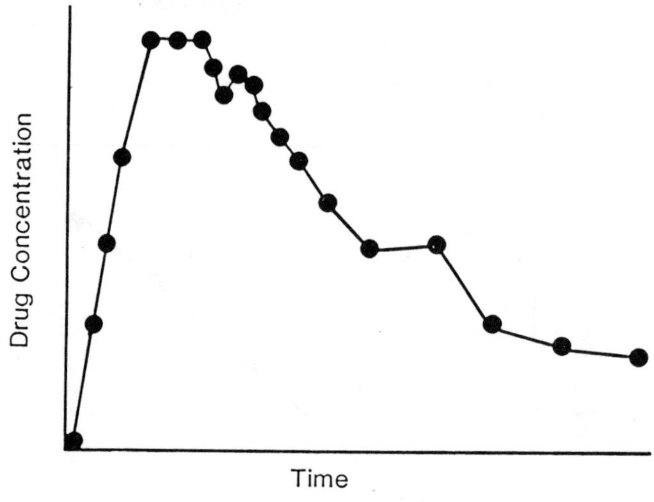

Figure 4.6 Profile of subject samples.

The concept of this module is to give the computing power of the system to the analyst in an interactive form so that, if required, many combinations of the data can be viewed and evaluated rapidly and the best fit chosen. Again, colour graphics aids and enhances this process. Similar programs could be used to display quality control values in the form of a Cusum chart for any particular test. There are many possibilities.

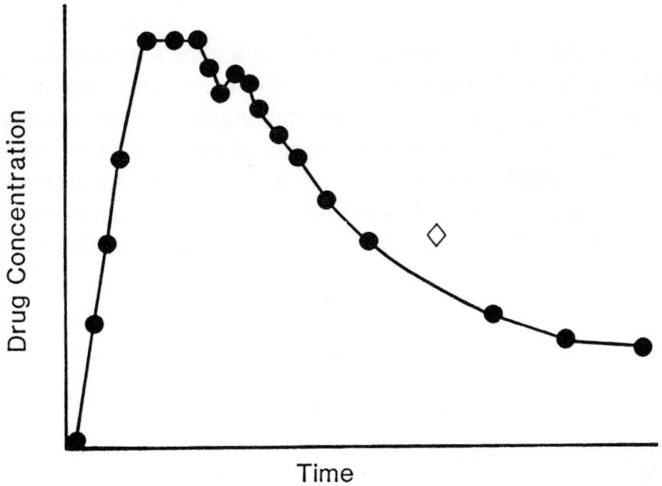

Figure 4.7 Interpretation of a profile of subject samples

4.2.4 Productivity

The productivity gains as a result of installing a LIMS are quite simply the sum of the benefits of storage, integrity and manipulation of data (Figure 4.8). It must be noted that at no stage does a LIMS aid the chemistry: its prime function is to help and speed the administration of the sample and the associated data. The database functions as the notebook, but the software modules surrounding it automatically update the records and dramatically reduce the need for administrative record-keeping or clerical duties. A LIMS will not speed the analytical process, it only reduces the administration to enable more time to be spent by staff on improving or monitoring the chemistry. Productivity increases achieved by an individual LIMS will be dependent on the aims and configuration of each system. According to Reber (9), no definitive study has been performed to assess the overall benefit of LIMS; but increases in productivity of 10-20% is the average estimate made by most laboratory managers who have installed such systems. A similar figure is given by Golden (10) who has assessed the impact of LIMS in both Research and Development (R&D) and QC environments. In the former, increases in productivity are the main benefit by removing the transcription and checking of data brought about by automatic data capture. This productivity is manifested by the speed at which a product can be perfected and brought to the market place. The rationale for a LIMS here is not only the improvement in productivity or in the speed at which specific analyses are completed but lies in the laboratory's ability to complete entire projects more quickly.

LIMS in a QC environment has the ability to quickly accept or reject raw material lots or manufactured goods, thus smaller raw material stocks can be held and finished goods released to be marketed sooner. These are areas of benefit in addition to the increases in productivity.

The use of bar coded labels generated from the author's system has saved staff in many departments from writing the labels by hand, enabling analysts to have uniformally labelled tubes and a speedy method for entering samples into the computer. If large numbers of labels are required then this is a very cost-effective method of producing them. Thus productivity increases from the LIMS can be seen in other areas such as those that supply the samples. Other areas where LIMS may be beneficial are the automatic generation of sample receipt forms or billing to customers.

Data Storage Benefits

+

Data Manipulation Benefits = Increased Productivity

+

Data Integrity Benefits

Figure 4.8 The relationship of productivity to other LIMS benefits

---

An aspect of LIMS included the integration of the system with word processing. This can be achieved by installing a word processing package within the LIMS. However, in the author's laboratory the LIMS had to be integrated with an existing corporate word processing system (Wang), the approach was therefore as shown in Figure 4.9. The LIMS can generate reports in predefined formats by writing data files to disk. These files are then transferred to the Wang word processing system by a second program via an IBM communications protocol (RJE 2780) for inclusion into the final report. Once the analytical portion of the final report is within the word processing system it is easily transferred via electronic mail to any part of the organization. Thus, speed, reliability and the ability to have the results in a suitable format for distribution are prime benefits of a LIMS.

Communication to other computer systems can be crucial for the success of a LIMS. Thus, alternative methods of transferring data or results should be considered such as a local area network (see Chapters 3 and 7) or Kermit, a software package (see Chapter 10).

Scheduling is possible with a LIMS, Berthrong (2) has described a system where this function receives a trigger from an external computer (such as an inventory controller or automated purchase requisition system) that a sample is due into the laboratory within the next six weeks. The minimal information supplied can be supplemented if required but it allows the laboratory management to see what workload will enter the laboratory during the next six weeks. It is left to management when, or if, to alter the priorities of individual samples.

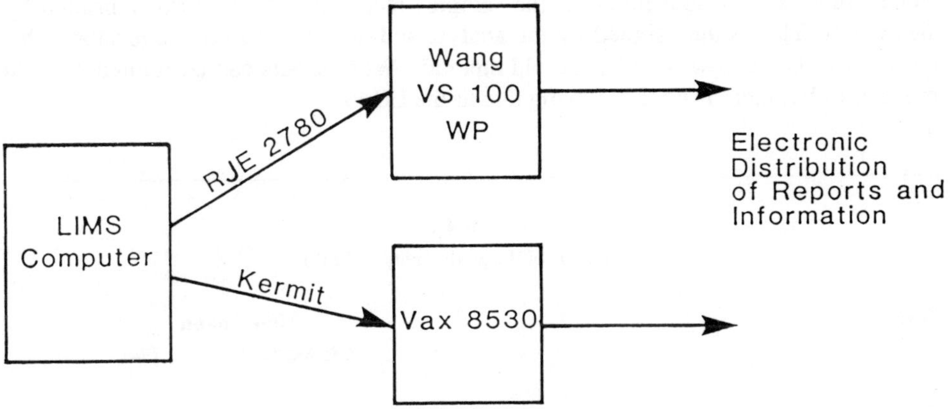

Figure 4.9 Transfer of results from LIMS to a word processing system and other computers.

Time saving is an important productivity gain and one which can often be quantified in terms of quicker release of product. Turnaround time, defined as that from when the sample enters the laboratory to when the result is released to the requestor, can be crucial, for instance in the hospital environment (Chapter 15) where the quick reporting of results can be life saving.

The overall productivity of a LIMS in the author's laboratory can be seen in Table 4.2 where a comparison is made between the procedures in place before (partly computerized) and after a LIMS was introduced. This work involves the analysis of 400 plasma samples from a clinical study in the pharmaceutical industry; the assay method uses HPLC and the throughput of the analytical method is 200 samples per week.

There is a small overhead in initially defining the work within the system: this was not necessary before. This time is quickly offset in the gains from uniformly labelling the small tubes with bar coded labels produced by the LIMS.

Samples entering the laboratory can be logged into either system. The old method required that each set of samples from an individual was entered into a computer data file. If there was a missing sample from a batch, the individual times of all the remaining samples had to be re-entered into the system. This was slow and tedious. In contrast, as the protocol had already been defined within the database, samples are entered by bar code. This is rapid, more so if laser-scanning readers are used, and error free as the sample identities are verified against entries already within the database. Simultaneously, if the times of the samples are different from those planned, the actual times can be entered by the analyst. Thus, whilst the two log-in procedures take the same time, the LIMS method is easier, less error prone and more information can be easily incorporated for the report stage. At the end of the LIMS procedure an acknowledgement of the samples received is automatically printed by the system This is then signed by the analyst and dispatched to the requestor. This form has to be hand-written by the old method. Work sheets can be printed by both procedures but there is no time saving due to the LIMS.

Table 4.2
Evaluation of the productivity of LIMS

Task	**Time Taken**	
	Old Method	*LIMS*
Sample Administration		
1. Define protocol	NA	0.5 h
2. Label tubes	4.0 h	2.0 h
3. Sample log-in	2.0 h	2.0 h
Analysis		
4. Print worksheets	0.5 h	0.5 h
5. Analysis of samples	10 D	10 D
Calculation of Results		
6. Interpretation of chromatograms	4.0 h	1.0 h
7. Transfer peak area ratio to computer	4.0 h	1.5 h
8. Interpretation and calculation	NA	2.0 h
9. Transcription checking	8.0 h	NA
Reporting Results		
10. Preparation of analytical report	2.0 D	2.0 h
11. Typing manuscript	2-10 D	NA
12. Transfer to word processing system	NA	1.0 h
13. Transcription checking	1-2 D	0.5 h

NA = Not applicable; Assumes 7 h working day

As can be see from Table 4.2 there is no difference in the time taken from the analysis of the samples by either method, demonstrating that a LIMS has no direct effect on the analytical procedures of a laboratory. The HPLC detector output in each case is fed into an integrator that stores it digitally on a floppy disk as both the raw and the reduced (peak areas or heights of each peak) data as well as printed output. Before LIMS, the analyst would transcribe the drug/internal standard peak ratios onto a sheet then enter these into a computer to construct a standard curve, and then calculate the analyte concentration in the samples. Transcription error checking must be undertaken throughout this process. In contrast, the reduced data files held on disk are transferred to the LIMS once the chromatograms are approved by the analyst. Programs in the LIMS display these files sequentially on the screen for the analyst to interpret. When complete the standard curve is displayed as shown in Figures 4.4 and 4.5, and then when acceptable the concentration in the unknown samples is calculated. At no time does the analyst enter any of the data to be manipulated or calculate any results; all keyboard entries by the scientific staff are concerned with the interpretation of the results. All calculations are undertaken by the computer. Thus, using validated programs, the emphasis shifts from transcription error checking to confirming the interpretation. The main gain is the virtual elimination of transcription error checking when calculating the results. Moreover, the tasks comprising this operation can easily be accommodated whilst samples are being analysed rather than waiting for the completion of all the assays.

Reporting results is an area where a LIMS excels: the time taken by the old method to prepare the report was about two weeks, depending on the workload of the typist. This can now be reduced to a day or less using the LIMS. The reason for this is the analyst can prepare the whole report using predefined forms which search the database and write the information to disk files for transfer to the word processing system. Instead of waiting for someone else to type the report, this task is devolved to the analyst, and knowing that the data are correct, he simply completes a few menus. There is no competition for the secretarial service and reporting is completed rapidly and efficiently. The time saved allows the analyst to do more productive tasks rather than checking manuscripts for typographical errors. The time taken with typing the report is that which has elapsed from when the manuscript is delivered to the typist to when it is returned to the analyst. This time is dependent on the workload of the typist and the priority of the report. Furthermore, it must be checked for errors and usually be returned for correction. The error checking that is undertaken with the LIMS is concerned with the format of the report and the correct presentation of the results.

4.2.5 Laboratory Management

The efficient use of laboratory resources is one way of judging the performance of an analytical laboratory. This should be covered under productivity benefits but because much needs to be done in this area it is prudent to discuss this under a separate heading to highlight the problem. The present capabilities of any LIMS are such that only rudimentary management and planning tasks are possible at present. Examples of this are simple calculations such as how many samples of a particular type were assayed in a month, or the calculation of a customer's analysis bills. These are of use as they free the analyst from the task of calculating the figures manually. For instance, laboratory supervisors can receive accurate up-to-date information about workload status and backlogs of samples currently undergoing analysis in the laboratory (6). Response to, or even anticipation of, events is possible to prevent backlogs of samples or to clear backlogs rapidly. Exception reporting, whereby only unusual actions that occurred in a laboratory are noted, is possible so that managerial time is spent only on those matters of current concern (11).

However, it must be appreciated that these are not the only tasks facing laboratory managers. They are also concerned with problems such as organizing future work schedules, planning how to use staff effectively and justifying resources. Planning future demand for services can be difficult and can only be based on the information available at the time a decision is made. This presents a dilemma: retrospective searches of the database are very effective and provide historical data easily. However, prospective extrapolation of data or calculating how the future workloads will affect resources is virtually non-existent with the present generation of LIMS. One method of overcoming the problem is to transfer a file resulting from a search of the database to a microcomputer for further processing by a spreadsheet program such as Lotus 1,2,3 (12). This involves the further purchase of a PC when a LIMS has already been installed, but may be the only practical means at present of gaining some limited management information.

How is this problem to be overcome? This depends upon the attitudes of both the LIMS suppliers and customers; if the suppliers do not perceive the need for these facilities they will not be added to the repertoire of LIMS. Therefore, it behoves the customers to demand these facilities of their supplier. But the main problem is that currently the customer is concerned about managing data and not the laboratory. A shift in attitudes and demands of the user is essential if these functions are to be added. The question then becomes how they should be implemented: should development be along the lines of the PC as described above or along alternative paths? One such has been suggested by Braithwaite (13) and it is the development of LIMS into an expert system able to help make decisions based on the current status of the laboratory.

The questions that would demonstrate a *true* Laboratory Information Management System might be, 'How will the laboratory cope with a new project?' or 'What resources do I need for the 3000 samples for the analysis of X that will be submitted during the next three months?' The LIMS should be able to calculate an answer for the manager to base a decision from the known workload and the current throughput of the assay in question. Priorities may also play a part in the process and a number of scenarios may be given. Managers would be able to plan effectively by using this information.

4.3 Constraints of a LIMS

When considering the benefits of a LIMS the potential user should also think of the constraints imposed by such a system. These are areas where care should be taken in design and evaluation of software. Although these areas will not be covered at length in this section of the book, it is important that they are given virtually the same amount of thought as the benefits in order to avoid rejection of the system by the users.

4.3.1 Error Correction

The majority of LIMS have a major disadvantage: error correction. This has been forced on suppliers through the need to make the system secure against tampering from unauthorized users. This aspect should be borne in mind when either planning or using a LIMS. According to Dessy (14) users should be able to remove themselves from situations they may have created that are incorrect, such as when entering the data, and there should be an easy mechanism to allow the user to re-enter the data. Once a result or piece of information has entered the database the users should specify a way of correcting the information/data before it is verified and cannot be changed.

4.3.2 Flexibility and Workability

The majority of LIMS publications stress the need for maximum flexibility so that the system can cope with the addition of new methods of working or analytical techniques that may arise in the future.

Software works well in routine situations; it is the unusual request or the procedure that has not been thought of that causes problems. Normally, a LIMS tends to be less flexible than the manual system that it replaced, which means that quite often work must be adapted to the system rather than the other way round. This can lead to problems and can cause frustration with the users. Even in the best designed systems this problem will inevitably occur; when it does there should be a relatively rapid procedure for overcoming it with as little disruption as possible.

If, in designing the system, the maximum flexibility is incorporated into the programs, it could lead to software that is cumbersome and slow to use and might be very expensive to write. Therefore, some compromises might be necessary.

4.3.3 Reliance on a LIMS

Once users are working with the LIMS they become reliant on it for their working, and if any component of the system is unreliable or malfunctions it can result in much lost work and user dissatisfaction. Great care should be taken in ensuring the reliability of the LIMS, outlined in the next two chapters are Graham Murkitt's suggestions to ensure the reliability of a system.

4.4 Justification of a LIMS

It is inevitable that the expenditure on a LIMS will have to be fully justified to a company's senior management in terms of the cost of installing and running the system weighed against the proposed benefits (see Chapter 2 by Peter Johnson). The purpose of justification is to demonstrate that the project has been well researched and has a reasonable chance of meeting the objectives set for it. Moreover, it is a means of avoiding spending large sums of money without an adequate return; these systems must justify themselves otherwise inefficient staff could remain inefficient. Individual corporations and organizations will differ in their approach, however, these are general guidelines that can be used and adapted where appropriate.

4.4.1 Systems Costs

The costs of the system purchase are relatively easy to calculate from the sum of the hardware and peripheral equipment together with that for the software component. Often one supplier will tender a complete package whereby there is one contact point for all the components of the LIMS; this is preferable and avoids the situation of one supplier blaming another for a problem.

The purchaser should also obtain quotations for maintenance of the equipment and the software. This will be a recurring expense over the life of the project and is dependent on factors such as the type of hardware used and the total cost of software purchased. The hardware maintenance contract has several alternatives, the first, and least expensive, is to have no contract. The user will telephone the service department and they will send an engineer when one is available, there will be no priority placed on this call and more important customers will have the use of the engineer first in this situation.

Given the reliance placed on a LIMS it is essential that a maintenance contract is taken out. This will give the user a guaranteed response time by which there will be an engineer on-site; the fault may not be fixed, but there will be someone in attendance. The costs of these contracts rise as the response time falls and care should be taken in their selection. For instance, if the computer captures chromatographic data, failure means that all data acquisition stops, therefore a short response contract is mandatory or there would be serious loss of throughput. However, with a LIMS where data capture is undertaken by distributed processors then a longer response contract would be acceptable. This is because acquisition could still be taking place while the computer was being repaired and data could be transferred later when the LIMS was operational again. A software maintenance contract to cover the cost of advice, solve problems and obtain upgrades is also advisable.

There may be additional costs, explained more fully in Chapter 12, to cover the expense of enhancing the software or to purchase extra hardware, mainly communications, memory or storage devices in later years. This does not usually form part of the equation for the formal justification of the system but should be remembered during the planning and budgeting stages of the project.

There will inevitably be the costs of a support team to run and maintain the system. At the very least this will be a part-time system manager whose duties will be outlined in the next chapter and may include a deputy to assist him in this task. These costs will recur during the lifetime of the project and will inevitably increase as they are direct labour costs.

4.4.2 Costing Existing Procedures

The next stage is to cost the existing procedures or working practice. This will usually be to determine how much time it takes to perform any particular task under normal circumstances. This is then converted into the cost of the operation using a value to represent the costs of salaries and other overheads for the laboratory.

4.4.3 Estimating Cost Savings of a LIMS

Areas where the LIMS will operate and improve laboratory operations and tasks should now be investigated to see how much time and money could be saved if the system were installed. If the proposed system is one of several within the organization, then examination of the records of these systems and their advantages will help enormously. However, if there are no others, then this process may be little more than guesswork. Many of the benefits of a LIMS have been listed in Section 4.2, however, there will be other areas unique to individual circumstances also for consideration. It is usually the areas of productivity that produce the greatest savings due to the system. This can be seen clearly in Section 4.2.4.

Sometimes staff saving may be an argument for installing a LIMS, but it is rare that substantial savings will be made as these will be offset by the need to recruit and train staff to support the system. It is safer to say that once the computer system has the resources to function effectively then it will be capable of absorbing more work without increases in staff being required.

The labour costs to calculate the savings will vary from country to country. For instance, in a country with very high labour costs it will be easy to justify a LIMS on the basis of productivity increases of about 5%. However, with very low labour costs it may not be possible to demonstrate any savings from the proposed purchase. Other factors should also be considered.

4.4.4 Intangible Benefits of a LIMS

The benefits of LIMS discussed above could have a monetary value placed on them by virtue of their increased efficiency (a so-called *tangible benefit*). There are many LIMS benefits that are difficult to equate to saving costs of time; these are the *intangible benefits*. They include data integrity, better reliability of results, release of results allowing decisions to be made sooner, increased accuracy, improved handling of data, and better motivation of staff by the elimination of tedious and dull work. These benefits are part of the system and part of the overall payback to the company. Whilst they will not contribute to any saving they might slow or reverse staff turnover and prevent the loss of skilled personnel from the laboratory, thus avoiding the hiring and training of their successors. At the very least, these areas should not be omitted from the justification simply because a value cannot be calculated for them.

However, as will be shown in Section 4.4.7 on economic analysis, it is essential in some instances to attempt to put a monetary value on these benefits to help the financial justification of the system.

4.4.5 Cost-benefit Analysis

Cost-benefit is a method of evaluating how productive a LIMS or any other project involving capital expenditure will be. The cost-benefit ratio is calculated simply as:

cost-benefit ratio = value of benefits value of all costs

A project should be cost effective as long as the total benefits exceed the total cost. Therefore, a ratio in excess of one shows the benefits outweigh the costs. These calculations can show the pay-back period, the time taken to recoup the initial investment and the anticipated savings over the life-time of the project. However, the approach is useful for simple projects but may be difficult to relate to a complex LIMS project with many goals. Therefore, it is important that the objectives of the system are well defined in order to gauge more accurately their impact on the laboratory.

Superficially, the cost-benefit ratio appears to provide an indication of the productivity of an investment. This is only true when considering a single investment and a single set of returns. When a LIMS must bear both the initial capital investment and the annual operating costs, such as maintenance and operation, the cost-benefit ratio fails to provide a clear picture of the value of the project as it usually under-rates the productivity of a project with high annual costs (15).

4.4.6 Internal Rate of Return

This has been proposed as a better method of measuring the feasibility of capital projects: the higher the rate, the better the project (15). This is a comparison of what the money would earn in interest if not used to purchase the LIMS, against the net savings produced by the system. In essence, would the company gain by putting the money into a savings account instead of spending it? The problem with this method of analysis is that ambiguous values for the internal rate of return can be calculated. The rate of interest charged on the money used to purchase the system can lead to problems and different results calculated by the cost-benefits ratio method. Overall, this method is more suited for projects in a commercial environment rather than a scientific one.

4.4.7 Net Present Value

This method of justification, also called the discounted cash flow, is concerned with maximising the value of a LIMS as measured by a parameter called net present value (NPV). This is defined as :

present value of all benefits - present value of all costs = NPV

For operation under a fixed budget, NPV is probably the best criterion for the evaluation of projects (15). This method focuses on the benefits of the system and can be used to assess productivity. It is useful when projects are competing for a limited capital budget as this method provides an assessment of the viability of the project within the overall financial resource. An NPV of zero means that the project has reached a break-even point; a greater value shows that profit can be generated by the LIMS but a negative number infers that the system would not be cost-effective.

An illustration of the use of NPV has been given by Golden (10) and is shown in Table 4.3. The assumptions involved are that the LIMS costs $250K and the capital cost is depreciated over five years; the system is installed in a QC laboratory with 60 staff whose costs, with overheads, is $3,600K. Money is borrowed at a 13% interest rate and the company is in a 50% US tax bracket. It is estimated that the system will produce productivity savings of 5 and 10% in the first two years of operation and 15% thereafter. The effect of the LIMS on speedier testing, more rapid claims against suppliers of inadequate raw materials, lower stock levels and lower insurance rates will be $25K and $50K in the first two years and $100K thereafter. The operating costs of the system are represented by an annual maintenance contract of $25K per annum plus one member of staff working full-time on the system for the first year with an additional cost of $35K. Using this method, the cost of the LIMS can be justified within two years.

However, examine the assumptions used in the example: this is a large laboratory with 60 staff and high overheads but there is no provision, apart from the first year, for a systems manager (Chapter 5) to maintain and run the LIMS. Additionally, this assumes a quality control environment where substantial savings on operating costs are available as lower stocks can be held at the site. These savings would not be available if the LIMS were installed in a research and development laboratory. There has also been no attempt to link the costs with inflation.

A spreadsheet, such as Lotus 1,2,3, can easily set up the equations used in Table 4.3 and be used to examine some of the assumptions used in the above example. No effect had been made for inflation, this has its main effects on the costs of the maintenance contract and the labour used to run the system. Inflation rates of 5 and 10% per annum had a minimal effect on the justification of the system. Slightly more effect was noticed when an additional cost of $35K was allotted for years 2-5 for a systems manager; even if this salary were adjusted for 10% inflation throughout the life of the project, this did not alter the payback of the LIMS from the second year. Removal of the revenue gains, which would not be available to a research and development laboratory, also had little effect on the overall outcome.

Table 4.3

Illustration of net present value for the economic justification of a LIMS

Cash Flow Item	Year 1	Year 2	Year 3	Year 4	Year 5
Cash Outflows					
Depreciation of capital	50	50	50	50	50
Interest on capital	32.5	26	19.5	13	6.5
Operating costs: labour	35	36.8	38.6	40.5	42.5
and maintenance	25	26.3	27.6	28.9	30.4
Tax on Revenue	12.5	25	50	50	50
Total cash outflow	155.0	164.0	185.7	182.5	179.4
Present Value Factor	0.885	0.783	0.693	0.613	0.543
Cash outflow present value	137.2	128.4	128.7	111.9	97.4
Cash Inflows					
Tax credit	41.3	13	9.8	6.5	3.3
Depreciation	47.5	47.5	47.5	47.5	47.5
Productivity	180	360	540	540	540
Revenue Gain	25	50	100	100	100
Total cash inflow	293.8	470.5	697.3	694.0	690.8
Present Value Factor	0.885	0.783	0.693	0.613	0.543
Cash inflow present value	260.0	368.5	483.2	425.6	374.9
Net Present Value	122.8	240.0	354.6	313.7	277.5
Cumulative NPV	122.8	362.8	717.4	1031.1	1308.6

Formulae:

Cash Flow Present Value for year n = Cash Flow (year n) x Present Value Interest Factor.

Present Value Interest Factor (PVIF) = 1/(1+I)n where n=year number and I=Interest.

Net Present Value (NPV) = Present Value Cash Inflow - Present Value Cash Outflow

Quoted from Golden(10)

This leaves the productivity savings to be examined: it is these large figures that enable the rapid payback of the system. If productivity was only 5% for each of the five years of the project, payback does not occur until the fourth year. If there were no productivity savings, the system would not be cost-effective in a research environment but would be in a QC area by virtue of the revenue gains.

If this example were a company in the United Kingdom, the tax rate is only 35% as opposed to 50% in the example quoted. If the LIMS is in an R&D environment then a tax credit for the whole amount is given in the first year. This is in contrast to a QC laboratory, where the company obtains a 25% credit of the full amount in the first year and 25% of the reduced balance in remaining years; under this system there is 1% of the original sum left after 17 years.

The above examples are for a large laboratory with high overheads where a LIMS would be relatively easy to justify. However, smaller laboratories or those with low overheads need to justify a LIMS with great care. As discussed above the intangible benefits of LIMS should have a monetary value placed on them. Wherever possible, these savings should be calculated, if they are estimated then it is better to be prudent and take a low value, otherwise expectation of the system may be over-estimated. As an example, the reduction in transcription error checking could be costed to show the value of a data integrity benefit.

4.5 Summary

The aim of this chapter has been to give the reader an overview of the benefits that are possible following the installation of a LIMS, these benefits are not claimed to be comprehensive as individual laboratories may configure or tailor a system to their own needs. Coupled with this has been an approach to the methods used to justify the purchase of a system. This should be useful in allowing scientific staff to anticipate the methods used to assess the viability of a LIMS project and to prepare their justification accordingly. For large projects of this nature care and planning is essential. Part of this process is manifested in the care required to justify the system, for without financial approval there will be no LIMS project.

Acknowledgement

The author wishes to thank Keith Button for the financial analyses in Section 4.4.7.

4.6 References

1. McLafferty, F.W., 'Trends in Analytical Instrumentation', *Science*, 226 (1984) 251-253

2. Berthrong, P.G., 'Computerization in a Pharmaceutical QC Laboratory', *American Laboratory*, 16 (1984) (February) 118-126

3. Berthrong, P.G. and Schaeffer, B.C., 'One Quality Control Lab's Experience with Automation and Data Management', in *Automation of Pharmaceutical Operations*, ed Fraade, D.J., (1983) pp137-142, Pharmaceutical Technology Publications, Springfield, Orgeon, USA

4. Gormley, G. and Berthrong, P.G., 'Automation of a Pharmaceutical QC Laboratory', *Computer Applications in the Laboratory*, 5 (1985) 282-285

5. Merrier, R.J. and Thompson, W.B., 'Information Management in the Undergraduate Instrumental Laboratory', *American Laboratory*, 15 (1983) (September) 56-66

6. Henderson, A.D., 'Beckman Software System at Smith Kline and French', *Analytical Proceedings*, 24 (1987) 112-114

7. McDowall, R.D., Pearce, J.C. and Murkitt, G.S., 'Laboratory Information Management Systems Part 2 Implementation', *Journal of Pharmaceutical and Biomedical Analysis*, in press

8. Kikkert, J.N. and Hanson, N., 'Quality Control in the Computerised Mineral, Metals and Chemical Laboratory', *37th Pittsburgh Conference Proceedings*, (1987) Abstract 1074

9. Reber, S., 'Laboratory Information Management Systems', *American Laboratory*, 15 (1983) (September) 78-85

10. Golden J.H., 'Economics of Laboratory Information Managements Systems', *Intelligent Instruments and Computers*, 3 (1985) (4) 13-18

11. Gibbon, G.A., 'The Impact of Laboratory Information Management Systems on Quality and Productivity', *37th Pittsburgh Conference Proceedings*, (1987) Abstract 933

12. Gillyon, E.C.P. and Horslen, N.C., 'Use of a Laboratory Information Management System (LIMS) in Conjunction with a Personal Computer running an Integrated Spreadsheet Package', *5th Pharmaceutical Technology Conference*, Harrogate, Yorks. Volume 2, 9th April (1986) pp370-378

13. Braithwaite, A., 'Decision Criteria for LIMS', *Laboratory Practice*, 35 (1986) 30-38

14. Dessy, R.E., 'Laboratory Information Management Systems Part 1', *Analytical Chemistry*, 55 (1983) 70A-79A

15. de Neufville, R and Stafford, J.H., 'Evaluation of Projects', in *Systems Analysis for Engineers and Managers*, 2nd edition, (1974), McGraw Hill, London pp159-183

Chapter 5

The LIMS Organization

G.S.Murkitt

5.1 Introduction

To ensure the successful installation and efficient running of a LIMS, an organization within the laboratory to support and maintain it is essential. This is because once operational, great reliance will be placed on the effective operation of the LIMS and therefore, the system itself must be highly reliable. A support organization will maintain the momentum and enthusiasm of the project through the implementation phase and into maturity.

This chapter contains the practical and relevant information on setting up an organization to get the maximum performance from a LIMS; it is based on experiences within a pharmaceutical bio-analysis laboratory. The areas discussed here are: the need for and role of a system manager, the essentials of user documentation, the training of users, some maintenance considerations, and the setting up of a LIMS user group. The information in this chapter should also be used in conjunction with that in the next, concerning implementation.

5.2 The LIMS System Manager

5.2.1 The Need for a System Manager

The person in charge of the LIMS is usually known as the system manager. It is imperative that this person be appointed as soon as possible after the decision to acquire a LIMS has been made. Ideally, he should be involved with the project from the start, being a thread of continuity from inception, via installation and validation to routine operational use. The LIMS manager is instrumental in ensuring the success of the system. To deny the system a manager is condemning it to failure.

The justification for a manager should be made at the same time as that for the system itself. This is obviously an additional cost to be taken into account, but the major reason is to ensure the efficient use of resources - allowing the users to use their LIMS.

The number and structure of LIMS support personnel will be dependent on many factors, such as the size and complexity of the computer system, the number of users, the importance of the work being conducted and the time when the system is being used, ie. normal working hours or round the clock operation. Even the simplest of systems should have a system manager. In these instances, the amount of work may not merit the role being full-time; possibly the person may have other responsibilities. Nevertheless there must be a person responsible for the maintenance of the LIMS to ensure that it operates at maximum efficiency.

The following attributes are important when considering the appointment of a system manager:

1. It is essential that the person is keen on computing, with a technical understanding of basic operating system commands.

2. The ability to write applications procedures and programs for the LIMS will circumvent the slow and expensive alternative of requesting such a change if the system has been purchased.

3. It is also of great importance that the system manager can understand the users' requirements and working practices; a scientist appointed from within the laboratory may be a solution.

4. The manager will need to converse with a wide range of people: he will need a friendly persuasion, realizing and accepting that criticism raised is directed at the system, and not pointed at them personally. To be effective the manager will require good interpersonal skills, one minute being firm with the supplier, yet the next being understanding to the users.

There will often be occasions when the manager is not present, for instance meetings, holidays or sickness. The appointment and training of a deputy is therefore an important consideration to provide cover for absence or helping at times of high demand. It is also worth noting that although all staff should be replaceable, the loss of the LIMS manager is not an enviable situation.

5.2.2 The Functions of a System Manager

The functions of the system manager are many and varied; they are summarized in Table 5.1.

In the author's opinion, the implementation period is the system manager's most intensive time. Upon receipt of the system, he or she should be expected to validate it by performing the appropriate tests to show the LIMS does what it is specified to do. This means devising tests, documenting both the anticipated and the actual test results and explaining any deviations. The execution of such validation protocols during acceptance testing should be the responsibility of the system manager as they can be quite complex. It follows that the running of re-validation protocols, ie. after system upgrades and bug fixes, is also an important function of this post.

Table 5.1

Some functions of a LIMS system manager

Documentation
- Writing of system procedures
- User documentation and its upkeep (new revisions, etc.)
- Acceptance testing and initial validation of the system
- Revalidation of software bug fixes
- Maintenance of training records
- Periodic revalidation of the whole system

Education and Training
- Education of users
- Training and retraining users
- Informing users of new upgrades, problems, etc.
- Member of the LIMS user group

Computer Housekeeping
- Troubleshooting and documenting problems
- Liaison with supplier and computer department
- Archive and retrieval of data
- Removal of unwanted files from system
- Deletion of unwanted data from database

The writing and upkeep of the system documentation, user training, basic system housekeeping and information dissemination are prime areas of responsibility for the LIMS manager and are discussed in greater detail throughout the rest of this chapter.

5.3 Documentation

Any procedure undertaken on the LIMS should be documented and this is presented fully in Chapter 25. The area discussed now is that of user documentation - how to use the LIMS software.

The LIMS will be new to all users and it is a daunting task to wade through the system unaided. Documentation eases this situation and helps overcome any initial fears about the LIMS. It is also a useful adjunct to training, for referral by new users when unfamiliar with an action and by experienced users for a specific task. Documentation is also an essential element of good laboratory practice (GLP).

The documentation of the LIMS software should be readable and aimed at the first time user. If a commercial system has been purchased it will come with instructions, however, these are rarely applicable to an individual situation especially if it has been customized. Therefore, the task of writing your own user documentation should be considered. To help, two useful guides to writing computer documentation have been published (1,2). Documentation in the author's laboratory is modular and self-contained, with standard abbreviations and style used throughout; each section is colour coded for easy reference, see Table 5.2. The aim is to produce material which can be referred to when appropriate, be understood and useful.

Table 5.2

Modular approach to a LIMS user guide

	Colour Coding
Introduction to LIMS	White/Black
LIMS Data Set Maintenance	White/Red
1. Protocol Management	Red/Black
1.1 Protocol entry	
1.2 Protocol modify	
1.3 Append to incomplete protocol	
1.4 Protocol display	
1.5 Protocol signoff	
2. Sample Receipt	Blue/Black
2.1 Barcode entry	
2.2 Manual entry	
3 Worksheet Generation	Beige/Black
5 Calibration and Results Logging	Yellow/Black
5.1 Data transfer from integrators to LIMS	
5.2 Results interpretation	
5.3 Balance results logging	
5.4 Manual results entry	
6. Results Validation	Green/Black
6.1 Quality control sample	
6.2 Calibration standards	
6.3 Subject samples	
7. Results Collation and Reporting	Grey/Black
7.1 Graphical overlays	
7.2 Report generation	

As the LIMS expands or new software upgrades become available then re-training or updating staff will be necessary. Additionally, the LIMS documentation should also be updated to reflect these changes. The modular nature of our documentation means that only those parts affected by change need to be replaced.

5.4 Education and Training

5.4.1 The Human Aspect

LIMS are complex systems which will only succeed if people want them to succeed.

A major reason that a LIMS can fail is because of people.

The implementation of a LIMS could quite naturally be met with resistance, due to the dislike of change and fears of an unknown way of working. The users should be identified and the effect of the system upon them examined. The successful implementation of a LIMS is dependent on winning the 'hearts and minds' of the users. There are major obstacles in the path to success but these can be overcome by education and training.

5.4.2 Education

Education of the users starts at the writing of the requirements specification and continues throughout the lifetime of the system. Informing the users of the potential gains in using the system should be the first step in this process; it should however, also include a realistic appraisal of the limitations of the system. Fears and barriers can be overcome by holding meetings in which the purpose of the LIMS can be explained with the advantages and disadvantages presented and discussed. This resultant increase in awareness can be further enhanced by encouraging people to become involved. Any suggestions forthcoming may have the added advantage of being useful as it is the users that will make the system a success or failure. It is also important to present the time-table for implementation so that users can see when they will be affected.

There may be apprehension at the prospect of using the new technology and the long learning curve which is necessary before they can reap the benefits of the system. This initial period is likely to be the most difficult and tense.

5.4.3 Training

Before the arrival of the system many commercial vendors run training courses for selected personnel and it is essential that both the system manager and deputy attend these. The purpose of this training is to give the system manager a good grounding in the essential operation of the system.

Initial impressions of a LIMS are very important, it is therefore imperative that the users are not exposed to an unready system. It is the author's opinion that the LIMS must not be released for general use until the major software bugs encountered have been removed, also, where necessary, a scheme to avoid problems already encountered should have been devised pending enhancements to solve the situation.

The best person to train the users is the systems manager, as this person will have the best understanding of the system. Training passed down the line, ie. from one person to another should be avoided. This is due to the complexity of the system, and hence the likelihood that a small piece of relevant information may be omitted or false impressions passed on, leading to unnecessary irritation and inconvenience.

It is not possible to train all potential users at once. First, it would result in poor and inefficient training due to the high trainee to trainer ratio, and second the swamping of the system by inexperienced users will undoubtedly lead to problems and the loss of system credibility. The training of an 'elite' has the advantages of creating an enhanced interest, whilst at the same time separating the users by a learning gap. If problems arise for the initial group it is possible to address these quickly and so avoid the exposure of many groups to the same problems.

Like all computer systems, the best way to learn how to use a LIMS is with 'hands-on experience'. Small, rather than large, groups of people should be trained, this allows individual tuition, if required. Well-worn paths should be demonstrated to ensure success, this may then be followed by working on schemes that mimic the trainees' intended work.

Training should take place at an allotted time and not fitted in around the routine work. This is essential, as otherwise it places the analytical staff in a position of learning new technology whilst still being expected to produce work on time, with the net result often being that both areas suffer.

Training should not be considered as a once only exercise. Retraining and updating will be necessary, especially when enhancements and software revisions are added to the system. This is particularly important in a system that is continually undergoing change.

5.5 Computer House-keeping and Problems

Equipment maintenance is an important consideration. GLP regulations require that all instrumentation used should be adequately inspected and maintained; in this respect a LIMS is no different from any other piece of equipment. However, there are other aspects of maintenance that are not appreciated until a LIMS has been installed. The purpose of this section is to highlight these less familiar aspects of computer housekeeping, so that these eventualities can be anticipated.

5.5.1 Database Back-up

This is a specialized job that should be left to the system manager or the computer department. Database back-up has two aspects for consideration, the first is the routine back-up for security purposes in case of disaster. These back-ups must be undertaken on a regular basis as the database can often be considered as raw data (3); a schedule for back-up can be seen in Chapter 7.

The second aspect is the archival of data that is not required on-line but must be retained for regulatory or historical reasons. Scientists should be contacted at regular intervals to ascertain if there is any data to be archived. This practice removes unwanted data from the system and prevents it from becoming full of unused material. The resulting tapes should be numbered uniquely and stored separately from routine back-up tapes to prevent any confusion.

5.5.2 File Clean-up and Disk Re-initialization

Regular tending of the system is a major function: old records should be deleted from the database once they are archived and the user accounts must be checked to ensure that files no longer used or required are removed. These tasks ensure that the computer can operate efficiently. If this is not undertaken the disks will become full and the system will grind to a halt. As a consequence of a routine file deletion schedule, the disks will become fragmented and it will be necessary to re-initialize the disk to compact the programs and files retained. A major problem with disk fragmentation is that the largest area or extent that is available for loading a program or file gradually diminishes until it is impossible to do this. Therefore, regular re-initialization of the disks will be required to prevent this problem. The intervals at which this is required will vary from system to system.

Complete? [ ]

Department of Drug Analysis SK&F Welwyn LIMS Log

Number [ ] Date [ ] Notifying analyst [ ]

Software module [ ]

Description of fault or problem

Action taken

Resolved OR Transfer to PE

Date By Whom SIR No [ ]

Figure 5.1 Problem Log

5.5.3 Problems

There should be a co-ordinated procedure for notification of system problems. These can include equipment malfunction, software problems and requests for enhancement. In the author's laboratory, as well as verbal communication, there is a standard form available for staff to notify the system manager of any problems with the LIMS (Figure 5.1). The form ensures that the situation is documented for both internal and external (ie. regulatory agencies) inspection, and makes it easy to identify faults and remedy them. The form is in three parts, one is returned to the user to show what action has been taken, another is passed to the computer department for information on how the system is operating and the last is retained by the system manager. Over a period of time this can reveal problem areas of the LIMS.

When any problem is passed to the supplier for solution, especially one involving software, an additional form (Figure 5.2) is used, to check that the problem has been successfully remedied. This is because software is not returned to its original state when a problem has been solved but is changed to a new state and revalidation will be required to show that the original problem has been fixed. This revalidation form outlines the tests to be undertaken to show the solution works, along with any expected test results. These are then compared to the actual results produced by the LIMS and when fixed the problem is signed off. In this way a complete appraisal documenting the history of the problem is available for inspection.

Department of Drug Analysis SK&F Welwyn
Re-validation Log

Number [ ] P-E SIR No [ ] Date of revalidation [ ]

Outline of test for revalidation

Test results (attach documentary evidence, if appropriate, to top copy)

Test Acceptable YES/NO Refer back to P-E YES/NO

Analyst ____________

Figure 5.2 Revalidation Form

5.6 LIMS User Groups

Involvement with both internal and external LIMS user groups is extremely important and useful.

5.6.1 Departmental LIMS User Group

The author's departmental LIMS user group is composed of system manager and deputy, along with representatives of the users and the laboratory management. It meets monthly and a typical agenda is shown in Table 5.3. In essence, they are formal meetings of interested parties with minutes kept as a record. The aim of these meetings is the improvement and expansion of the system, disseminating relevant information and canvassing opinion.

Table 5.3

A typical LIMS user group agenda

1. Review of last minutes
2. Action points
3. Review of outstanding problems
   a) New problems since last meeting
   b) Progress on existing problems
4. Enhancements
   a) Those requested since last meeting
   b) Progress on existing enhancements
5. Status of documentation
6. Users' forum
7. Any other business
8. Date of the next meeting

By the means of this forum, there is a method of channelling information and views between the users, the system manager and management. All users can be informed of potential problems, and new procedures can be explained and discussed. The forum also allows for updating users on the progress in solving problems they have encountered and improvements requested. Any problems or concerns of the users can be raised and recorded. All decisions affecting the system can be made with the confidence that the users' needs and comments have been taken into account.

5.6.2 External LIMS User Groups

Similar benefits are possible with a user group affiliated to one company's product. More informal in nature, these meetings can be held up to four times a year. They are a forum allowing fruitful discussions between users of similar systems and the vendor. The appreciation of mutual problems, and awareness of new features are the major benefits of attending such meetings. The requests for new options are strengthened considerably when requested by many users. Long term development, both in the direction and enhancement of the system may be discussed. One manufacturer asks representatives to vote on a wish list of improvements, and the item with the most votes is implemented.

5.7 Summary

In order to obtain the maximum benefit from a LIMS, an organization to support and maintain the system is essential. The elements of this structure are a system manager, the efficient training of users, together with good user documentation and a procedure for the notification, documentation and solution of problems. The establishment of a departmental LIMS user group is beneficial for monitoring progress and for the future development of the system.

5.8 References

1. Weiss, E.H., *How to Write a Usable User Manual*, ISI Press, Philadelphia, (1985)

2. Brockman, R.J., *Writing Better Computer User Documentation, from Paper to On-line*, John Wiley and Sons, New York, (1985)

3. Taylor, D.W., 'Inspector of Computer-Supported Toxicological Data Submitted to the FDA', *Drug Information Journal*, 18 (1984) 189-194

Chapter 6

Implementation

G.S.Murkitt

6.1 Introduction

The purpose of this chapter is to help the staff and management of a laboratory contemplating the introduction of a LIMS, by discussing the pertinent areas and some approaches useful to ensure a successful implementation.

Implementation covers the period from the ordering of the system until its satisfactory operation, supported by the appropriate training and documentation. The implementation of a LIMS has the potential to be the most exciting and rewarding period of its lifespan, yet at the same time equally gruelling and harrowing. It is a major operation, which requires much planning and effort by the LIMS project team, discussed further in Chapter 12. Furthermore, the organization outlined in the last chapter will aid this process.

6.2 A Strategy for Success

Some factors for the success and failure of a LIMS are listed in Table 6.1. The major point is that the LIMS project team should be aware of the causes of failure and take active steps to avoid them, whilst incorporating as many as possible of the success factors into an implementation plan.

Two of the most important factors for success are that the project must be seen to be worthwhile by all concerned and that, as discussed in Chapter 2, there should be a strong management commitment to the proposed system. The aims of the LIMS should be achievable within a reasonable time-frame; this is especially important in a laboratory that is continually changing or evolving, as the requirements might have changed when the system is eventually operational. The system aims should be achieved by thorough attention to detail in the design stage which includes the incorporation of the needs and ideas of the users.

At the start of the project, the objectives of the system should be recorded as well as the criteria by which they will be judged. A continuing programme of education and training will help to ensure that the users are motivated to use the system. As operation of the system progresses, an expertise in the system should be built up that can be used to improve the programs and add different functions if required. Once the system has been installed it should be thoroughly tested and validated prior to its routine use, in order to avoid losing credibility with the users. Throughout the life of the project, tact and diplomacy will be needed to ensure that the LIMS does not become the centre of a power structure (1).

Table 6.1

Factors for the success and failure of a LIMS

Factors for Success

A worthwhile project.
Good management commitment.
A strong, technically competent project leader, respected by all.
Thorough attention to detail in the design of the system.
Thorough system testing and training/education of users.
Ego-less decisions and solutions that do not utilize the ability of a LIMS to a centre of power structures(1).
Analysis of the needs of users.
The development of expertise in the installed system for future expansion and development.
Do not be overambitious.
Record agreed objectives with predefined success criteria.

Factors for Failure

Complex or grandiose beginnings.
Approached in a sceptical 'evaluation' style.
The replacement of an existing system without adding new functions.
The solving of problems put off until later.
Inadequate support.
Users' expectations not met.
Non-adaption or inflexible working.

Grandiose ideas that are impossible to achieve and the company that wants the latest technological wonder are two major factors for failure. This is followed closely by a lack of adequate support, which may manifest itself in the political factions amongst users or the LIMS project team, or a lack of resources for the project. The inability to add further functions easily, or the inability to adapt to new work, allows the users to view the LIMS as inflexible, which can mean the loss of credibility. If the expectations of the users are not met this will be another reason for failure. It is important, therefore, that the education process, as outlined in Chapter 5, keeps expectations close to reality.

6.3 Planning for Implementation

6.3.1 Planning Considerations

The planning for the implementation of a LIMS should begin from the initial conception of the project. This means deciding in advance what needs to be achieved and then how it is to be done. Planning requires predicting how long it will take to achieve a certain task and considerations should include:

- Personnel needs.
- Resource management: both human and material.
- The hardware and software testing requirements.
- Requirements for interfacing with current computer systems.
- Changes to existing manual procedures.
- Identifing any jobs or tasks which will be superseded or created.
- Ensuring the motivation of users.
- Preparing the relevant documentation for users and operating staff.
- Purchasing and siting of peripherals and wiring.

Included in this plan will be the overall time-scale for the system development. Obviously for a package that requires little or no customization this will be relatively short compared to one that requires software to be written. The plan should include the delivery and staging of the computer hardware, and its environmental requirements.

Training of key personnel to use the system must also be covered. To be meaningful, this must be undertaken after the system has been installed at the users' site and acceptance tests begun. It is no use training potential users who then have to wait for their LIMS to be delivered so that they can reinforce their original training which they may by now have forgotten.

The key to successful planning is to reduce the implementation process into manageable tasks which can be easily defined and to which realistic time-scales can be assigned. Progress should be regularly reviewed and tasks rescheduled if necessary. All parties should be kept informed and regularly updated. It is crucial that the project team controls the implementation, rather than let the implementation control them.

6.3.2 Planning Aids

The various tasks that need to be performed during an implementation plan can be visualized by either a bar chart or a network diagram. A bar chart (Figure 6.1) consists simply of a horizontal time axis with the tasks listed vertically. A shaded box or an arrow is drawn on the chart for each task covering the time that each is scheduled to begin and end. The bar chart has the advantage that the time-scales of each activity are shown clearly but it suffers from the disadvantage that it does not show the inter-relationships and dependency of one task to another.

Task	Time Scale (weeks)													
	1	2	3	4	5	6	7	8	9	10	11	12	13	14
Hardware														
Site Meeting	↔													
Site Survey			↔											
Agree Inst. Date				↔										
Installation											↔	↔		
Cabling									↔	↔	↔			
Site Preparation							↔	↔	↔	↔				
Hardware Testing													↔	
Training Key Users										↔				
Training Key Users														↔

Figure 6.1 Implementation bar chart

To overcome this, network diagrams can be used as shown in Figure 6.2, although here the time-scale is not illustrated. Care must be taken before attempting the construction of a network, as the following definitions are necessary:

1. Each objective in the plan is specified as an event.

2. The time and resources needed so that progress can be made to the next event is known as an activity.

3. The longest time through the network is known as the critical path and is the rate limiting step for the project.

4. The excess time that can be added to any activity before it lies on the critical path is the float.

There are microcomputer programs available that use either method to plan a project, these have the advantage that information can be quickly updated, the time schedules recalculated or the graphs re-drawn relatively rapidly.

The important point is that the tasks, whichever way they are visualized, have the resources available when they need them so that they do not become rate limiting for the project as a whole.

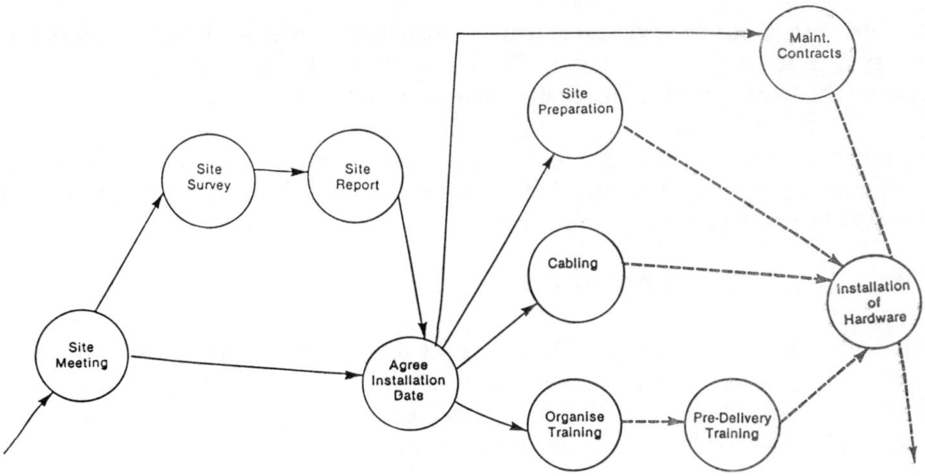

Figure 6.2 A Network diagram

6.3.3 Physical Aspects

Consideration should be given to the sites where the various pieces of hardware are to be located, and to any special requirements for those areas. The computer will generally be placed in a position along with other site computers in areas specifically designed for them. Such specifications of the computer site will include air-conditioning to control temperature and humidity, false flooring to allow for the extensive cabling, anti-static mats to prevent damage due to static electricity, uninterruptable power supply (UPS) to ensure continued operation in times of power cuts and surges, along with the security considerations, ie. limited access to the system console and the tapes maintained for historical and regulatory needs. Vendors will in most cases make an inspection of the site and advise on the requirements needed for their particular system.

The users will interact with the system through a variety of peripheral devices which will be accessed frequently during the working day. It is important that they are located near to the working environment, but these devices should be in a separate 'dry' area, to prevent the possible spread of biological or chemical contamination. However, some peripherals such as bar code readers or terminals attached to balances will have to be located in the 'wet' areas; all personnel should treat these devices as 'dirty' and take the appropriate precautions.

6.4 Styles of Implementation

Once the system has been validated and the documentation is available, the users can be exposed to the new system. There are primarily three ways this can be accomplished: total immersion, parallel operation, and selected use.

Regardless of which method is chosen, the learning curve is usually long, requiring much training and understanding, but the benefits of a well designed system should always outweigh the time spent on training.

6.4.1. Total Immersion

This is the introduction of the LIMS on a set date, with everybody using the new system, ie. a finish to the previous mode of working. It is a bold manner of implementation and requires 100% confidence gained from previously validating the LIMS, as well as total commitment on the behalf of the laboratory management and users.

This approach is merited when the new and old working practices differ radically. Implementation in this manner is best suited to periods of calm, allowing prompt attention to problems as and when they arise, with support staff to deal with them.

6.4.2 Parallel Operation

As the name suggests, this involves running both the old and new systems to allow cross-checking and confirmation of data. It should be borne in mind that this will only be applicable to certain tasks when the data from both systems are comparable. If the data do differ you have the added problem of resolving which is correct. The advantage of this approach is that if the LIMS has problems you can still produce results with the old procedures. Due to the burden on resources of this style of implementation, ie. duplicating the work, a time limit should be set after which only the LIMS should be used.

6.4.3 Selected Use

In this style of implementation, only a selected group will work on the LIMS. This cautious approach, nonetheless, is a real test of the system with the advantage of allowing prompt attention to any problems as they arise. If time and resources permit, known data can be fed into the new system and the results compared with the original to show the operation of the new system.

6.5 Post Implementation

Building credibility of the system in users' eyes involves providing simple services that work well initially with the more difficult tasks following later (1). Enhancements and upgrades are possible in an environment that is constructive and encourages suggestions. If the users do not demand change from day one then they are not accepting the system, they are putting up with it. Schneider (2) on the subject of implementation has said '...the project team must be highly visible and responsive to the requests and needs of the laboratory personnel during this time ...then...be prepared for requests for new and increased functions for the LIMS. It is a natural outgrowth of the acceptance of the system by its users'.

6.6 Summary

The components of a successful implementation are: the formation of a project group responsible for the planning, monitoring and control of this phase of the LIMS, winning over the users by creating an environment conducive to open discussion, exchange of views and providing accurate and relevant information. The implementation should be managed, so that problems that appear are localized, minor and can be promptly remedied.

As credibility is most easily lost during this phase of a LIMS, the highly visible support of management throughout this time is vital to the success of the project.

6.7 References

1. Dessy, R.E., 'Managing the Electronic Laboratory Part 1', *Analytical Chemistry*, 56 (1984) 725A -731A

2. Dessy, R.E., 'Laboratory Information Management Systems Part 1', *Analytical Chemistry*, 55 (1983) 70A - 80A.

Chapter 7

Computer Hardware

M.Bertram

7.1 Introduction

The selection of computer hardware for a LIMS is something that cannot be treated in isolation. The constraints of software, the existing expertise of staff, any corporate computing policy and existing hardware, especially that which is to be connected to the LIMS, must be taken into account. This complicates the decision making process. This chapter concentrates on the selection criteria for hardware alone, rather than the interaction with other factors which tend to be dependent on individual organizations.

The problem of obsolescence also has to be considered. It is extremely frustrating to order hardware only to find a couple of months later that a new range of equipment has been announced that gives more power for the same price with reduced maintenance costs. There is no solution to this except to talk to sales people, existing users (especially large users and those associated with user groups) and to be aware of comments in the computer press. If the hardware supplier considers the potential customer to be a valued account then a non-disclosure briefing may be arranged where the future product directions will be discussed.

Commercial LIMS normally only run on one or two manufacturers' equipment. If an in-house development is planned then there are usually fewer constraints on the choice of the hardware.

7.2 Processors

7.2.1 Centralized Systems

The choice of a main processor vendor may revolve around hardware or operating systems considerations, corporate policy, or the availability of LIMS software running on the processor. Once this decision is made then a choice of the size of the processor has to be made, depending on price against performance.

The sizing of a machine depends on the volume and nature of the workload put upon it but the decision can only be made in the knowledge of how the processor works in the field. Advice should therefore be taken from both the hardware and/or LIMS supplier and from other users. The latter is important as some sales people do have a tendency to underspecify the size of a processor in order to gain the initial business, with the belief that if the machine is underpowered then the user is much more likely to upgrade it rather than discard it and start again from scratch.

The use of benchmarks is one way to try to assess machine performance, but it is extremely difficult, if not impossible, to simulate the complex workloads expected with a LIMS and the author is not aware of any successful attempts to run benchmarks in this context.

The question of premature obsolescence was mentioned in the introduction. This is especially a problem with processors and PCs where the machine life-cycles are becoming shorter and shorter, and a range of computers may only be marketed for one or two years.

Normally, you have to accept the inevitability that a system will have some downtime even if the maintenance and support is totally satisfactory. If only very small amounts of downtime are acceptable then special hardware configurations of normal equipment can be set up. In these systems there is normally resilience to the failure of one component. For example, the data disk drive may be duplicated and data written to both disks; this is called *shadowing* and is illustrated in Figure 7.1. If one disk fails then the system can continue working using the other disk.

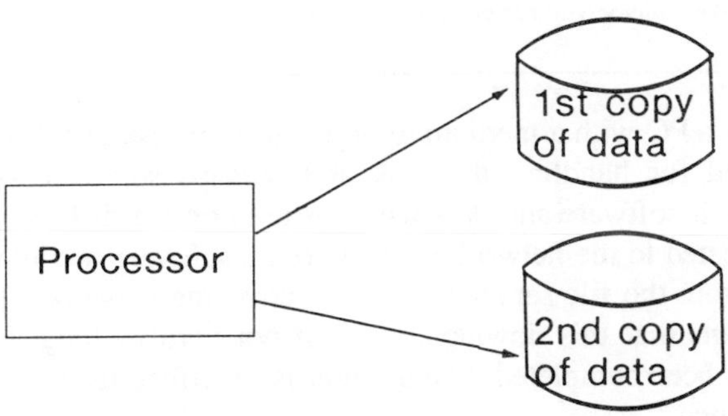

Figure 7.1 Hardware configuration for data shadowing

As an alternative, hardware can be purchased from a specialist supplier such as Tandem or Stratus. The equation to be considered here is the cost of the hardware against the benefit of the lowered downtime.

7.2.2 Networked Microcomputers

The selection of processors used to be only a question of choice of supplier and sizing but the development of networked microcomputer systems has added a new dimension to be considered.

Figure 7.2 illustrates the essential components of a networked microcomputer system. It consists of a local area network connecting a number of microcomputers (PCs). Local area networks (LANS) can be described as hardware and software for connecting information processing equipment in a limited geographic area (ie. an office, building, complex of buildings, or campus) using reliable high-speed communications. The configuration of this network may be in the form of a bus, a ring or a star, as illustrated in Figure 7.3

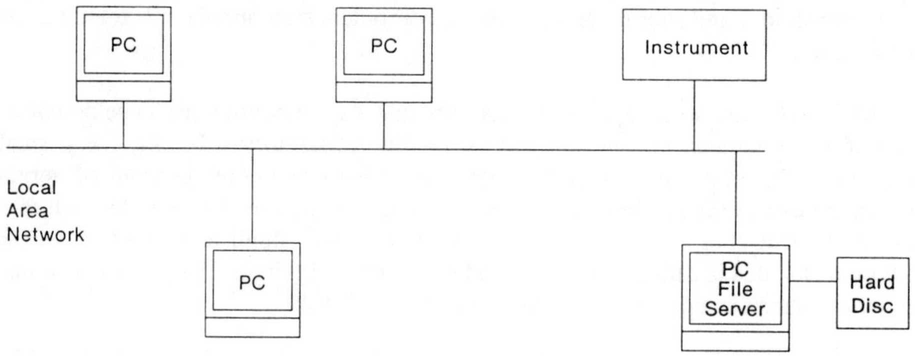

Figure 7.2. A networked microcomputer system

The file server is a PC with a medium to high capacity disk attached to it. This PC is often solely used for handling the disk and perhaps some back-up medium (eg. streamer tape). All software and data will be stored on the disk. When a user on one of the PCs connected to the network wants to run a software module or access data it is downloaded from the file server to the PC across the network. The high speed of the network means that this downloading will not involve long delays. Once this downloading has been completed then the user is, in effect, using the PC as if it were a stand-alone device.

Analytical instruments are connected to PCs for data capture; after capture the data is downloaded to the file server. Either each instrument can be connected to a separate PC, or hardware and software are available to enable a number of instruments to be connected to one PC. The rate of data generation must be taken into account when configuring the instrument to PC links.

The individual PCs may have printers connected directly to themselves or a printer server may be used. This allows a higher priced, high quality printer, often a laser printer, to be accessible by all users of the system. It often consists of a PC with a printer attached and the other PCs in the network send all their requests for printing to it.

There are a number of advantages to the networked microcomputer approach: good response times are possible with no variable time-sharing delays, as are to be expected with the traditional approach of a terminal connected to a multi-user central processor. These advantages are especially noticeable in graphics work.

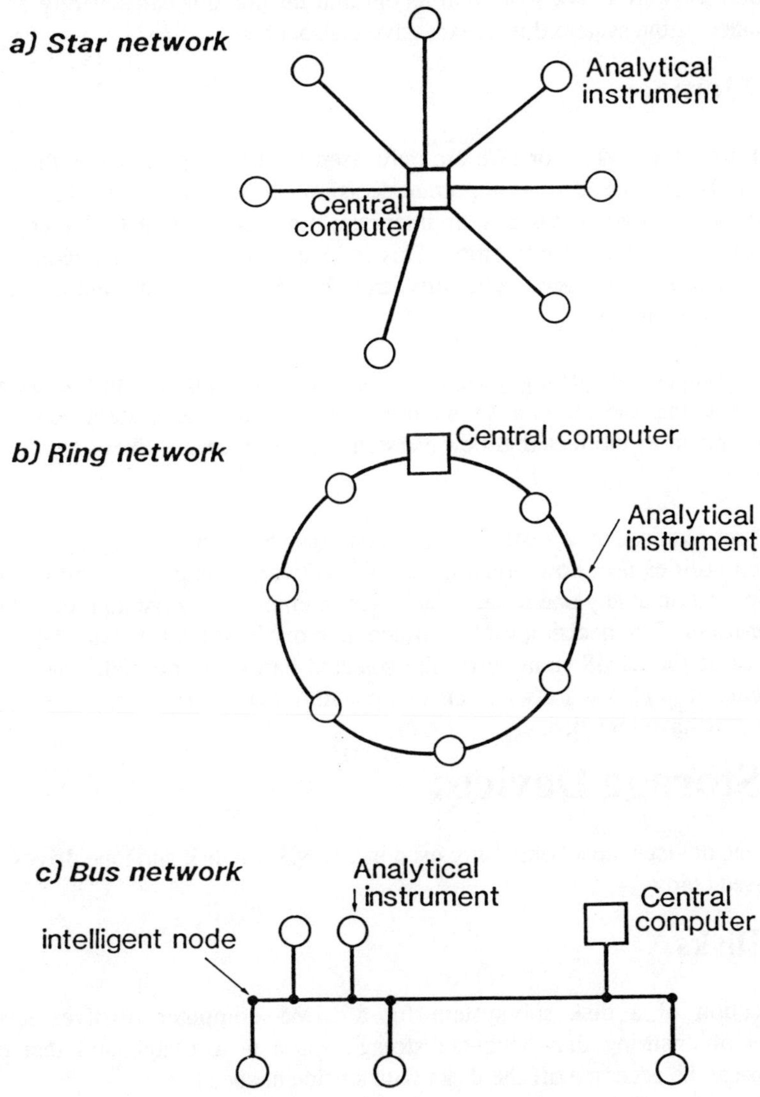

Figure 7.3. Local area network (LAN) topologies.

Growth is by the addition of extra PCs rather than by having to upgrade a central processor when it becomes overloaded. This means that the cost increments are much more gradual. The disadvantage of a network approach is that the disk capacities available are generally smaller than those available with a central processor system, and the rate that information can be accessed off the disks is also slower. This is especially important as the number of PCs attached to the network grows, ten PCs is often considered as a practical maximum number using one file server. If the application software is not written in an optimal fashion this can severely degrade the performance of the system due to excessive disk access.

7.2.3 Memory

Whether a central system or PCs are purchased for LIMS processing, the amount of memory to be purchased is an important decision. New generations of memory chips are appearing at two to three year intervals, providing a four-fold increase in the amount of data stored on one chip. This is leading to a continued reduction in the price of memory, increased reliability and the availability of greater amounts of memory for processors.

Software authors are writing programs that take advantage of this extra memory. When estimating the amount of memory required for the system some attention should be given to the fact that new versions of software are likely to require more memory.

The amount required by a system depends on how much the operating system and its associated utilities use, how much the LIMS software occupies and how many users will be logged on at any one time. Insufficient memory gives rise to reduced response times for users. The decision on how much memory is best left to somebody who has experience of the LIMS running on the selected hardware, normally the supplier of the system. It is also worthwhile checking with other LIMS users to see how much memory they have installed.

7.3 Storage Devices

The storage devices most commonly used in a LIMS are disk and tape drives and each is discussed below.

7.3.1 Disks

The selection of a disk sub-system for a LIMS computer revolves around the questions of ensuring that adequate storage space is available and that data and programs can be accessed off the disks with sufficient speed.

The amount of data held by the system will depend on the rate that data is captured by the system and how long it is retained on-line before it is archived and deleted. (The question of data retention policy is discussed by Don Mattes in Chapter 24) Additional storage capacity may be required for audit trails and transaction logs.

In addition to data, sufficient space is needed to hold the operating system and associated utilities (eg. compilers and editors) and the LIMS programs themselves. The data storage requirements for data and programs need to be combined, and a reasonable contingency added, to give the final requirement for presentation to the hardware supplier.

The maximum possible disk configuration for the hardware should also be obtained in terms of number of disks and their capacity. If this maximum is close to the final requirement then serious consideration should be given to other hardware configurations that allow more disk capacity. Do not believe a computer salesman who says extra capacity will be available in the future; it may not happen.

As well as having sufficient storage capacity, it is important that the disk sub-system is configured so that the speed that data can be written to and read from the disks (the I/O throughput) is adequate. This throughput is determined by the performance of the disk drives as well as the linkages between the central processor and the drives. The type of I/O workload also needs to be considered, for example, how does the accessing of operating system files interact with the data files and any transaction files? If the configuration is not optimum then the response seen by the user of the system may be limited by the speed that data can be written or read from the disks.

The normal way that I/O bottlenecks are overcome is by spreading the data files across a number of disks, for example, keeping the operating system files on a separate disk from data and applications programs. The question of disk sub-system configuration is a complex one and the advice of the hardware supplier should be taken. It is also advisable to check with other users of the equipment to see their configuration and what problems they have experienced.

Disk drives support either removable or fixed packs. The advantages of removable packs are that they can be changed quickly, can be used for back-up (but this could be an expensive option), archiving and the loading of software. The disadvantages are that they are of lower reliability and generally of lower density than fixed disks. The trend in the computer industry is towards fixed disks, especially the sealed unit or Winchester drives. Rapid advances have been seen in the capacity of these disks especially for microcomputers.

Optical disks, after many years of promise but little else, are now more available. They have very large capacities at relatively cheap prices. Two types can be purchased at present. The CD-ROM (compact disk read-only memory) is a read-only device which can be used in conjunction with commercially available databases which are relatively static. WORM devices (write once, read many times) are very attractive for holding data with long retention times, ie. archiving, and may become an alternative means for taking back-ups. This is especially interesting in the context of GLP requirements.

Given these considerations, the WORM devices are more likely to be of interest in the context of LIMS.

7.3.2 Tape Drives

Tape drives are normally used for three main purposes: back-up, archiving, and loading software including operating systems. They can either be of the traditional reel-to-reel type or they can be of cassette format. When a drive is purchased for back-up purposes it should have adequate performance in order to back up the disks within a reasonable time. Tape drives are available in 800, 1600 and 6250 bits per inch (bpi) density. Modern large disks frequently require 6250 bpi drives in order to keep the number of tapes required to store all the data to a small number, and to keep the back-up time to a reasonable length.

The specification of tape drives given by manufacturers is usually inadequate to calculate how long it will take to back-up a disk. It is advisable to ask the manufacturer how long he expects it to take, as well as talking to a user of the equipment, if one is easily contactable.

Streamer tape drives have become increasingly available. These drives are optimized for continuous reading and writing, as required during back-ups, but are of lower performance as stop-start devices, for reading selected files from a tape. This reduced performance is balanced by lower cost.

If the drive is being used to load software, you should check whether the LIMS software is available from the supplier in the required format or can easily be converted to that format.

The selection of a tape drive for back-up purposes depends on the disk configuration chosen, the time available for back-ups, and the type of back-ups. An outline plan for taking back-ups should be developed before a drive is selected. The type of questions addressed should be: when are they going to take place - in the evening, lunchtime, weekend or during the day? Are complete back-ups going to be taken each time or should incremental back-ups be taken? (An incremental back-up scheme is where a full back-up is taken only, say, once a week and then only those files which have changed are backed up in between times. Table 7.1 shows an example tape back-up scheme). How long should tapes be kept before they are re-used? How many copies of back-up tapes should be taken and where are they going to be kept?

Many of these questions can be answered by reference to the policies and procedures of a central computing department as well as by interpretation of GLP requirements, if appropriate.

Table 7.1

Example tape back-up schedule

Weekly Incremental Back-up Schedule		3 Monthly (assuming 4 week month)	
Day	*Back-up tape*	*Week*	*Back-up Tapes*
Friday	Full #1	1	Full #1
Monday	Incremental #1	2	Full #2
Tuesday	Incremental #2	3	Full #3
Wednesday	Incremental #3	4	Full #4
Thursday	Incremental #4		Offsite monthly #1
Friday	Full #2	5	Full #1
Monday	Incremental #1	6	Full #2
Tuesday	Incremental #2	7	Full #3
Wednesday	Incremental #3	8	Full #4
Thursday	Incremental #4	9	Offsite monthly #2
.	.	.	
.	.	.	
.	.	.	
		12	Offsite monthly #3
		.	
		.	
		.	
		16	Offsite monthly #1

7.4 Communications

The hardware needed for both internal and external communications should be considered as part of the requirement for a LIMS. This will be discussed in four parts: terminal to processor communications, computer to computer communications, computer to instrument communications and, finally, external communications.

7.4.1 Terminal to Processor Communications

The simplest terminal to processor communications consists of direct wiring. The main problem with this is lack of flexibility, in that the terminal can only communicate to other computers through the computer it is directly connected to, which is not always possible or very efficient.

There are two main ways to gain extra flexibility. The first is to install a terminal switching system available on the market from vendors such as Gandalf and Micom. In these systems the user, when wishing to connect to a computer, first interacts with the switch and selects a computer to attach to. Once the connection has been made the switch acts passively, not interacting with the communications, except to close down the connection when the session is concluded by either the user or the computer.

The second means of flexible terminal to computer communications is through a local area network (LAN) as described in Section 7.2.2 on networked microcomputers, except that both the terminal and computer are connected to the LAN. The interfaces to the network have some intelligence giving rise to extra power relative to a switching system. An example of this extra functionality is the ability to run multiple simultaneous sessions on either the same or different computers. The addition of an extra computer to the network only involves one connection, while with a switched system the number of connections has to be at least equal to the maximum number of simultaneous users expected. As usual there is a disadvantage to the attractive approach of installing a LAN and that is cost, although this is reducing.

7.4.2 Computer to Computer Communications

There is often the requirement to communicate between the LIMS and other computers; this aspect is essential to integrate the system into the organization. The LIMS vendor will be able to specify what hardware and software is required for such communication, but check if any development is required to make a workable system, and ensure that the system is robust. A number of communication systems require considerable continuing support to keep them operational.

7.4.3 Computer to Instrument Communications

Once it is decided which data are going to be acquired by the LIMS, then the means by which it is achieved will normally be specified by the prospective vendors as part of their proposals. The question of whether one way or two way communication between the computer and the instrument or experiment is required should be addressed. In the case of one way communication the computer only acquires data. In two way communication the computer can control the instrument by downloading worklists, methods or raw data for reprocessing. Two way communication is not common at present but its use is likely to grow as instruments become more intelligent. Greater discussion of instrument communication is given in Chapter 10.

7.4.4 External Communications

External communications also should be considered. Data may need to be acquired from or sent to external computers. This communication may be by means of dialled connections, leased lines or public networks (such as British Telecom's PSS). The choice between these methods needs to be made in terms of data volumes, transfer rates and cost. A central computing department or the hardware vendor can help with the choice.

7.5 Peripherals

7.5.1 Terminals

The days of an unintelligent terminal connected directly to a computer are coming to a close. It is being replaced by the work station connected to a network. At the lower price/capability end of the spectrum the work station will be a PC. The *de facto* standard at present is the IBM compatible PC running MS-DOS. The factors leading to this change are the steadily reducing price differential between terminals and PCs and the growing need for users to access multiple computer systems.

The LIMS suppliers are well aware of this trend and are adding the ability to link to work stations and transfer data between their systems and the work stations, or are supporting the work station-only environment in terms of PC networks.

The present day purchaser of terminals or PCs for a LIMS has a choice, but a choice which is simple if the protection of the investment for the future is important: a PC every time. For the rest of this section when a terminal is mentioned it also implies a PC.

The numbers of terminals to be purchased depends on the number of users, how frequently they will access the system and for how long. In deciding upon numbers some feeling should also be obtained for the type of work to be undertaken on the terminal: is it just to be used for reporting, or is it to be used to control or acquire data from instruments or experiments? In the latter case contention for its use will have the effect of lowering the utilization of expensive equipment.

Other questions to be thought about include graphics and colour facilities. The information handled by a LIMS is often best portrayed in graphical form and a fair proportion, if not all, the terminals purchased for a LIMS should have graphics capability. The interpretation of complex graphics is enhanced through the use of colour. The use of colour for presenting only text is not cost effective. Much of the advantage of presenting information in graphical format and with colour on a screen is lost if it cannot be converted to hard copy. The purchase of these terminals should therefore be connected to the purchase of printers and plotters with the relevant capabilities.

A final factor to be taken into consideration in purchasing terminals is the question of whether they can access other computers on the site. As far as possible, terminals should be purchased with the capability to access other relevant systems. There is little more frustrating to the user than having to use one terminal for one system and a second one for another, never mind the waste of valuable bench or desk space.

7.5.2 Printers

The dividing line between printers and plotters has changed over recent years and has become blurred. Originally, printers only supported a limited character set and any graphics printing had to be done by means of plotters. Modern printers generally handle a much wider range of characters and can often also print graphics. The present dividing line between the two types of devices is now much more a question of colour, with plotters providing multi-colour output by means of pens, and this is the differentiation used here and in the following sections. In fact, this division is becoming further blurred with the arrival of multi-colour ink jet printer-plotters and the use of multi-colour ribbons on impact printers.

The requirement for printers can be split into two. First printers need to be attached to terminals which will print relatively small amounts of output directly from the terminal. Second, printers are required to be connected directly to the processor, for larger throughput printing.

The first requirement is normally met by cheap dot matrix printers while the second is being met more and more by the xerographic based laser printer.

The first question to be asked is whether graphics as well as text based output is required for a printer. Then a list of printers with the required capabilities and supported by the hardware and software needs to be put together. The actual choice of printer can then be determined by factors such as print quality, throughput, price, operating noise level and whether it will handle cut and/or continuous stationery. The quoted throughput of printers cannot always be obtained in normal running conditions due to factors such as set-up time. Also, the throughput of laser printers can be substantially reduced when printing graphics rather than text. The running cost of the printer should also be ascertained; some laser printers have surprisingly high costs per page of output.

A recent development is that of print servers as described in Section 7.2.2. They often support extra printing functionality, such as the mixing of text, graphics and images, which is required for desk-top publishing.

7.5.3 Plotters

Plotters are used for multi-colour graphics output. Like printers they can either be associated with a terminal or directly connected to a processor. The dominant plotter technology is presently pen-based. The choice of a plotter is restricted by the choice of hardware and software. Other factors in the selection are the number of colours available, plotting speed, maximum size of plots and whether paper handling is manual or automatic.

7.6 Maintenance

The maintenance and support required for both hardware and software needs to be investigated. The cost of hardware maintenance is usually in the range of 5% - 10% of the purchase price per annum while software maintenance costs are in the region of 10% - 15% of the current software price per annum. The traditional approach has been to contract with the hardware supplier to mend any inoperable hardware *in situ*. With the lowering cost of peripheral items such as terminals or PCs it has become more attractive to purchase spare items or boards, swop out broken items and send them away for repair. This approach is one way of lowering maintenance costs.

Hardware manufacturers are also attempting to reduce the price of maintenance by the use of remote preventative maintenance and problem solving. Here a remote centre dials up the computer at regular intervals, or when problems have been reported, and interrogates the database of performance statistics that the computer has collected. The centre can then diagnose potential and actual problems and schedule a maintenance engineer to come and correct them. A number of suppliers offer reduced maintenance costs if this service is taken up.

7.7 Summary

This chapter has discussed the choice of hardware components for a LIMS under a number of categories. It is essential that the proposed hardware configuration is discussed critically within the users' organization, with the vendor and with existing users, as good planning before the ordering and installation of the hardware will help to ensure that your LIMS is a success.

Chapter 8

The User Interface

Ross Burgess

8.1 The Need for User-friendly Systems

Computers today offer a great deal of power compared to those of only a few years ago: part of this extra power can and should be utilized to meet the user more than halfway, by providing outputs in a form that is easy to understand, and accepting inputs in a form that the user is happy with. Thus a LIMS, like any type of computer system, needs a good user interface if it is to be used effectively.

There has been much research into the theory of user interfaces, but the theory has not always been put to good effect. This is partly because computer systems available today are inherently complex to use. As an analogy, consider the telephone. No instruction manual is required for a telephone, because the way of using it is almost self-evident. There is not even a switch to turn it on, other than the obvious action of picking up the receiver. The process of dialling is simple because the numbers on the dial correspond to the numbers which the user has in mind when making a call. However, telephones are becoming more complex. Modern automatic systems allow the user to call the last number again without re-dialling; to instruct the system to keep ringing an engaged extension until it is free; to announce that the user is moving temporarily to another extension, and have the calls redirected. There is only one thing wrong with all these facilities: it is very difficult to remember how to use them. Probably the facilities are obtained by keying in a numeric code. For a full-time telephonist, this is a very good arrangement; but for someone who needs a particular facility only once a week or so, the codes become impossible to remember. What started as a simple system has become complex and difficult to use: it needs further technical developments for it to be made user-friendly once again.

The same is true for computer systems. An integrated system with database manager, spreadsheet, and word processor can be a very useful productivity aid, but all too often such a system is under-utilized because there are too many commands and it is difficult to remember them all.

The key to an effective user interface is that it should be appropriate to the different types of people who will be making use of the computer, and their different needs. Fundamentally, there are three kinds of computer users:

1. Programmers.
2. Operators.
3. True end-users.

For both programmers and operators, the computer itself and its software is the focus of their work, but for the true end-user, for example a chemist or analyst, the computer is merely a means to an end.

Many of the most powerful computer programs in existence, for instance the operating systems that help the computers themselves to organize their work, use complex command languages, and are therefore often criticized for being *not user-friendly* . But operating systems are not designed for ordinary users, they are designed for use by programmers or operators. The programmer should ensure that the user never has to see the operating system, let alone know its command language. The system should load the appropriate package as soon as the user signs on, and the package itself should provide a user-friendly way of using the few operating system facilities likely to be required.

As computers have become smaller and cheaper, the distinction between the user and the operator has tended to become blurred. This is sometimes expressed by saying that microcomputers do not need operators; it might be truer to say that the user of a microcomputer often has to become an operator as well, changing disks, doing back-ups, and coping with system errors. For personal computing this may be acceptable; but for a system holding important data such as the results of experiments, it is still important to separate the user and operator functions. Perhaps operators as such may not be required, but at least the system should (as far as possible) hide the nuts and bolts of its operations from the users, leaving them to get on with running the application, and the system manager to cope with the needs of the computer.

8.2 Some Types of Human-Computer Interaction

Different computer programs vary considerably in the way they are used and interacted with. This is due to a number of different factors:

1. The programs may have been designed with different screens, keyboards, and other equipment in mind.

2. Software producers vary considerably in the importance they attach to the user interface in the design of the system.

3. Different programmers may make arbitrary decisions as to how to structure the user interface.

4. In older programs, lack of memory space may have limited the size of program that could be accommodated.

If the memory available for the program is limited, the user interface is likely to be given a relatively low priority. Moreover, if the design of a user interface is left to programmers, they may fail to understand the needs of those who are much less familiar with computers than themselves.

It is insufficiently recognised that the design of a user interface is a very specialized task. Users buying a standard software package should expect the user interface to have been done properly, but this is unfortunately not always the case. If the package is designed for a specific purpose, the authors may have given a great deal of attention to the way the application itself is achieved, but little thought to the user interface. In a more widespread type of application, such as word processors and database managers, market pressure is likely to have made the suppliers give much more thought to the user interface, and indeed the packaging of the product generally.

8.2.1 What Constitutes a User-friendly System?

In considering the user-friendliness of a computer system, there are a few rules and guidelines that need to be borne in mind.

One generally accepted rule is that, at least for the unskilled user, interaction with the computer should be *mode-independent*. Some programs break this rule quite extensively. For instance certain widely used text editing programs use the ordinary alphabetic keys of the keyboard as control keys when in *command mode*. Thus, if you hit the'X' key, this will sometimes put an X in your text in the normal way, and sometimes delete the character that the cursor is positioned on, depending whether you are in command mode or insertion mode. Not surprisingly, new users find they are constantly doing the wrong thing by mistake. Indeed a new user getting to grips with such a program can be recognised by the sound of muffled cursing, mixed with the repeated bleeps that the terminal makes as the system rejects erroneous commands.

This sort of user interface is no doubt good for programmers and computer buffs, who can become very proficient in making the product sing and dance, and often take pride in their mastery of such a recalcitrant package. However, it should not be given to managers who want to compose an internal memo, or it will confirm all their worst beliefs about computers. Today most VDU terminals and personal computers have a range of programmable function keys, so that it should be possible to use some for ordinary input, others for control purposes, and thus avoid the need to change *modes* at all.

8.3 Menus and Commands

Whenever humans communicate with a computer, they are using some sort of language that the computer is programmed to understand. By analogy with communicating with another human being from a foreign country, there are two ways that the conversation could proceed. One way would be to learn the foreign language in order to make yourself understood. This will take time and effort, and there is a good chance of making mistakes. Nonetheless, this is the way that many computer systems behaved in the past. A more reasonable approach, since the conversation will cover only a restricted subject matter, is for the computer, the foreigner in the analogy, to be primed with questions to ask, and the ability to make sense of the answers.

There are thus two main ways of giving instructions to a computer: by typing commands, or by answering questions, that is, selecting items from a menu. Later in this chapter an alternative approach, analogous perhaps to sign language, will be discussed.

The choice between a command-type interface and a menu interface is a fairly fundamental one in the design of any computer software, but one that should not be made too hastily. Generally, the best answer may be to have not one or the other exclusively, but a mixture of both, or perhaps an option which users themselves can make depending on their familiarity with the system. New or occasional users will normally be happier with a menu (provided that the menu items are unambiguous and easy to understand) while more experienced users may prefer a command approach, which can be quicker to use.

8.3.1 Types of Menu

The traditional menu on older computer systems comprised a full screen, one line per item, often numbered. There were instructions to type in the number corresponding to the selection required, and then hit the Return key, as shown in Figure 8.1. Aspects of this are discussed further in the next chapter on databases.

More modern user interfaces feature different types of menu, which do not take up the full screen, and are in some ways easier to use. These days it is common to provide several methods of selecting items from menus. The most straightforward is by *pick and point*. There is a menu cursor, often shown by the entire menu item being highlighted on the screen; the user can move the cursor up and down until the required item is reached, and then hit the Return key to confirm the selection. Most systems have several means of selection, by keying in a number corresponding to the menu item, or the first letter of the option in question. Occasionally, you come across software which has only pick and point, or only selection by initial letter, but this can generally be put down to poor system design.

The Lotus 1-2-3 spreadsheet package has made the so-called *ring menus* popular, see Figure 8.2. If the user hits a special 'menu' key (the '/' key in the case of Lotus) a single line menu is displayed at the top of the screen showing the various options as one word items. One of the items is highlighted, and this item is explained in more detail on the line below. The user can move the highlighting from one item to the next, and hit Return when satisfied, or type in the initial letter of the command word.

The experienced user can often find menus tiresome, and prefers a quicker route. Some software therefore has a slight in-built delay in displaying the menu when the menu key is hit; users who already know what option they are going to use can hit the key in question immediately, and the program will then not bother to display the menu.

Database Maintenance Functions

Record Maintenance

A – Add a Record
D – Delete a Record
M – Modify a Record

Type in your choice by letter: ____

Figure 8.1 A traditional menu

Many applications require a series of menus, one following on from another, so that starting from the main opening menu a further menu is selected, and so on until the actual program required is reached. The various menus can generally be regarded as a tree, in which users can start at the root, and work their way along one of the branches. For instance, a top-level menu might include the item 'print'. Having selected this the user might have a further menu to determine the document to be printed, another menu to choose between draft and letter quality printing, and so on. Having completed the work at the lowest level, the user may then need to work all the way up again. Good application software will allow the user to escape from the current menu immediately, back up to the top level, provided this would not by-pass any important safeguards.

```
Archive   Cut   Delete   Edit   File   Move   Spell   eXit
Do spelling check on the current file

of this style of user interface.

The "Lotus" style menu features one line at the top of the screen
giving the various options which can be selected, often in
alphabetical order.

You can select an option by typing its initial letter, or by moving
the cursor. Often there is no conventional cursor, but the entire
option is highlighted or shown in a different colour.

The next line gives a more detailed explanation of the option
currently selected.

This type of user interface is particularly valuable where the
```

Figure 8.2 A Lotus-style menu shown in a word processing application

A common complaint about computer systems is that they ask for information which they already have available to them. For instance, if the user is using a word processing program, and wishes to use the 'edit' command to load in a document created previously, the program will often demand the name of the file. But of course the computer knows what documents there are on disk, and the software could very easily display the list on the screen, and invite the user to choose one or the other by means of cursor selection. Some packages show the names of files already on disk, but still require the user to type in the name of the particular file in full.

8.4 Some Particular Problems

8.4.1 The Return Key

The use of the Return key sometimes gives rise to difficulties. Its original function, of course, was derived from the typewriter, and has two elements: repositioning the cursor at the left-hand side of the page ('carriage return') and moving it down to the next line ('line feed'). It retains this function in many computer applications, although in word processing for instance, the cursor may move automatically to the start of the next line without the need to press Return.

A secondary function that the Return key has acquired is that of indicating that input is complete and ready for processing. In this context it is often referred to as the'Enter' key. Sometimes the two uses conflict, for instance if a series of menu options are arranged as a column, some software will interpret hitting Return to mean that the user wants to select the item in question, other software (less commonly) thinks they want to move down to the next item. It is worth mentioning that some keyboards have both a Return and an Enter key (sometimes with identical functions, sometimes not) while some have a special Do or Execute key.

8.4.2 Dealing with Text

Text input is a key part of the user interface to computers. Nearly every computer function needs some amount of inputting of text by the user, whether it is the name of a command to the operating system, customer name and address details in an order entry program, or the description of a sample in a LIMS. Unfortunately, there are not yet universally accepted standards for the most basic functions such as correcting the last character typed, moving the cursor backwards and forwards, deleting the character that the cursor is now on, changing your mind and clearing the entire field or confirming that the input is correct. For instance many packages use the Backspace key to delete characters to the left of the cursor, and the Delete key to delete to the right; however, one of the most popular word processing packages for personal computers uses the backspace to move the cursor left without deleting, and the Delete key to delete to the left.

8.4.3 Dealing with Errors

Any computer system needs to deal with mistakes by the user, and return error messages if the action requested is impossible or inconsistent, or perhaps just cannot be performed at the time. Many computer programs are unhelpful when errors arise, and give obscure messages, with little advice for what to do next. The reason for this is sometimes just carelessness on thc part of system designers, particularly since error handling is often treated as an afterthought in the design of a system. Sometimes there is limited space available for the table of error messages, and the designers have doubled up, making error messages more general purpose, and hence less specific and helpful. A typical example is the error message, 'Incorrect number of parameters'. Even assuming that the user knows what a 'parameter' is, this is not a very helpful message. Does it mean too many parameters, or too few? After all, the system must know how many parameters it was expecting, and the message should be more specific.

The ideal approach for errors is no doubt to prevent them from occurring. For instance, selecting one item from a menu, the user may be told that the particular facility cannot be used at this point for some reason. A far better method would be to omit the item from the menu at times when it was unavailable. An alternative approach, used in a number of current systems, is to show in faint lettering the menu items which are currently unavailable, and to ignore any attempts to select them. More generally, a sound principle of program design is to consider the various errors that could occur, and design the logic of the user interface as far as possible so that such errors just do not arise.

Many computer programs are very unforgiving in their attitude to input errors. For instance if there is no distinction between upper and lower case in a particular context, there is no reason (other than laziness on the part of the programmer) for the program to insist that one rather than the other must be used. If the computer asks a question to which the answer is yes or no, it should accept 'Y' or 'y', 'N' or 'n'. Similarly, in programs designed for the naive user, it is as well to treat 'O' (the letter) and '0' (zero) as equivalent versions of the same thing, similarly 'I' and '1', unless of course there is a real distinction to be made.

8.4.4 Help Facilities

Many software packages incorporate 'help' facilities, invoked by pressing a special key (perhaps the question mark, or function key F1). The best help facilities are *context sensitive* - in other words the program displays a help message specific to the part of the application which the user has reached, and makes suggestions as to what options are available at this point.

Like many other aspects of a good user interface, a help facility has a cost in terms of the disk space that must be provided for the help text, and the logic that must be included in the program to retrieve the appropriate piece of text.

8.4.5 Response Times

The speed of a computer system's response can be an important factor in determining how user-friendly it appears to be. It is irritating to type a command and then have to wait a long time for the computer to respond. It is still worse if the response time is very variable. This can easily happen with multi-user systems, particularly those which aim to be 'fair' to all their users, and therefore react to increased load on the system by reducing the time allocated to all the tasks, and thus causing the service to all the users to slow down. In certain applications the variability of the response has been found to be so distracting that it was necessary to put in artificial delays, to slow down the response when the system is lightly loaded, so that the difference when the system is more heavily loaded does not seem too great.

It has been found experimentally, that delays are more tolerable at certain points in the course of an interaction with the computer. Say the user is inputting a complicated transaction, consisting of many fields of data. He will not want to be held up when moving from one field to another; however, when the entire transaction has been entered, and the user finally hits the Return key to signify that it is complete, a psychological event called *closure* occurs: the user sits back for a moment or two, to draw breath (figuratively speaking) and may need to turn to the next document, or decide what to do next. A longer pause on the part of the computer: perhaps one or two seconds, is quite acceptable at this point. Most likely the user will appreciate that this is a point at which the computer has a considerable amount of processing to do.

An inappropriate response time may actually confuse the user as to what is going on. In particular, if a command is going to take an appreciable time to execute, it should at least respond immediately with some message such as 'processing your request - please wait', otherwise the user may think that the program expects some more input, and start pressing additional keys, or repeatedly hitting Return, which in its turn may cause problems. The 'please wait' message should be prominently displayed, and perhaps flashing, which at least gives the impression that something is happening, although too much flashing can cause eye strain.

8.4.6 Where am I Now?

An important element of a user interface is that the user should not be left in doubt about what is going on. The worst offenders are those programs which move from one mode to another, without giving the user any visible indication that they have done so.

An important element in user interfaces is the mental picture that the user has of what the computer is actually doing. A typist using a word processor may have only the haziest idea of what is going on inside the computer. This may or may not matter, depending on whether it leads to inappropriate behaviour. For instance, a user who does not understand the difference between a document being held in memory and saved to disk may turn the computer off, or exit from the application, without having saved the file he has been working on.

A common symptom of poor understanding of a computer system is that the user can cope perfectly well while it is working normally, but panics as soon as an error condition occurs. Panic generally takes the form of pressing every key in sight, rebooting the machine, or even turning it off. Such a response can clearly be very inappropriate at times.

8.5 The WIMP Approach

Many of today's ideas on the user interface come from the work of Xerox's Palo Alto Research Centre (PARC) which developed what has been called WIMP (window/icon/mouse/pointer). The WIMP type of presentation takes the pick and point technique one stage further. Rather than moving a text cursor about using the arrow keys, there is a special input device such as a 'mouse' (a little box on wheels or a rolling ball, connected by a cord to the computer). A pointer moves about on the screen in time with the movements of the mouse on the bench. Rather than having to type commands, or select an option from a menu, a number of little pictures ('icons') are shown on the screen (Figure 8.3). The user can select items to be actioned by moving the pointer over the icon in question and pressing one of the buttons on the mouse. Other features of this type of presentation include 'pop-up' or 'pull-down' menus, in little windows of their own, which appear only when requested.

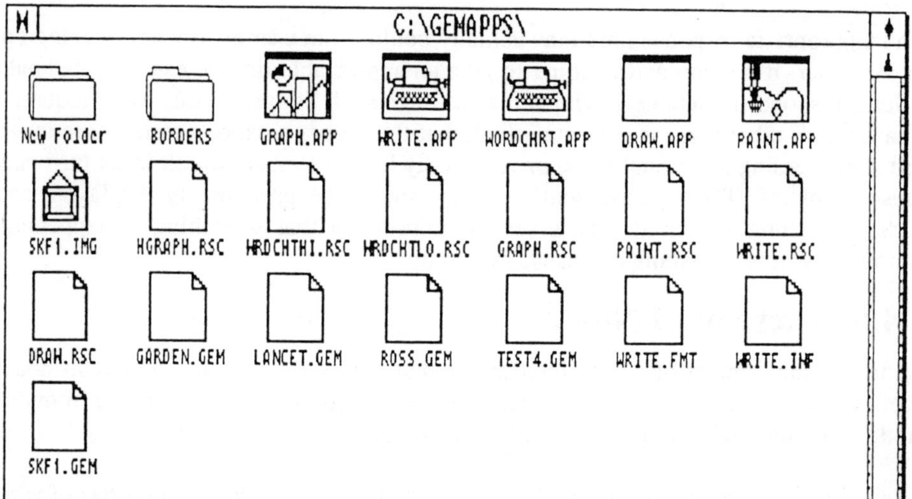

Figure 8.3 Icons (as used in the GEM package supplied by Digital Research Ltd)

The WIMP approach also affects the appearance of the display screen: rather than simple lines of text, it is a graphical display, divided up into a number of windows as shown in Figure 8.4. Some systems have the windows side by side, others have them partly overlapping, similar to papers lying around on a desk; each window represents an object such as a program or task in progress, or a document being processed. The mouse allows the user to move around easily from one window to another, and to manipulate objects via their icons.

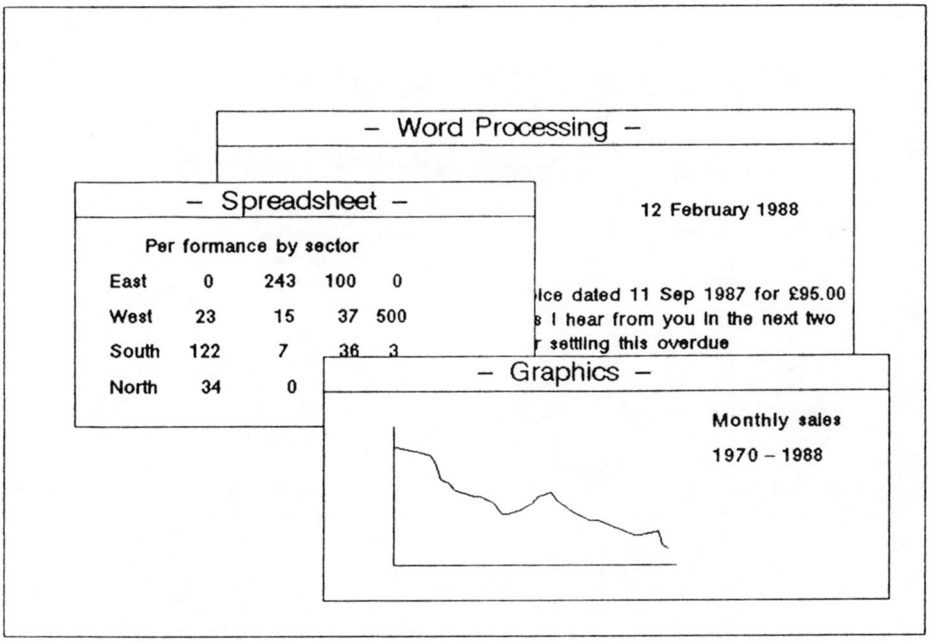

Figure 8.4 The use of windows on the screen

The use of windows on the screen is particularly appropriate where there are several entirely different contexts or *modes* which the user may need to switch between.

The WIMP approach, although devised by Xerox, owes its popularity to Apple, whose Macintosh personal computer has made this type of presentation very popular. It has since begun to appear on many other types of computer system. Although not used in conjunction with LIMS, products using a WIMP user interface are now appearing for scientific applications, examples being the Beckman Gold, Dynamic Solutions' Maxima and Nelson Analytical's TurboChrom.

The WIMP approach lends itself particularly to high-resolution graphical screens, but it is certainly possible to use a text-only display (providing it has at least line drawing facilities). However, the essence of windowing is that the windows themselves should be clearly distinguished from their surroundings: if the normal screen display uses lines and boxes, a window may not stand out very well.

8.5.1 Using the Mouse

To previous generations of computer users, used to the keyboard, a mouse as shown in Figure 8.5, for input may seem a frivolous adjunct to the serious business of computing. But it has a definite part to play in modern computer systems.

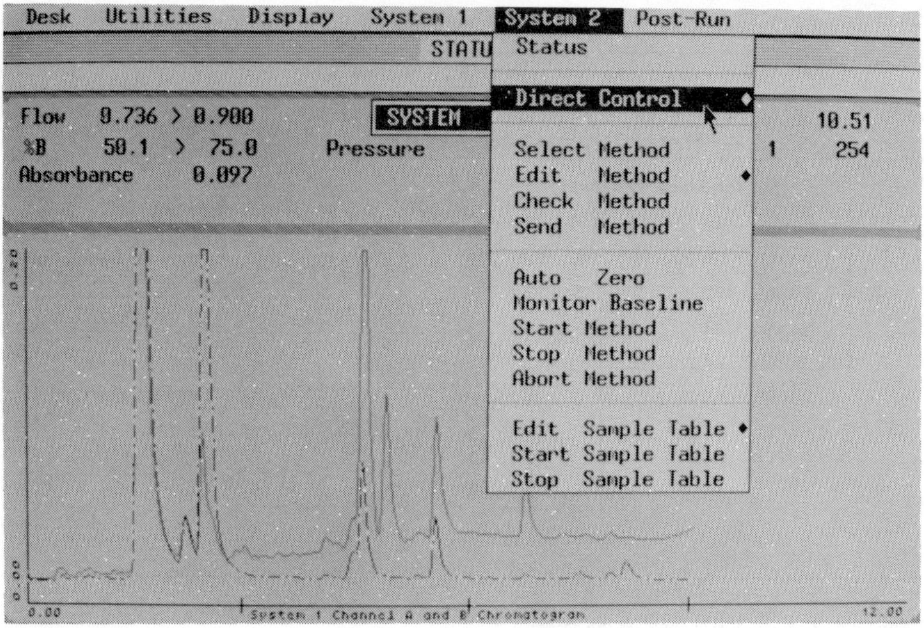

Figure 8.5 A mouse used for program selection (courtesy Beckman Instruments)

A mouse is much more than just a replacement for cursor arrow keys, for several reasons:

1. It can move in small increments, not just a character at a time; it is therefore ideal for all cases where fine resolution of movement is needed, and can even be used (by those with some dexterity) to produce drawings, given suitable software.

2. It can move very quickly from one part of the screen to another, and is therefore suitable where the sequence of items to be dealt with is difficult to predict or to organise in advance.

3. It naturally allows movement to all parts of the screen, whereas cursor keys, tabs, and so on typically allow movement only to specific fields or within a particular window; this can be either an advantage or a disadvantage, depending on the context.

The use of a mouse does, however, depend very much on an instant response from the computer. In particular it would be difficult to use on a heavily-loaded multi-user system, as the cursor on the screen would always be lagging behind the movement of the hand.

There are now fairly well established conventions for the use of a mouse. Given a screen display showing various disk files represented as icons, there are several things that can be achieved with the icons displayed. First, of course, the user needs to position the mouse cursor over the icon in question. Then he can *click* the icon to select it, using a single press on the left hand button; he can *drag* it, by holding the mouse button down and moving the mouse: this moves not only the cursor, but also an image of the icon itself; or he can *open* the icon with a double click. The effect of the various actions depends on what sort of object is being manipulated. For instance, double clicking on an icon representing a program will probably start it running; double clicking on an icon representing a document may show its contents on the screen, perhaps by calling an editor or word processing program.

Dragging an icon has the effect that when the button is released the system makes a new copy of the object. For instance, there might be two windows open representing different disks, and a file could be dragged from one to the other, to make a copy on the other disk.

Another technique with the mouse is to select items at once by *shift-clicking* them, in other words holding the Shift key on the keyboard while moving the cursor around, and clicking particular items with the mouse button. On releasing the Shift key, all the items in question will have been selected for use.

8.5.2 Putting Things Right

Among the most important features of a satisfactory user interface is the ability to *undo* a command made in error. It is unfortunately all too common to delete a piece of text, or erase an entire file, and then realize that the wrong name was typed and deleted the very thing that should have been saved. An undo key could be most welcome in such a situation, and is offered by an increasing number of software packages. An alternative approach, to cover the specific case of deleting the wrong thing, is the WIMP convention of deleting by dragging the item across the screen to dump it in a *waste paper basket* icon. This makes it very difficult to delete an item without meaning to, and there is often the ability, just as in real life, to retrieve items from the waste bin, provided it has not yet been emptied by a separate utility.

The WIMP type of approach can certainly reduce the likelihood of certain types of errors. For instance you may be in the habit of copying your working files to a different diskette for safe keeping. If one day by mistake COPY B:* A: was typed instead of COPY A:* B: you could find the entire day's work was lost. Using the WIMP type approach, of actually dragging the icon representing the data from one place to another, there is much less danger of making this mistake.

8.6 Coping with Interruptions

A truly user-friendly computer system would allow interruptions to the flow of work; for instance it could allow you to break off from entering data into the system, consult a database or do some calculations, and then resume the input program where it had been left off. Today's computer systems are beginning to offer this sort of facility, but implemented in various ways. One of these is to have multiple windows on the screen, each representing a different program; an other is to have *multiple virtual terminals*, in other words the entire screen is given up to one program, but at the press of a button the display can be instantly replaced by a different program.

8.7 Using Colour

More and more terminals are appearing with colour screens, and they are particularly popular on personal computers. Colour allows the software designer to achieve a much greater degree of emphasis and distinction between different parts of the display, but of course it should not be over-used, for practical as well as aesthetic reasons. Moreover, it is dangerous to rely totally on colour: you need to remember not only the significant minority of the male population with defective colour vision, but also the much larger proportion of the VDU population with monochrome only, some but not all, able to represent colours by different shades of grey.

8.8 Other Forms of Input

The keyboard and the mouse are not the only means of getting input into a computer system. Sometimes it may be desirable to capture data from a printed source, for instance using a technique such as OCR (optical character recognition) which scans a typed or printed sheet optically and converts the printed characters back into computer-processable information. Frequently, the information in question will have been produced by a computer system in the first place, so re-entering it may not seem very sensible. However, such techniques come into their own when the printed information in question is attached to a physical object such as a laboratory specimen. The aim in such cases is twofold: to reduce the effort and time required to collect the data, and to reduce the likelihood of error. For this type of application, a special technology has developed, that of bar coding, which is specifically designed for computer input.

8.8.1 Bar Codes and Bar Code Readers

Bar codes are a familiar sight on grocery items, allowing data capture to a level of detail which would have been quite impracticable using traditional methods, because of the need to process individual items extremely quickly. In the case of laboratory systems, however, the need is not so much for speed, as for the elimination of human error.

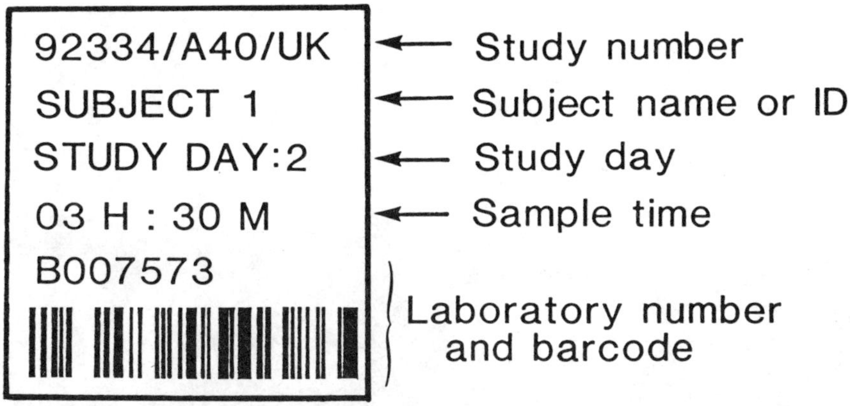

Figure 8.6 A bar code label

A bar code is essentially a sequence of black and white stripes as shown in Figure 8.6, some thick and some thin, which can be interpreted by a computer system according to an established coding scheme, which relates different patterns to numeric or alphabetic characters. In a LIMS, the encoded data would typically be a sample number or the equivalent.

Bar code readers can take several forms. The most common is a pen with a light source at the tip which is moved by hand across the code, shown above in Figure 8.7. If the code is read and decoded successfully, the terminal emits a bleep to inform the operator. Clearly a certain basic dexterity is required; in practice, as with using a mouse, new operators have some difficulty at first, but their success rate rises rapidly with practice. Generally, the reader is physically attached to the computer system or terminal, but portable battery-powered readers are also available, so that the data can be collected off-line (for instance while checking the stock in a store room) and fed into the computer subsequently.

Fixed beam readers are an alternative to the hand wands; in this case it is the bar code (attached to the sample) that is moved past the reading head. This can be a more appropriate solution if the items need to be manually handled, as in a supermarket checkout. For large numbers of samples, the use of laser scanners should be considered. These are high speed devices that can read the label when it crosses the light path at any angle.

8.8.2 Voice Input and Output

Voice input and output will certainly be important elements in the user interface of the future, but as yet the technology has had only modest success.

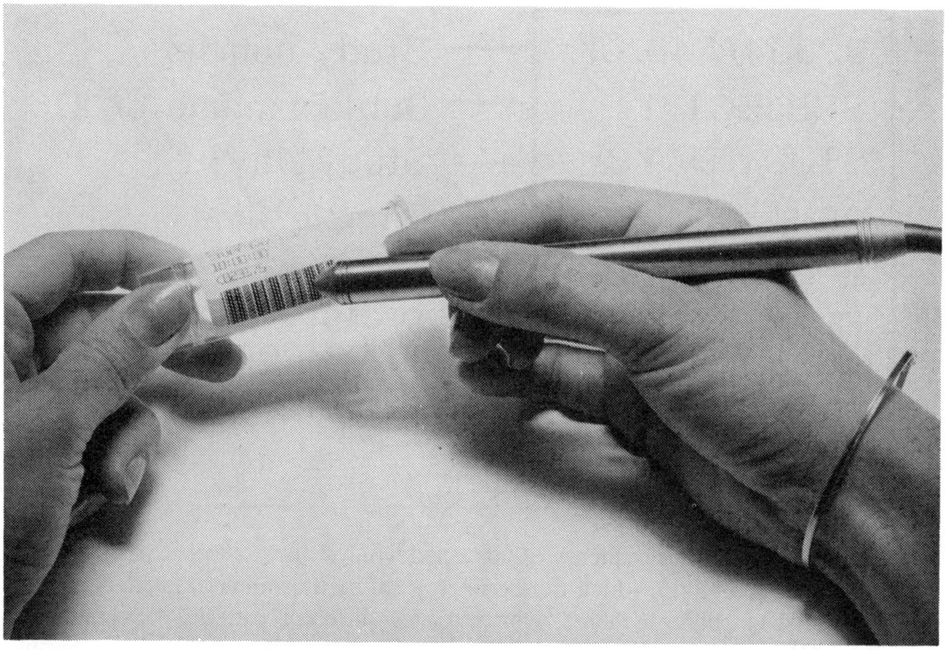

Figure 8.7 A bar code reader

Speech recognition could ultimately revolutionize computing and make information technology far more accessible to non-specialist users. A type of speech recognition was demonstrated as long ago as 1939, but many problems still remain. Today three types of voice input can be identified:

1. Voice recording for subsequent playback, without any attempt to interpret it.

2. Limited speech recognition, with a small vocabulary, used to give commands to a computer; this is now established technology, and available on home micros.

3. Full speech recognition, for example dictating text into a word processor; this is still at the leading edge of technology.

Many voice recognition systems require each user to train the system by dictating every word in the vocabulary. This is perfectly reasonable for most current applications of speech input, but would be impossible for full dictation, which might require a 20,000 word vocabulary. Some recent systems enable the user to train the computer by dictating a short standard text.

In the laboratory, voice input of a rather basic kind can be extremely useful, for instance as a way of inputting data while keeping both hands free, perhaps in order to handle some wet or dangerous material. Clearly, the process may lead to some errors if the computer misinterprets what is said to it; but this can happen with any type of input. One feature of voice input is that when the system makes a mistake it produces a word so totally out of context that it is not too difficult to detect.

Speech generation from text is less difficult in principle than speech recognition, but the quality of the speech generated is still not entirely satisfactory. The difficulties here are more to do with the acceptability of speech as an output medium, and the possibilities of disturbing, or being overheard by, other users. Probably for most types of computer interaction for some time to come, a visual display will be preferred.

8.9 Summary

The way that the users interact with a LIMS is vital to its acceptance. Care and forethought in the design of the user interface can enhance any program. Although little, other than commands and menus, has been developed for use with LIMS to date, many possibilities are available if the users require them and the manufacturers develop them.

Chapter 9

Databases

Adrian Hampshire

9.1 Introduction

The purpose of this chapter is to give the team responsible for selecting and implementing a LIMS some insight into the Database Management System (DBMS). In particular, what features to look for in assessing a DBMS, how it can work for them and, in outline, what types are available. The chapter is written primarily with the user of the system in mind rather than for the benefit of computer staff.

Today's computer systems have the ability to store very large amounts of data. The availability of this data for fast and selective retrieval allows the scientist to undertake storage and analysis of amounts of laboratory information that would, in the past, have proved impossible. Current database and disk technology allows the maintenance of a database holding upwards of 1,000 million characters. It is obvious that management of such volumes of data requires powerful software. It is also self evident that this data represents a very significant asset for a company, requiring close and careful management. Only with the development of Database Management Systems has it become possible for an organization to realize the full value of its data.

Past attitudes to the storage of data by computers have resulted in the development of isolated systems which have largely ignored the value of the data to other parts of the same organization. For example, GLP requires that *curricula vitae* are maintained which demonstrate that a scientist is competent to carry out a particular function. It would seem reasonable to keep such records within a LIMS. However, it is likely that the Personnel Department is already holding the same information using its own computer system. A more effective solution that avoids many of the problems associated with keeping multiple copies of the same data up to date would be for a LIMS to have access to the required Personnel records.

Obviously, the above example requires a system that 'sees' beyond the boundaries of the laboratory information and understands data held by other arms of the same organization. The difficulties associated with the development of applications software aimed at solving a particular problem also show within a LIMS as well as between a LIMS and the rest of the organization.

For example, consider a collection of application programs each aimed at handling a particular analytical test. In order that the data are analysed and presented in the correct context it is often necessary to store data such as time of sampling, and information about the sample or donor, in addition to the actual result. Computer software systems developed to handle such analytical results often stored this additional information for their own use and disregarded the fact that other systems on the same machine were storing identical information for their own purposes. Thus duplication of data flourished, computer resources were wasted, and the goal of data as a shareable resource within an organization was made unattainable.

Moreover, problems arose when any of this duplicated information had to be changed. For instance, if a sample volume was found to be incorrectly measured, then all systems that held that information would have to be reviewed and corrected.

The obvious step was to make sure that all information required for the whole analytical data handling system was held in a common store accessible by all parts of the system. Figure 9.1 depicts a computer system developed without the use of a database management system. Many data files are created by individual programs.

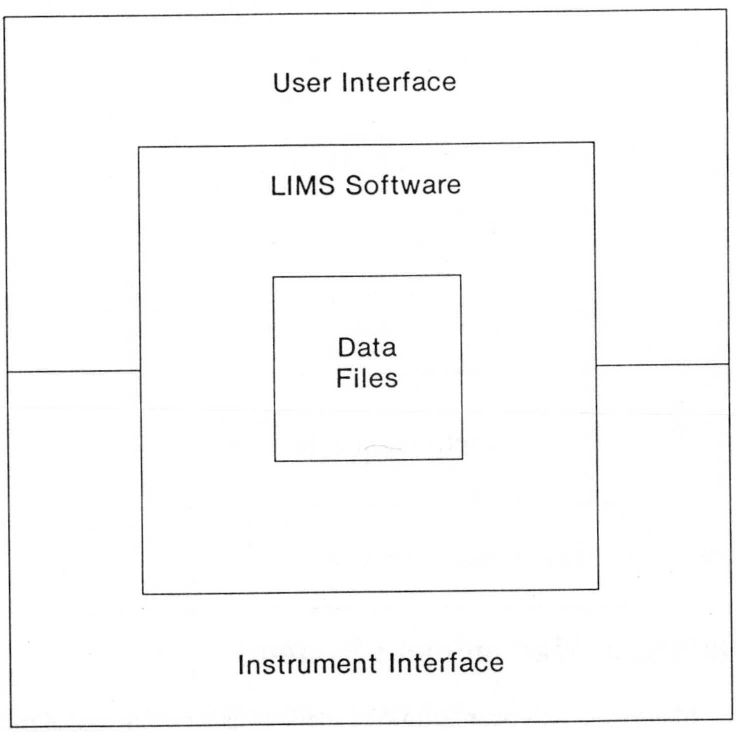

Figure 9.1 The interaction of data files and software without a database

The problem with such monolithic file systems was found in their inability to cope with the changing demands of the users of the systems for new analytical tests and different ways of reporting the data. At this point the technology of the DBMS was introduced. Figure 9.2 shows that the database provides an interface between the actual data files and the software of a LIMS. The responsibility for dealing directly with the data is that of the database, and the LIMS software only handles data through it. A DBMS enables the sharing of data across apparently independent functions in an organization. This is far more difficult to achieve without a common means of managing the data. Modern databases provide more than just a complex data file handling capability as many now offer a query facility for the user to search and retrieve data and a report writer to format the output.

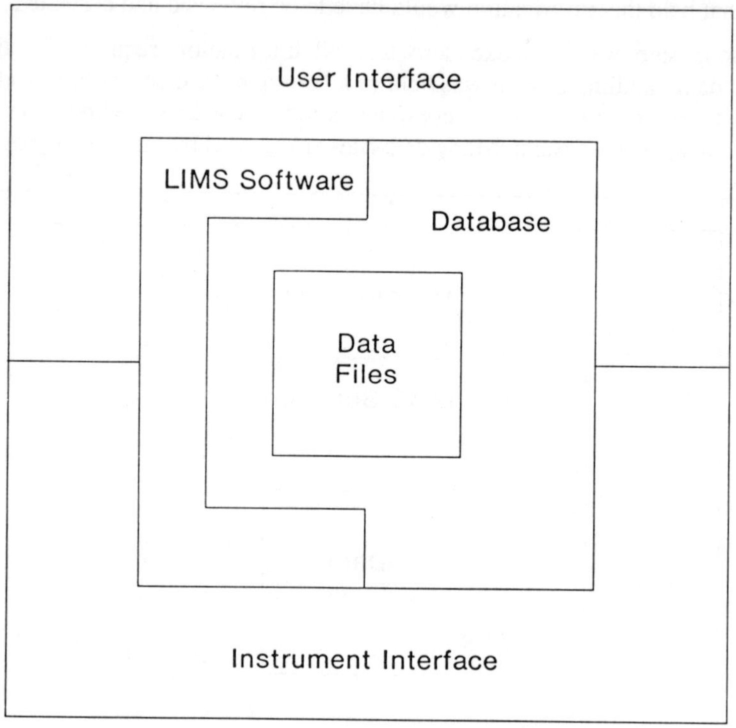

Figure 9.2 The interaction of the database with LIMS software.

9.1.1 Database Management Systems

A Database Management System (DBMS) is defined as a system that allows multiple independent users to have concurrent access to a central repository of information (1).

The key concepts embodied in this definition are:

1. *Multiple independent users* For an organization to realize the full value of the information it holds, it must be available to many users throughout that organization. If the users come from different functional areas of an

organization then the 'view' they each require of the data held may vary considerably. The scientist at the bench will be most interested in the analytical tests under their control. The manager may be more interested in the throughput of samples and the tests undertaken. The Patents Department may want to examine all the relevant data held for a particular potential product. Thus the data must be held in a manner that maintains independence from the various applications that may wish to use it, but also makes its easy for new applications to be developed which look at the data in different ways.

Most importantly, a major purpose of the DBMS is to maintain a single occurrence of any one data item: no duplication, no redundancy.

2. *Concurrent access* For the data to be properly available to the organization, more than one user at any one time must have access. Note that control of access to the data is of vital importance; any system that fails to provide comprehensive security facilities should be rejected at an early stage.

   In any LIMS the data are not static but changes frequently. How is access to the data controlled in this environment? If two users wish to read the same analytical result then the DBMS should support this. However, if one of the users is reading the result in order to change it (eg. apply a correction to an incorrectly recorded result) should the second user be allowed to read the result which is undergoing change? There are various strategies for handling this situation; their effect should be carefully considered when selecting the database.

3. *Central repository* To be available to the multiple users, the data must be stored in a central repository. This does not mean that the data must all be held centrally in one large computer system but rather that any user may access the data required wherever it is held. Many of today's databases are able to hold the data distributed across a number of computers (distributed database). It is the DBMS that presents the data to the users as if it were held in a single system.

4. *Information* Information is anything that users wish to store that is relevant to them achieving the goals of their job and providing information to other users to achieve their goals. The consequence of this wide definition of information is that the data the DBMS is expected to store can be very complex in nature. Information is often complex in nature and if the DBMS is unable to cope with the required complexity then its benefit is reduced.

In summary, a database enables the storage, retrieval and reporting (ie. management) of logically related data in an effective and efficient fashion. It avoids unnecessary duplication of the data and makes it available for use by a number of applications. Data are stored independently of the application that requires them and, thus, do not constrain development of new applications to fit any pattern established by past developments.

In determining whether a particular DBMS is suitable for use in a LIMS, primary consideration is given here to what the user of the system should expect to find. The computer systems group will have many additional points to consider and some indication of these is also given.

9.2 Assessing a DBMS

In order to assess the qualities of any DBMS that is offered as part of a LIMS package, seven fundamental questions need to be answered:

1. What can I store in the database?
2. How do I 'talk' to the DBMS?
3. How do I get my data into the database?
4. How do I access the data?
5. How can I report the data I have retrieved?
6. How is security and integrity of the data ensured?
7. What support does the supplier offer in additional services?

These questions can now be expanded to provide an assessment of each of the DBMS products that may be proposed whilst selecting a LIMS. The points are summarised in Table 9.1 as a possible checklist for database evaluation which can be tailored to an organization's requirements.

9.2.1 Data Storage - *(What can I store in the database?)*

Data comes in many forms and databases vary in their ability to handle this variety. Some of the forms encountered are identified below and range from standard data types that you would expect any DBMS to support, to more complex types which will not be common.

- *Alpha Data* constructed from the characters A to Z, usually both upper and lower case, although this should be checked.

- *Alphanumeric Data* constructed from the characters A to Z and digits 0 to 9. Punctuation characters and other printing characters are often included under this data type.

- *Integer* Whole numbers, both negative and positive, having no decimal part.

- *Fixed decimal* Numbers, both negative and positive, having a decimal portion which is of predefined size when the number is output.

- *Floating point* Numbers, both negative and positive, having two parts, a fixed point part and an exponent part, eg. 1.23 10E-3 which is a common method for representing the number 0.00123.

- *Boolean* Data that can have one of only two values, *TRUE* or *FALSE*.

It is an essential prerequisite that any DBMS should support all of the above data types. Lack of support for any one may cause serious problems in the use of the LIMS.

Less common data storage formats are:

- *Date and time* Some systems may handle values for date and time in special ways. Depending on how the date and time is held internally by the DBMS it may be possible to carry out arithmetic on dates and time values. For example, it may be possible to ask the system to retrieve all results entered into the database since TODAY - 5, ie. 5 days ago.

  All DBMSs should offer a variety of formats for displaying or printing the date. The different conventions for displaying the date between Europe and the USA are a constant source of confusion. The following are formats that might be expected:

  DD/MM/YY eg. 23/6/87 (23rd June 1987)
  MM/DD/YY eg. 6/5/85 (6th May 1985)
  DDMMMYY eg. 12OCT86 (12th October 1986)
  DD-MMM-YYYY eg. 25-JAN-1976 (25th January 1976)
  YYMMDD eg. 860905 (5th September 1986)

- *Arrays* If available, arrays can offer a way to introduce repeating groups of data in a controlled fashion. Strict data analysis techniques would lead to designing a database for storing three replicated results as:

  Sample No, Replicate No, Replicate Value
  Sample No, Replicate No, Replicate Value
  Sample No, Replicate No, Replicate Value

  An array would store this data as:

  Sample No, Replicate Value Array

  Where 'Replicate Value Array' is an array of three elements, each of which holds the value of the replicate and the replicate number is implicit in the number of the array element.

- *Text* While most DBMSs will support the storage of textual information as alpha or alphanumeric data, the storage of such data will (in most cases) be restricted to fields of fixed and predetermined length. In designing the database the analyst will have decided, for example, that 30 characters will be sufficient to store the name of a test. Whilst many fields of textual information can have predetermined maximum lengths, certain data requires a far more flexible method of storage to meet the needs of a LIMS. If, for example, scientists are expected to record significant observations on the results of a series of tests, they cannot necessarily be constrained to limiting their observations to fit in the fixed length field; they should be allowed to enter text without unnecessary constraints.

Note: It is not adequate for a supplier to claim that text can be stored in an unrestricted fashion if this is actually achieved by defining fixed length fields of a very large maximum size. In many cases this would result in each occurrence of that field being allocated space to the maximum size of the field, even if this meant that most of that space was occupied by blanks.

If the DBMS offers support for text handling, you might also look for useful additional facilities such as spelling checkers and thesauri.

- *Missing values* Numerical data can have three standard values: a positive value, zero, and a negative value. In many instances the fact that a value is absent, eg. not recorded, also has significance. This fact is recorded by assigning to the datum point the value of 'missing'. The difference in averages obtained from three replicates of a test result can be seen if one datum point that is absent is taken to be zero rather than missing.

Replicate results	Replicate results
24	24
0	missing
26	26
--	--
Ave 16.67	Ave 25

9.2.2 Interacting with the DBMS *(How do I talk to the DBMS?)*

The previous chapter discussed the user interface; the comments in this section are specifically directed at databases, however, they are also intended to amplify some of the points concerning how the users may perceive a system. Experience indicates that as users become more familiar with a system they tend:

1. To want to reduce to a minimum the number of keystrokes required to provide the information required.
2. To want to reduce the amount of prompting provided by the system in eliciting that information.

Thus, although a menu-driven system satisfies the first point it will be disliked by more experienced users because of the large amount of prompting given by the system to elicit a single character response.

Some points for consideration when interacting with a DBMS are:

Command If communication with the database is through a command language, then the content and structure of that language must be assessed. There must be a command for each action: a command language that groups distinct functions as options within a single command may lead to unnecessary confusion among the users.

The syntax rules for the various commands should be consistent, once the user understands the way one command is structured then other commands should fit the same pattern. For example, given the following two commands: *DISPLAY* which displays the contents of a record on a device which is usually the user's terminal, and *PRINT* which directs a copy of a record to a device which is usually a hard copy printer, and given that both commands allow the user to supply a device name on which to *DISPLAY* or *PRINT* the record, then the manner in which the DBMS command language allows these devices to be specified should be consistent. If the *DISPLAY* command has the syntax:

DISPLAY device name

and the PRINT command has the syntax:

PRINT/OUTPUT=device name

then it is obvious that the user will easily become confused.

The above example is worryingly common.

A balance must also be struck between understandability and wordiness in a command language. Thus, if the command necessary to recall information from the database is RETRIEVE RECORD then though it is clear to the novice user what the purpose of the command is, the experienced user will balk at typing in the 15 characters required to issue the command. A common and acceptable way often used to give the best of both worlds is to allow use of the command in an abbreviated form. Many systems allow the user to abbreviate the command any amount as long as the system can still distinguish it from any other.

However, some DBMSs make the mistake of allowing only one abbreviated form of a command. The consequence to the user is similar to introducing a second set of commands to learn that perform the same action as the original command.

Menus As discussed in Chapter 8, are often used to simplify the user interface. The ability for experienced users to discard the use of menus altogether and employ a command language instead should be seen as a mark in favour of the DBMS under consideration.

Where menus are employed look for the following:

In a complex tree structure of menus the system should always let the user know where they are.

Menus that use letters rather than numbers for their selection are preferable, especially if the option letter bears some relation to the action it selects.

Systems that mix letter and number selection should be viewed with suspicion.

The way of leaving a menu and returning to a higher level should be consistent. Some systems go back up a level if the user enters no option at a particular menu, others have a specific option to return to the previous level. There are arguments in favour of both: fewest number of keystrokes versus elimination of unintended return to higher level by accidental pressing of the return key.

Strip menus are usually displayed at the top or bottom of the screen and often related to a particular line of special function keys on the keyboard. The user interacts with the system by pressing the required function key as indicated by the strip menu. The system changes the strip menu to a different set of actions attributed to each function key as required. Such systems can find wide acceptance because the strip menu can be rapidly changed and the user has only to press one key to invoke an action. However, they can fail if the keyboard offers too few special function keys and thus the selection of the required action needs too many intermediate strip menus. A minimum of eight function keys is required.

- *Help* It is important that the user is given all help necessary in the use of the system. This assistance is of three types: on-line help, computer based instruction (CBI), and documentation.

A user may well get to a point where he or she is lost as to what to do next. He may have forgotten the proper syntax of the command or he may be unclear about what commands are available; he needs 'help'. Reference to the manuals is a possible course but the system also should be able to provide assistance.

On-line help facilities should be available that can reasonably answer most of the user's queries. The aim should not be to provide an on-line version of the reference manual, as finding the required information in a manual can often prove a major problem. The 'help' information should be in the form of short descriptions of the commands available and their syntax. Ideally, examples should be available to illustrate the use of the commands.The user acceptance of the system will be greatly enhanced if the on-line help is given in context. For example, if the user is using the report writer element of the DBMS he should not receive 'help' information about deleting records but only about report writer topics.

On-line tutorials Many computer software packages offer on-line tutorials, sometimes referred to as self-teach or computer-based instruction (CBI) packages. These courses provide the user with a self-paced tutorial in the use of the particular package. Examples range in complexity from simple text displays which progress at a rate determined by the user, to interactive systems that provide fully documented examples and exercises in the use of the system.

Whilst CBI courses cannot fully replace proper instruction courses they are a useful adjunct to the training offered by most DBMS suppliers.

- *Documentation* To write documentation that provides the reader with the required information whilst maintaining comprehensiveness and clarity requires particular skills that all too often appear to be missing in the software suppliers' staff. No DBMS evaluation should be considered complete without a critical examination of the documentation. Look for the following items of documentation:

  Reference Manual this should be a comprehensive listing of all commands available in the system, grouped in alphabetic order rather than by function. Each entry should provide a full description of the command, its correct syntax, the messages it may return and annotated examples of its use.

  User Guide this document should be a description, organized by functional area, of the use of the commands described in the Reference Manual. From this document it should be possible for the user to understand how the various commands interrelate.

  A Primer should be written for the benefit of the novice user. Its purpose is to introduce the new user to the concepts of the system and how they fit together.

  Quick Reference Guide Any form of aide memoire is useful. A quick reference card should provide brief descriptions of each command with the valid syntax.

  Additional documentation relating to design and administration of databases will be required by the computing staff.

9.2.3 User and Instrument Interface *(How do I get data into the database?)*

One of the major tasks of a LIMS is the collection of data from instruments and scientists. The collection of data from instruments is usually handled by specific applications programs within the LIMS software. However, collection of data from the scientist may make use of data input facilities provided by the DBMS.

Data input by VDU is likely to be one of two main types, either prompted input or screen format input.

- *Prompted input* This refers to the 'question and answer' interaction that is carried out between computer and user where the computer asks for a particular data item and the user supplies the required information.

- *Screen input* Here the computer 'paints' a form on the VDU screen which resembles a printed form. The user can move a cursor around this form filling data at required points and only dispatching the whole form to the computer when satisfied with the answers.

In reviewing screen input facilities consider, at least, the following:

- Can the user ask for 'help' for the particular field that he is completing?

- If validation rules are applied to the data, when is any invalidity signalled? When the users move the cursor to the next field or only when they dispatch the whole form for further processing? The former is preferable.

- Does the system highlight the current field for data entry, eg. by putting the field into reverse video?

- Will the system keep header information such as subject information displayed on the screen whilst allowing the user to enter multiple lines of individual information such as time and sample information (ie. selective scrolling of the screen)?

For both types of data input you should also investigate the following:

- *Calculated fields* Can the software readily calculate the result of applying a complex formula involving data held in one or more fields and place the result in another field?

- *Default fields* If the user chooses not to supply a particular data point can the system apply a standard (default) answer? Is this default 'factory-set' or can it be altered by the user, or does it allow the last entered value for that field to be taken as the new default?

- *Table look-up* Can the user select the answer to a particular prompt for data from a displayed table of alternatives?

- *Force case change* It may be essential to the system that certain data be supplied in upper case characters only. If so, changing the case of the data given by the user should be handled automatically by the system.

- *Required input* Where the system must have a particular data point it should be a function of the data input facility to ensure this is given. It may not be able to ensure that the answer given is meaningful but it should be able to make sure that some answer is provided.

- *Validation* The data input facility should undertake validation of the replies it receives. Validation should include:

  Range checking Greater than, less than, less than or equal to, greater than or equal to, equal to, etc.
  Look-up list Error should be signalled if the value supplied is not found in a given look-up list of allowable values.
  Look-up relation Error should be signalled if the value supplied is not found as a result of a search of another part of the database.
  Calculated expression If the results of a calculation on the data supplied falls outside a defined range or a single value an error is signalled.
  Date and time stamps The data input facility should be able to mark the exact date and time when a particular data point was supplied.

9.2.4 Searching the Database *(How do I retrieve the data?)*

It is important that the user can retrieve the information required in whatever combination needed and that, subject to suitable security considerations, the system should place as few constraints as possible on doing this. The combinations of data types that a user may wish to associate are so varied that any system in which such relationships must be predefined is likely to prove very limiting. For example, whilst a systems designer can readily determine that a user will want to look at all the test results for a particular sample, he may not so easily spot the usefulness of a report of the test results from a samples trend.

It is highly desirable that these associations can be made when required. Of course, permanent or static relationships should also be possible as these can often be more efficiently implemented than dynamic associations.

When looking at the data retrieval capabilities of a database, the following should be covered:

- *Selection* In retrieving data from a DBMS it is general practice to form collections or 'sets' that meet particular selection requirements eg. in a fictional DBMS the command:

  FIND AGE GT 43

  will construct a set of records from the database which obey the requirement that the AGE field (subject age) is greater than 43.

All DBMSs should offer the following range of operators as standard:

GE - Greater than or equal to
GT - Greater than
EQ - Equal to
LT - Less than
LE - Less than or equal to

In addition, the following logical operators should always be available to allow the logical combination of sets:

AND - eg. SPECIES EQ RAT AND MOUSE
OR - eg. DATE GE 01-JAN-1985 OR LT 01-JAN-1987
NOT - eg. TEST EQ (LC03 OR LC07)AND NOT LC44

The DBMS may offer other logical operators for combining sets which may be useful in a LIMS. For example:

BEFORE and AFTER for handling dates
BETWEEN which is equivalent to GT AND LT

It should always be possible to build these operators into complex record selection expressions, for example:

FIND TEST EQ SP01 AND INSTRUMENT EQ SKF0032 AND (OPERATOR EQ JONES OR SMITH)

ie. find all those records of the test called SP01 which were undertaken on the instrument numbered SKF0032 by either Jones or Smith.

In supplying an operand, ie. a field value, the user may not wish to supply an exact value but a 'wild-card' which represents any combination of one or more characters instead. For example:

FIND SUBJECT EQ JO*

The wild-card is the * and means we wish to retrieve all records that have a subject name starting with the characters JO and followed by any other characters. Thus JONES, JOHNSON and JOSEPHS would be accepted whilst JAMES, JERMYN and JACKS would be rejected.

When it comes to retrieving the exact records wanted, two distinct strategies are available.

Strategy A allows the creation of a set of records using a particular record selection expression. This set is further reduced by applying to it a second selection and so on until the collection is refined to the final required set. For example:

```
FIND TEST EQ SP03
Set #1 = 115 records found
FIND #1 AND SUBJECT EQ JONES
Set #2 = 25 records found
FIND #2 AND DATE = 12-AUG-1986
Set #3 = 3 records found
```

Strategy B allows the build up of an increasingly complex record selection expression with which to retrieve the required records in a single command. For example:

```
FIND TEST EQ SP03
Set #1 = 115 records found
FIND TEST EQ SP03 AND SUBJECT EQ JONES
Set #2 = 25 records found
FIND TEST EQ SP03 AND SUBJECT EQ JONES AND DATE =
12-AUG-1986
Set #3 = 3 records found
```

Whilst either strategy results in the same collection, the relative efficiencies may be significantly different. With Strategy A the system only ever retrieves records with TEST EQ SP03 once and thereafter uses this collection as the base for searching for records with SUBJECT EQ JONES rather than returning to the whole database. Strategy B, however, always looks at the whole database applying more and more complex record selections to retrieve the required records. In the example of Strategy B, records which obey the requirement that TEST EQ SP03 are actually retrieved three times and those with SUBJECT EQ JONES twice, in getting to the final collection.

Both strategies are in regular use. Strategy A offers greater flexibility whilst Strategy B is more economical with computer resources and forms the basis of SQL, IBM's structured query language which is becoming an industry standard in database record retrieval languages.

Saving selected set Retrieving the targe records often requires complex record selection expressions. To have to derive the required expression each time the data is required would be tiresome. The DBMS should offer some way of storing either the data retrieved by a particular selection expression, the selection expression itself or, preferably, both.

- *Procedures* Often there is a need to place a logical process around a set of records selection expressions. The combination of logic steps and selection steps are often termed procedures. An example section of a procedure might read:

```
:
:
ASK REPLY 'Select either of the following species - RAT/MOUSE'
IF REPLY = 'RAT' THEN
FIND SPECIES = 'RAT'
ELSE
FIND SPECIES = 'MOUSE'
END IF
:
:
```

  An editing facility to help store and maintain these procedures should be expected from the DBMS.

  Look for the ability to pass parameters to a procedure, provision of suitable handling of errors (Are the error messages understandable?) and ability for one procedure to call another. This last facility is called *nesting* and often there is a limit to the number of levels of nesting. A limit of less than four might prove restricting.

- *Default Fields* When searching the database must the fields to be searched always be entered or does the system allow the setting up of a list of default fields which are automatically searched if no field is specified?

- *Variables* It is often most useful to hold a single data value in temporary storage for the duration of a session. This value may be derived from one or more fields in records retrieved during the session either as the direct value by calculation or, entered by the user. For example, if a procedure was executed which retrieved and reported values for any test undertaken after a certain date, then if that date is stored as a variable it could be retrieved when the procedure was run, to eliminate entering the same information.

- *Mechanisms of Retrieval* It should not be necessary for the user of the DBMS to know how the system actually achieves the data retrieval. All that the user should be aware of is that the system can successfully recover the required data in a timely fashion. It is perhaps unfortunate that in many databases the mechanism has a direct influence on the efficiency of the operation and the user must be aware of and understand it.

- For example, many systems use the principle of keys or indices, much like the index of a book, to achieve faster location of the data. The information stored may be indexed under a number of levels of keys, eg. primary, secondary, or not indexed at all. In some systems the user will need to be aware how the information is keyed so as to achieve the fastest retrievals. In some systems the user may not be able to directly retrieve non-keyed information at all but will have to get the data via retrieval on a keyed field.

- A DBMS that requires the user to understand the mechanism of retrieval should, at least, receive a more critical examination than one where this is not necessary.

9.2.5 Reporting the Data *(How do I report the data I have retrieved?)*

Every DBMS should provide the facility to produce comprehensive reports containing the data that has been retrieved. Such a facility is usually called a 'Report Writer'. Defining reports is usually achieved by one of three methods:

1. *Command language* The Report Writer may have a specific command language like the main DBMS software. If this is so, you should examine the language and check that it is consistent with the main DBMS language. Differences, for instance, in syntax rules may lead to confusion among the users.

2. *By Forms* The Report Writer may present a form on the terminal which the user completes; giving the details of what is wanted and how it should be formatted.

3. *Interactive* Some DBMSs provide a Report Writer which displays the result of each command issued. The starting point is a very simple report format with perhaps standard titles, column headings and layout automatically worked out by the software. The user then progressively refines the layout and content of the report observing the effect of each command he issues on the report generated. This method, whilst always showing the user the effect of his commands, can prove tedious.

Having defined the report it is essential that the report format can be saved for future use.

The Report Writer should be able to send its reports to the screen, to a printer or to a permanent file on disk. Sorting data within a report should also be possible. The Report Writer should be able to handle sorting, in ascending or descending order, of fields containing alphanumeric, numeric or date information. It should also be able to sort on more than one field at once.

(a) TEST SP01 RESULTS PAGE: 1 (b)
DATE: 12-AUG (c)

(d)	BATCH NO	TEST DATE	RESULT	
	SQ1/C	10-AUG	0.017	
	(e)	(e)	0.015	
	(e)	11-AUG	0.020	
			0.021	
		12-AUG	0.025	
			0.027	
		AVERAGE:	0.0208	(f)
	SQ2/A	10-AUG	0.011	
			0.011	
		11-AUG	0.018	
			0.019	
		12-AUG	0.023	
			0.025	
		AVERAGE:	0.0178	

Note: The above results are subject to final release (g)

(a) Report title
(b) Automatic page numbering
(c) Automatic dating
(d) Column headers
(e) Suppression of related information
(f) Logical break with simple statistics
(g) Page or report footer

Figure 9.3 A report

A major function of the Report Writer must be to provide an intelligent, automatic formatting of the data. Thus, column titles, pagination, spreading columns to fit the page should be done by the software. However, it is essential that the user can override these automatic placements and impose his own requirements.

The user should also be able to add headers and footers to the report and to each page, justify fields to left, right or centre, insert logical breaks to distinguish groupings of data, suppress unnecessary repetition of data and undertake some simple statistics. Figure 9.3 illustrates some of these features.

Whilst most Report Writers can offer reporting of the data as shown in Figure 9.3, few can offer the facility to cross tabulate the data as in the Figure 9.4. Cross tabulated format is often very useful in reporting scientific data. Any Report Writer offering this capability as standard must be viewed more favourably.

TEST SP01 RESULTS — PAGE: 1, DATE: 12-AUG

BATCH NO			BATCH NO		
SQ1/C			SQ2/A		
10-AUG	11-AUG	12-AUG	10-AUG	11-AUG	12-AUG
0.017	0.020	0.025	0.011	0.018	0.023
0.015	0.021	0.027	0.011	0.019	0.025
AVERAGE: 0.0208				0.0178	

Note: The above results are subject to final release

Figure 9.4 A cross tabulated report

Reports should not be restricted just to text. Graphs form a major part of reporting scientific data. When looking at the graphing capabilities of a Report Writer assess the quality of the output. If it is of sufficiently high quality it may then be included in a formal report without further work, thus saving time and effort. Look for the following graph types as standard: bar and pie charts, scatter diagrams and line plots.

- Is the graph linked to the statistics of the report to allow you, for example, to add standard error bars around a mean point?
- Is the graph linked dynamically to the data so that changing a data value reflects immediately in the graph rather than requiring the recompilation of the graph?
- Have you control over format and text placement as in the text part of the Report Writer?
- Is the method of defining a graphical report similar and consistent with the way text reports are constructed? If not, then the users will encounter more problems than necessary.
- Can graph definitions be stored and edited like text report definitions?

9.2.6 Database Integrity *(How do I ensure security of the data?)*

Having spent much time and effort in acquiring the data the user will want to be sure that the data is stored safely from deliberate illegal access or accidental loss. Whilst many considerations of integrity will be fielded by the systems group, the user should be aware of what issues are being considered.

- *Recovery from system failure* Journalling or keeping to a transaction log is a technique for recording all transactions which alter the database content. Along with back-up copies of the database a journal can be used to restore the content of the database after a major problem so that, at best, only the record currently being entered or modified during the crash is lost.

- *Tracing database changes* Audit trails are used to record the changing of any data in the database. They are of vital importance where the complete life history of a data point has to be recorded. Whilst many DBMSs have the ability to record what was changed, by whom and when, few allow the user to enter any detail of the reason for change. In many industries where regulatory authorities require the organization to provide documentation of any changes to original or derived data this lack of provision of a 'reason for change' entry in the audit trail is a major shortcoming.

- *Security of access* Security of the data from unauthorized access must be carefully considered. Having such a highly valued asset as the data normally held in a LIMS, the company will want to ensure that it is held securely. The DBMS should provide security checks at all of the following levels of data:

  Database.
  Record within the database.
  Field within record.
  Procedures, graphs and reports.
  Forms.
  Database definitions.

  Protection can be achieved by the proper use of passwords and user identification. Look for the security facilities offered and consider whether you can convince the potential users of the system to accept and abide by these security measures. Without their acceptance the most complex security facilities can be compromised.

- *Security monitoring* The DBMS can monitor breaches or attempted breaches of security. Without the information provided by a comprehensive security monitor the user will not know of any potential or successful breaches of the system.

9.2.7 Support *(What can I expect the supplier to offer in services?)*

In some cases the suppliers of the database and the LIMS will be the same. If, however, the LIMS supplier has chosen a DBMS from another source and has avoided the trap of re-inventing the wheel, then the user should assess the supplier of the proposed DBMS. Primary questions are whether or not they are likely to remain in business throughout the life of the LIMS and whether they will continue to support the product for the same period.

The supplier should offer comprehensive support in the use of the product. Look for:

- *'Hot-line'* The hot-line is a telephone number you can use to get immediate assistance with problems encountered in using the DBMS.

- *'Bug' reporting* All software packages of the size of a DBMS will inevitably have a small percentage of errors in the software. There should be channels for reporting these to the supplier and also to find out if problems encountered by other sites will affect you. The supplier should report to all sites any proposals for correcting the problems and interim solutions if they are available.

- *Start-up services* To help you get going with the product the supplier should provide start-up services to 'hand hold' you in the early usage of the product.

- *On-site training* Training is always better done using the database as configured for actual use in the laboratory's LIMS, as specific examples are easy to understand. The supplier should be able to offer training carried out on-site using the LIMS set-up.

- *Consultancy* The supplier should offer consultancy in using the product. After any initial set-up services the purchaser will probably have to pay for any consultancy undertaken by the supplier.

9.2.8 Miscellaneous

Finally, there are a number of general points that should be considered when assessing a DBMS.

- *Concurrent databases* As well as the LIMS database the software may be responsible for other databases such as one for chemical structures. The user may wish to relate information in the LIMS database with chemical structure and other information. This requires more than one database to be 'open' or available concurrently. If any limit is placed on the concurrency then its effect should be assessed.

- *Indexing* The DBMS will probably provide the ability to create an index of the contents of one or more fields in order to optimize data retrieval speeds. Users should establish how many indices can be formed and whether any limit is likely to affect the use of the system even though the actual mechanism of data retrieval should properly be transparent to them.

Background mode operation The DBMS should be able to function in background or 'batch mode'. If the user carries out a transaction with a computer by issuing a command, waiting for the result, issuing another command and waiting again then the user is operating in 'interactive mode'. If a file containing a list of commands (a job) is created which is then submitted to the computer for execution and the results are only viewed when the whole job has been completed this is termed batch or background mode as the operator is free, during the execution of the job, to issue other commands to the computer in interactive mode. Obviously, the ability to carry out more than one job at once is a great advantage and should be looked for in any DBMS.

Database restructuring It is inevitable that the database design will have to be changed over time either as a result of errors found, inefficiencies, or of changing requirements for data storage. You should carefully assess what effect modifications to the design will have on the LIMS.

Can the user add and remove particular fields?

Can the length of fields, their data type, name and validation requirements be changed by the user?

Can the expressions by which certain field values are calculated be changed?

Can the screen form definitions be changed easily?

How do all these changes affect the existing database? Must all existing data be transferred into the new format (rebuild the database)? If this is the case the user is likely to have to restrict access to the system during the rebuild which may take several hours or days. On the other hand, does the DBMS allow alterations without disturbing access to the current data?

Linkage to other utilities A DBMS should never exist in isolation from all other functions being carried out on that computer or other computers linked to it. If an electronic mail system is used for instance, staff will probably want to move data from the DBMS into such a system for dispatch to colleagues.

If a word processing system is in use then the transfer of extracted data from the database to include in constructing word processed reports is vital. Data transfer to spreadsheet packages or other applications running on the system may also be important. Therefore, communication between the database and these other software programs should be assessed.

Table 9.1

A suggested checklist for assessing a database for a LIMS

1. Data storage types (9.2.1)
- Alpha
- Alphanumeric
- Integer
- Fixed decimal numbers
- Floating point numbers
- Boolean
- Date and time
  - European
  - US
  - US military
- Arrays
- Text
- Handling of missing values

2. Interacting with the DBMS (9.2.2)

1. Command
- Single command for each action
- Consistent syntax
- Abbreviations possible?

2. Menus
- Position in tree
- Mix letters and numbers
- Use with command language
- Consistency of operation
- Strip menu with sufficient function keys
- Pop up menu
- Light pen
- Mouse
- Joystick
- Windows
- On-line help
- On-line tutorials

3. Documentation (9.2.2)
- Reference manual
- User guide
- Primer
- Quick reference guide

4. Data entry (9.2.3)
- Prompted input
- Screen input
- Calculated fields
- Default fields
- Table look-up
- Force case change
- Required input
- Concurrent data access

5. Validation of input (9.2.3)
- Range checking
- Look-up list
- Calculated expression
- Look-up relation
- Date and time stamps
- Command language

6. Searching the database (9.2.4)
- Selection options
- Logical operations
- Predefined searches
- Default fields
- Variables
- Procedures
- Nesting depth
- Wild cards
- Mechanisms of retrieval

7. Reporting the data (9.2.5)
- Command language
- Consistency with DBMS software
- Forms
- Interactive report writer
- Saving report criteria
- Add header/footer
- Sorting data
- Insert logical breaks
- Cross tabulate data
- Graphical output
- Bar charts
- Pie charts
- Scatter diagrams
- Line plots
- Linkage to statistics
- Dynamic linkage to data
- Format and text placement
- Consistency with report writer
- Storage of graphical definitions

8. Database integrity (9.2.6)
- Recovery from system failure
- Audit trail
- Ease of searching audit file
- Security of access
- Security monitoring

9. Support (9.2.7)
- Hot-line
- Bug reporting
- Start-up service
- Consultancy
- On-site training

10. General points (9.2.8)
- Distributed database capability
- Indexing
- Background operation
- Ease of restructuring DMBS
- Linkage to other utilities
- Understandable error messages
- Communication to other computers and packages

9.3 DBMS Types

There are three main DBMS types which are based on the three major methods for modelling data on computers. The types are the relational, network and hierarchical models. The characteristics of each will be briefly examined.

9.3.1 Relational Model

The relational DBMS (RDBMS) model is, perhaps, the most easy to understand, especially for the scientist, as a relation is no more than a table containing data. Each horizontal row is an instance of a record; each column heading is an attribute.

The example of two relations that might be used in a LIMS is shown in Figure 9.5 where the test relationships are displayed and those for the sample in Figure 9.6.

Relation: TEST

Test Name	Test Ref	Calib. Ref	Result Units

Figure 9.5 The relationships for 'test'

Relation: SAMPLE

Samp. Ref	Samp. Time	Samp. Vol	Test Ref	....

Figure 9.6 The relationships for 'sample'

The query types that the relational model offers are:

- *Projection*
  eg. PROJECT TEST NAME
  would form a list of the names of all the tests.
- *Selection*
  eg. SELECT SAMPLE REF = AH101/C
  would form a collection containing any records in the sample set with the sample reference of AH101/C.
- *Join*
  eg. JOIN TEST AND SAMPLE OVER TEST REF
  would result in a new relation of the form shown in Figure 9.7.

Relation: TEST and SAMPLE

Samp. Ref.	Samp. Time	Samp. Vol	Test Ref	Test name	Calib. Ref	....

Figure 9.7. The relationship between sample and test

- *Union* The union query allows two collections to be combined if they have the same attributes. Thus, if we created a collection containing all tests carried out on a Sample XYZ and a second collection containing all tests carried out on Sample ABC then the union would be all records containing test information for Samples XYZ or ABC.
- *Difference* The difference query when applied to two sets creates a new collection of the records which can be found in one set but not in another.

Relational database management systems therefore present a limited number of relatively simple concepts for the user to grasp. When originally proposed by Codd (2) in 1970 they were regarded as likely to result in performance problems due to the requirement for substantial computer resources in order to to function correctly. This has not been borne out by experience and many popular implementations of the relational DBMS are now available. Examples are: ORACLE, INGRES, RDB, BASIS-DM, INFORMIX, SYSTEM-R, SQL/DS and DB2

9.3.2 Network Model

The network model has some of its roots in the presumption of inefficiency of the relational model. The CODASYL Data Base Task Group proposed the network model in 1971 (3) it found wide acceptance for a time and many products based on it are still in use despite the rising popularity of the relational model. This is because the investment in network model databases is so great that few users can afford the high cost of changing.

The main concept of the network model is that of linking two record types together. The relationship between the record types may be many-to-one, ie. a sample may be subjected to many tests or one-to-one, ie. a subject has only one father.

The association or linking of records for a set of test records with the sample record is called the linking of set members with set owner. (Note that this is a mathematically incorrect use of the term 'set' which leads to confusion.) The following examples show diagrammatically, two instances of the set of sample tests.

Owner: Sample AH101/C	
Member:	Test ABC
	Test DEF
	Test XYZ
Owner: Sample BC231/D	
Member:	Test GHI
	Test XYZ

'Navigating' sets to be able to retrieve the data can be complex and does much to explain the fading popularity of the network model. However, because linking sets and members is achieved in the computer by relatively efficient and well understood means the implementation of network systems is less problematical than relational systems. Examples are: DBMS-11 (VAX-DBMS), IDMS.

9.3.3 Hierarchical Model

The organization of the hierarchical model differs from the network model in that instead of records joined together by a named link, the structure is in the form of a tree.

For example: a sample is the head of a tree containing a number of test detail records, and each test detail record is linked to a number of replicates of the test data measurement.

SAMPLE LC01A2
TEST ABC
Replicate 1
Replicate 2
Replicate 3

TEST DEF
Replicate 1
Replicate 2
Replicate 3

Each tree is called a physical database record type. The collection of all the subordinates of a given root is called a physical database record instance. In our example, the SAMPLE physical database record type is only one of a number of record types that might exist in the LIMS database. Each record type is called a segment.

The hierarchical model is considered easier to understand than the network model, however, it may not be as easy to grasp as the relational model. The hierarchical model has achieved the widest acceptance of the three models. This is only now reducing in favour of the Relational model. Examples of a hierarchical database are: IMS, FOCUS, RAMIS.

The network and hierarchical models can each be manipulated to emulate each other's modelling strategy and both can be made to look like a relational model. In assessing a DBMS it important to know what the underlying modelling technique is so that you may judge whether or not the database will be able to cope with the flexibility you require in a LIMS.

The user should be careful to spot those databases that use one modelling system to masquerade as another, as this is more than likely to present limitations in the use of the system in handling the required data, and can only rarely be achieved with a reasonable degree of efficiency.

9.4 Summary

This chapter is intended to give the scientist an overview of a database and what is needed in a LIMS evaluation.

9.5 References

1. Engles, R.W., 'A Tutorial on Data Base Organization', *Annual Review in Automatic Programming,* eds Halpern and McGee, Pergamon Press (1974)

2. Codd, E.F., 'A Relational Model of Data for Large Shared Data Banks', *Communications of the Association of Computing Machinery,* 13, (1970) 377-387

3. CODASYL Systems Committee, 'Feature Analysis of Generalised Data Base Management Systems', Technical Report, *Association of Computing Machinery,* New York, (1971)

Additional Reading.

Churchman, C.W., *The Systems Approach,* Dell, New York, (1969)

Date, C.J., *An Introduction to Database Systems Volume 1 Fourth Ed,* Addison-Wesley (1986)

King, J.M., *Evaluating Database Management Systems,* Nostrand Reinhold, New York, (1981)

Martin, J.M., *Computer Data-Base Organization,* Prentice-Hall,New Jersey, (1977)

Sibley, E.H., 'How to Select and Evaluate a DBMS', *Journal of Information Systems Management,* 2, (1986) (2) 40-49

Chapter 10

Interfacing and Communications with Analytical Instruments

J.C.Pearce

10.1 Introduction

A major function of a LIMS is to integrate sample information with the corresponding analytical results. This is achieved reliably and automatically by interfacing or connecting the analytical instrument to a LIMS computer. However, there are many factors to consider. This chapter will look at the major problems and their possible solutions. When considering the interfacing of analytical instruments, the questions to ask are:

1. What are the advantages of a connection to a LIMS?
2. What would be achieved?
3. Will the benefits outweigh the technical difficulties?

10.1.1 Interfacing - The Advantages

The advantages of interfacing laboratory instruments and passing results directly into a LIMS are primarily concerned with the philosophy that the manual input of data becomes redundant and the inevitable transcription errors that accompany it are eliminated. There then exists a direct correlation between the sample identity and the corresponding analytical result - an essential commodity. As seen in Chapter 4, the greatest benefits of a LIMS are gained by the laboratory that transfers its data automatically.

It is also worth noting that interfacing of instruments to a LIMS offers the additional benefit that all data will be collected and manipulated in a uniform manner (1). This has the advantage that the derived results will be consistent and allow easy review, comparison and conformity to regulatory agency guidelines. In addition, the electronic transfer, manipulation and interpretation of data is both rapid and reliable.

10.1.2 Interfacing - Is it Worthwhile?

Before an instrument can be interfaced to a LIMS, or any computer, the conversion of an analogue signal to a series of digital numbers that is representative of it is necessary. Analogue to digital conversion is explained in greater detail in the next chapter. Once in a digital form, data can be manipulated, reduced and transmitted easily by computers. Digital data can be sent to a LIMS directly or via a work station or microcomputer, or the whole process can be undertaken by an intelligent instrument. The various ways are shown in Figure 3.1.

Interfacing is the term given to the connection of an instrument to a computer and it involves two stages. The first is the physical connection between the two processors and the second is the transfer and interpretation of the transmitted data. As discussed in Chapter 1, the question of what data to send to a LIMS is crucial. For our purposes here, it will be assumed that data transfer involves some form of reduced data: results or summary tables in either binary or ASCII files.

The usefulness of interfacing is dependent on the frequency of transfer and the use to which the data are put within a LIMS. To look at this question in more depth, consider the author's laboratory. There are a number of microprocessor controlled instruments, eg. thin-layer radio-chromatographic detectors, which would be relatively easy to interface to a LIMS. However, as the data are displayed, evaluated, reduced and reported locally by the analyst, it is difficult to envisage any further benefits being offered by interfacing. If the data were to be used further, for instance in calculations, then the user would interface the instrument. However, if data were only transferred infrequently, then the cost might not be justified.

For some items of analytical equipment, eg. spectrometers, it is not only of little use but also not cost effective to attempt a direct connection, as the software for dedicated instruments offers greater flexibility than that currently available for a LIMS. The original signals can be efficiently reduced and processed locally and then a summary file can be sent for inclusion in the database. Where interfacing is not practicable, often a manual entry in the LIMS giving the reference number and location of the spectrum plus a few lines of free comment may suffice. Interfacing this type of equipment is an area for development in the future.

10.1.3 Interfacing - The Problems

The underlying problem is that there is no accepted standard format for the transfer of data to and from analytical instruments (2). This is manifested in a number of ways: the different sockets used for output, the wiring standards, the control codes used for transmission and the file format all differ, even from the same manufacturer.

Another important factor is the age of the instruments. It is rare that an automated laboratory can be built from first principles (3), so potential users will be faced with the task of interfacing their existing instruments to the system. If they are older than five to seven years, then it will be more difficult to connect them and it may be more cost effective to replace them. The capital cost of making such a decision may seem daunting, but the purchase of a data station or an intelligent instrument will certainly make the task of interfacing easier. Also, the labour cost for the alteration is usually underestimated and the time taken to complete the exercise must also be taken into account.

Interfacing requires that the hardware and the software function correctly. Making the hardware connection could, in theory, be as simple as attaching a suitable cable between the LIMS and the instrument. However, this is rarely true in practice. Compatibility problems may arise because the communications settings on each processor are not the same, factors here are the baud rate, parity bit setting and the number of stop bits. Also, there can be confusion about which processor is the sender and which is the receiver. The introduction of breakout boxes, to visualize the way the signals are sent, has done much to overcome this type of problem. In essence, this equipment allows communication parameters to be easily changed and established until the right combination is reached. These parameters can then be installed upon the system in a permanent form.

Once the hardware is configured correctly, there may still be a problem as the data transmitted cannot be interpreted by the LIMS. The most probable explanation is that the format of the data is not compatible with the software of the recipient. There is no standard format for data file content and potentially each different instrument in a laboratory may require custom software to be written for it. Moreover, it may be necessary to reformat the file or rely on software within the LIMS to interpret it.

10.1.4 Interfacing - The Solutions?

Perhaps the most significant constraint on the development of the automated laboratory in recent years has been the inability to connect existing analytical instruments to a LIMS computer in a simple and reliable manner. The same argument is also true when we consider the requirements for a communication between a LIMS and other computers. The good news for the analyst is that recently a variety of approaches have been developed to try to overcome this hardship. Before looking at these, communications protocols are be discussed.

10.2 Communications Protocols

There are several protocols that can be used for interfacing equipment; the remarks here will be limited to the two major ones used in the laboratory - RS-232C and IEEE-488.

10.2.1 RS-232C

The RS-232C protocol is the most common method used for data transfer between analytical instruments and a computer. However, it is only a 'Recommended Standard' which manufacturers can implement in different ways; this is a standard without standards. In its simplest form, it can consist of two 25 pin D-shaped plugs connected by a twin twisted wire. Traditionally, pins 2 and 3 of the plug are designated transmit and receive respectively with one further connection at pin 7 for signal ground (Figure 10.1). Additional lines can be used for more sophisticated communication, but unfortunately there is no standard for assigning these and different manufacturers may (and do) adopt different variations which complicates the situation further.

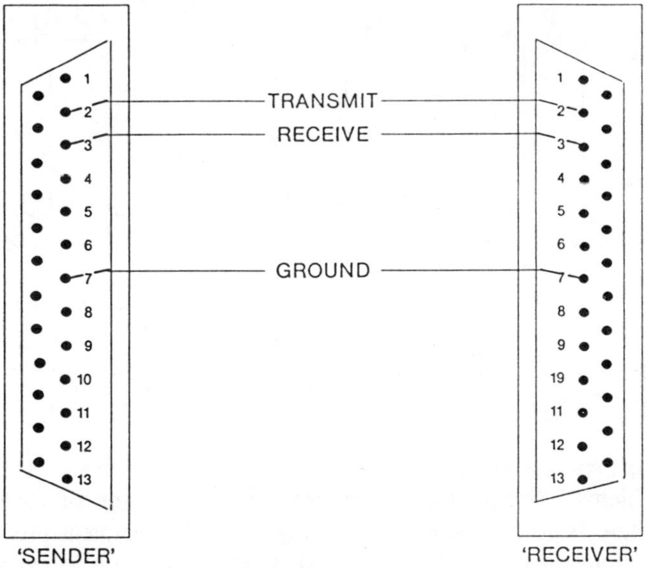

Figure 10.1 RS-232C minimum pin connections

Data, either numbers or letters, are passed between an instrument and a computer as binary encoded characters, the accepted format for which is the American Standard Code for Information Interchange (ASCII). All numbers, letters, as well as characters for the computer to use are represented by a series of seven or eight bit 1s or 0s called a 'byte'. These are transmitted along the line as a series of voltage pulses, a value of -5 to -15v represents a 1 and one of +5 to +15v represents an 0.

To inform the receiving processor that information is to be sent, a start bit is transmitted; similarly, there is also a stop bit inserted to signify the end of the character. Rudimentary error checking is sometimes used by the receiver to check whether the character sent was correctly received; a parity bit is sent as the last bit of the character string, this is chosen so that the number of 1 bits in the signal is always either odd or even. When the character is received by the recipient processor then it sums the number of 1 bits and compares this with the parity bit; if they are the same, the sent and received signal may have been the same. The transfer of a character as a byte is shown in Figure 10.2, the signal has a start bit, a parity bit and two stop bits. This will be followed by the next character to be sent which will also have a start bit to indicate that transmission is to begin. For further information on this subject, the reader is referred to the excellent article by Dessy (4).

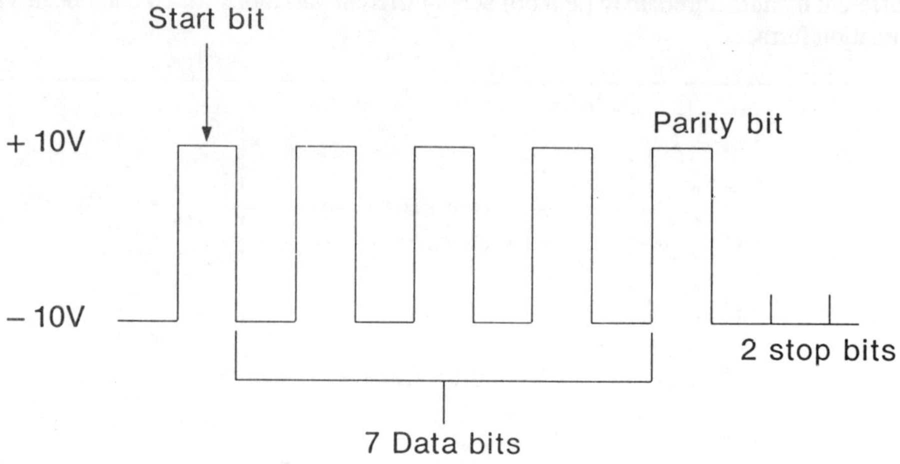

Figure 10.2 Representation of a character transferred by RS-232C

Once the wires, sockets and speeds of transmission have been selected there are still potential problems. When data are transmitted from many of today's analytical instruments there is no provision for feedback from the recipient processor. In this instance, the recipient processor must make special provision to avoid missing any data, this may involve dedicating a line for a fixed time to ensure capturing the data. This in turn, means the computer will not be able to perform other tasks as effectively if it is prepared for data transmission. Care must be taken when interfacing instruments of this type.

10.2.2 IEEE-488

The IEEE-488 protocol was developed by Hewlett Packard (HP) in the early 1970s: and unlike the RS-232C serial interface it follows a more rigidly defined standard, including the wires and the connectors to be used, because the Institute of Electrical and Electronic Engineers (IEEE) adopted it in 1975 (5). This protocol is also known as the HPIB (Hewlett Packard Interface Bus) or the GPIB (General Purpose Interface Bus).

Instead of sending one character at a time down a single line there are eight parallel data lines and each bit of the character is sent simultaneously to the receiver. As a consequence, data transfer is very much faster compared to the RS-232C. In addition to data transmission there are five lines dedicated for control of equipment, see Figure 10.3. This communication standard has proved popular with instrument manufacturers where the distance between modules is within the maximum distance of four metres. However, when data are transferred over greater distances the cost of parallel interfacing becomes prohibitive. In this case, either the serial approach is used initially or the parallel output is converted to a serial signal for transmission.

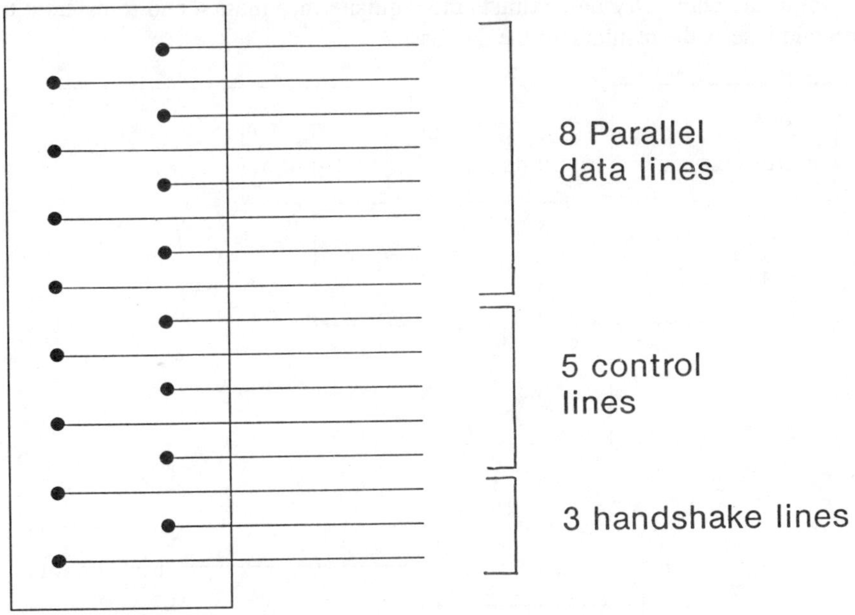

Figure 10.3 The IEEE-488 parallel interface.

10.3 The Manufacturers' Approach

The approach taken by three analytical equipment companies has been selected for appraisal. The need to interface not only their own but other manufacturers' equipment to a LIMS has been the driving force for these developments. Other companies may adopt different approaches towards the problem, but it is not within the scope of this chapter to review them all.

This section is intended to cover non chromatographic instruments as chromatographic data acquisition is the subject of the next chapter.

10.3.1 Beckman

The approach taken by Beckman to interface non chromatographic instruments is to use RS-232C output and link this either via their Mk5 Digimetry instrument interface coupler or directly to the computer as shown in Figure 10.4.

Each Digimetry unit can connect up to eight instruments. In addition to data transfer, it can also act as a terminal for the LIMS, accept sample identification via bar codes, offer instrument controls as well as perform real-time data acquisition. This unit also has a keyboard and display panel through which the analyst can pass information to and from the database. The user can write a custom program to acquire data and pass it to the central computer with the Laboratory Interface Language (LIL) which is derived from Pascal. Once the data are acquired from the instrument, the LIL program in the Digimetry unit extracts the requisite information and transfers it to the LIMS and inserts the results into the database.

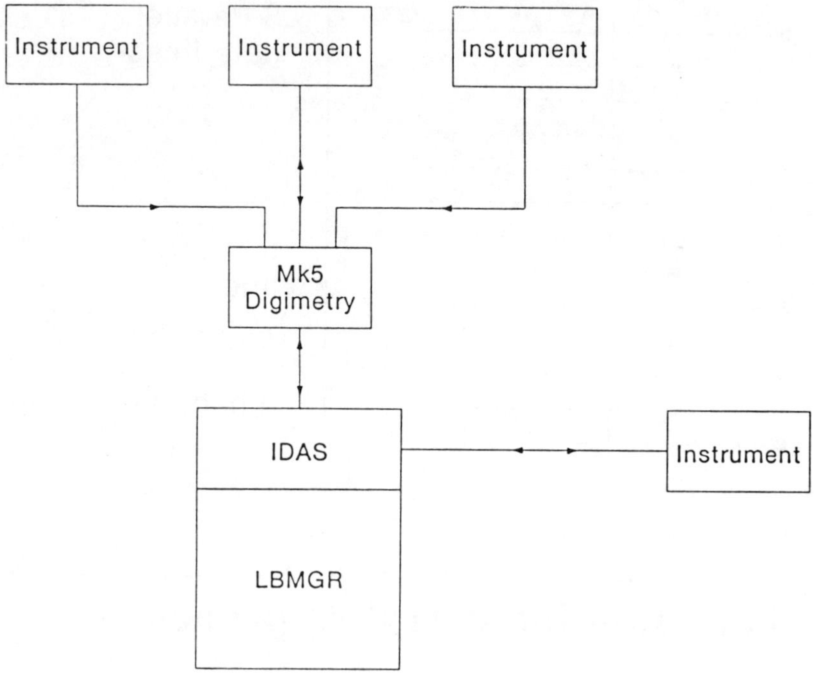

Figure 10.4 The Beckman approach to instrument interfacing

Communication with the database is useful as it allows decisions to be taken based on the answer received, eg. if the analyst needs to know if the last determination was within specification this can be relayed to the display panel on the Digimetry unit.

Within the computer another program, called Instrument Data Acquisition System (IDAS), can be used if the communication facilities of the Digimetry unit are not required. Here, as shown in Figure 10.4, files are transferred directly to the computer for processing and specific user programs written in LIL are used to extract the information required and incorporate it into the database.

10.3.2 Hewlett-Packard

The problem of interfacing both their own and other manufacturers' analytical instruments to Labsam has been accomplished in a number of ways as shown in Figure 10.5. In most of these instances both single or bi-directional communication is possible. Instruments are interfaced to the computer via RS-232C communication. A program known as the On-Line Handler is used to set up the computer to acquire the data from an instrument by correctly setting up the communications line, then the files are passed to programs which extract the required information. These programs can be user written or come from a central library, usually one program is needed for each item of equipment. The extracted information is passed to the database by the On-line Handler.

Another approach is to use a program called Logger which directs incoming files into a temporary storage area, then a second program, using Labtools is automatically triggered. Essentially Labtools is a set of routines that allows users to access certain files to abstract information and then enter it into the database.

Data are passed from Chemstations using a utility called Quickcom. Once stored inside the HP 1000, the files can be accessed using either Labtools or the LAS programs to extract and transfer data to the database. It is also possible to make the Chemstation an interactive terminal for Labsam.

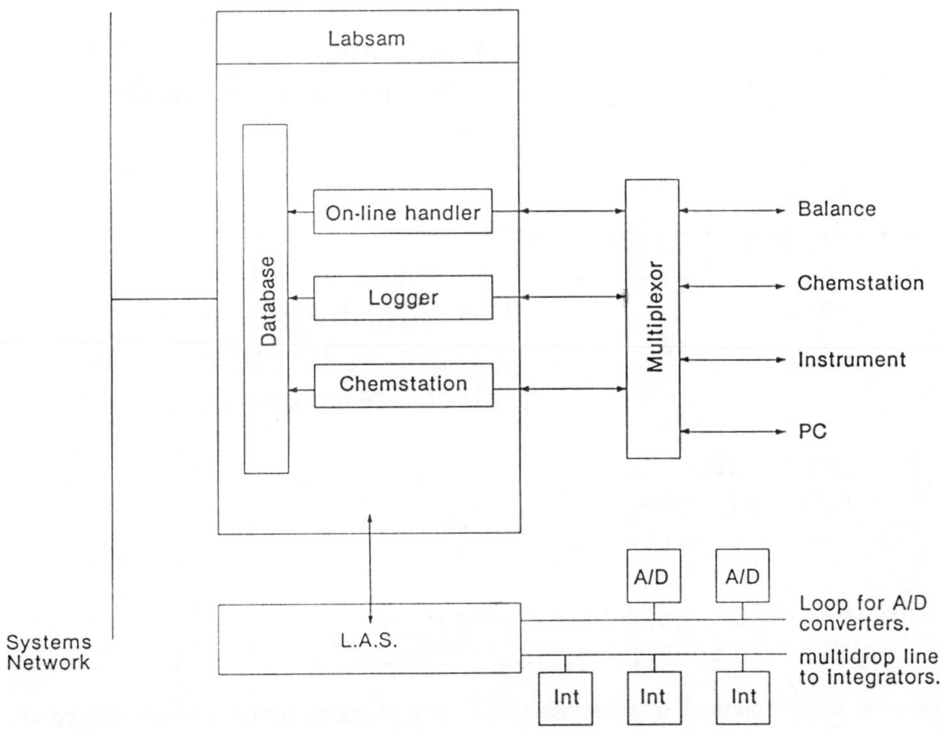

Figure 10.5 The Hewlett-Packard approach to instrument interfacing

10.3.3 Perkin-Elmer

Perkin-Elmer have for some time put emphasis on standard methods of communications, as evidenced with their 'Everywhere' concept. Ideally, any manufacturers equipment should be interfaced via an RS-232C protocol for data and file transmission to a LIMS computer.

The transfer and interpretation of the data file is shown in Figure 10.6. The way this functions is dependent on the architecture of the instrument involved. In order to cope with the widest range of instruments, there are two main methods of connection.

With the first, instruments with an RS-232C output can be connected directly to a LIMS. The output file, typically a report, is interpreted by a template within the LIMS 2000 software and extracts the appropriate data. Information is abstracted from the files by means of user defined electronic templates that highlight the information to be abstracted. If an instrument has a parallel output then this must be converted to a serial signal before transmission to the LIMS.

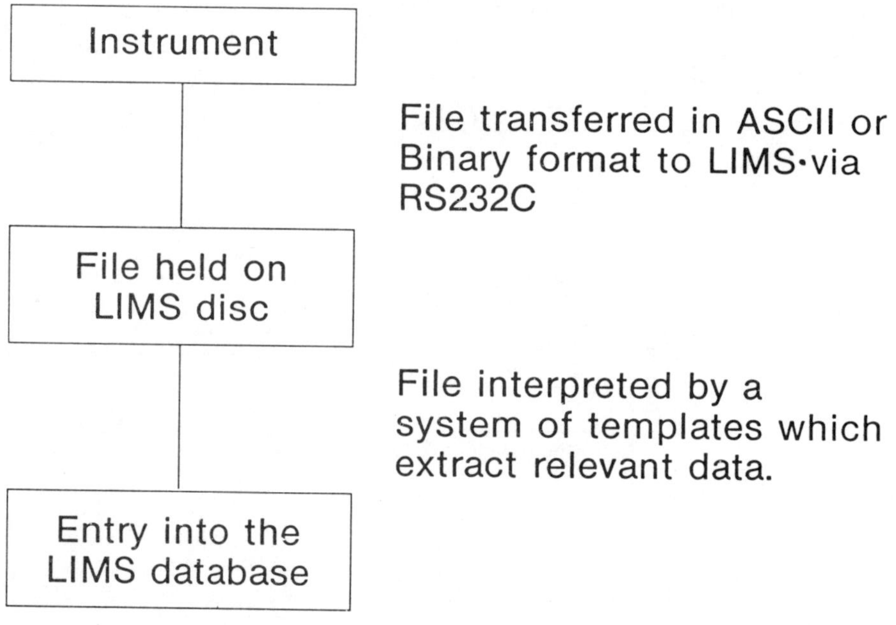

Figure 10.6 The Perkin-Elmer approach to instrument interfacing

With the second method, instruments which use a PC as a controller are connected via the micro to the LIMS. Any file on the PC disk which contains results can be transferred to the central computer by a P-E utility called TOLIMS which resides on the PC. Once within LIMS the same templates are used to extract the required information from the file.

10.4 Software Utilities

The previous section discussed what some instrument manufacturers are doing to overcome the lack of communication standards in passing files from instruments to LIMS and *vice-versa*. A question arises, can the task be done by a software utility?

10.4.1 Kermit

This is a public domain communications program which was developed at Columbia University during the 1960s. Kermit is now available in a range of computer languages for over 200 different computers, all of which can speak to and understand one another. Each computer/operating system combination requires a different version of Kermit. The required version can be obtained on tape or disk for a nominal charge; copying and further development of the program is encouraged.

The problems of data transfer are concerned with three main areas: the physical media to be used, the file structure and its content and error checking. Kermit uses the lowest common denominator to overcome these problems by using simple RS-232C terminal connections, allowing communications between most computer systems. There are no special hardware requirements necessary which is why Kermit is attractive to use. It was developed originally to overcome a local problem of transferring files between computers and was designed to be used over telephone lines for sending packets of data in either direction. Error checking routines are included to ensure that the data transmitted are correct and an immediate acknowledgement is sent from the recieving computer after each packet has been correctly received.

The initial action of the sender and receiver prior to exchanging information is for each to configure the other by sending the required parameters that will be used during transmission. In addition, Kermit can offer server mode which allows control of all functions from one end of the line once the two versions are linked.

Kermit is currently being used in the author's establishment to transfer data files from a LIMS to a DEC VAX and IBM PCs for further manipulation and reporting. The purpose of this operation is to make use of software packages on the VAX which are not available on the LIMS. The use of Kermit in this way has proved to be both reliable and highly cost effective. Once the file has been transferred to a LIMS, procedures and utilities are still required to open the files and extract the specified information, Kermit aids only the data transfer process.

10.5 Local Area Networks (LANS)

The subject of local area networks has been covered in Chapters 3 and 7 by Alan Braithwaite and Mike Bertram respectively. However, it is worth noting that a networking approach, where applicable, may overcome some of the interfacing problems outlined earlier. Laboratory instruments can be linked through appropriate interfaces to personal computers or a minicomputer, which are in turn connected to the main network. The LIMS computer and other computing facilities are also connected to this network to give the totally integrated package. The PCs themselves may also operate as part of a smaller separate network in which each is fed by a file-server.

10.6 The LIMS - Robot Interface

The demands in recent years for analysts to increase productivity has led to considerable interest in the use of flexible robotics systems (6). Many companies have invested in robotics and if this trend continues there will be as much a need to integrate a robot into a LIMS environment as any other analytical instrument. Progress has, however, been slow in this area and for good reason. Robot manufacturers have adopted their own programming languages which have been designed to make life easy for the analyst. These languages such as EASYLAB used by the Zymark Corporation, use simple procedures where the low level instructions of the computer are transparent to the user. It is ideal for independent application but attempts to interface such a system have until recently proved very difficult.

However, Abbott (7), has described the application of laboratory robotics for optimizing the physical testing of absorbent structures used in the pulp and paper industry. The system consisted of specialised testing equipment working in conjunction with robots, from which data was collected by a data acquisition system and transferred to a PC before transmission to a LIMS.

Hamilton (8) has described a system for the determination of wet and dry cell mass in fermentation processes. Use is made of the RS-232C interface of the PC used to control the robot, prompt users and direct files to a database on an HP 1000 computer. Samples can be positively identified by the use of bar coded sample tubes and a laser scanner. This allows the fermentation samples to be placed in a rack in any order and undergo measurement; when complete a list is printed by the PC giving the location and order of analysis.

These two examples, although representing a significant step towards a totally integrated robot/LIMS unit incorporate little or no feedback whereby the robot can make decisions based on the results received by the database. It is possible that this goal will not be achieved until artificial intelligence becomes part of the system (9).

10.7 Future Developments

Failure on the part of the instrument manufacturers to agree standards for the format and transmission of data has, in the opinion of the author, been a major constraint on the road to total automation within the analytical laboratory. This has been true for both direct transfer of data to LIMS and the networking of analytical instruments. Individual manufacturers have made some progress in this area. However, the much needed standardization, both in data format and communications, appears to be some way off.

Hopefully, developments from the instrument vendors towards this goal will appear within the next few years as the number of frustrated LIMS users increases and pressure is brought to bear. Only then will analysts feel confident that they can reap the full benefit of a totally integrated LIMS.

The drive for compatibility will be accompanied by the utilization of networks which allow distributed processing power capable of two-way communication with both intelligent instruments and the LIMS database. In this way, analytical instruments may be directly controlled by a LIMS, though in such a way that the performance of the central computer is not compromised.

10.8 References

1. Berthrong, P.G. and Mansfield, P.B., 'Automation Systems for Small Laboratories: Part Two', *American Laboratory*, 18 (1986) 82-91

2. Borman, S.A., 'Laboratory Computer Networks', *Analytical Chemistry*, 56 (1984) 408A-413A

3. Squirrell, D.C.M., 'The Why, When and Wherefore of Automatic Analysis', *Analytical Proceedings*, 22 (1985) 79-80

4. Dessy, R.E., 'The PC Connection Part I', *Analytical Chemistry*, 58 (1986) 678A-683A

5. Smardzewski, R.R., Microprocessor Programming and Application for Scientists and Engineers, Elsevier, Amsterdam, (1984)

6. Arndt, R.W., 'Automating the Status Quo', *Chemistry in Britain*, 22 (1986) 974

7. Abbott, J.C., 'Optimising Physical Testing of Absorbent Structures Using Laboratory Robots, Micro and Mini Computers and Information Management Systems', *International Dissolving Pulps Conference*, February 1987, pp. 237-245

8. Hamilton, S.D., 'Robotic Assays for Fermentation Products', in *Advances in Laboratory Automation Robotics*, eds Strimaitus, J.R. and Hawk, G.L., Zymark Corporation, Hopkinton, MA, (1987) pp. 1-21

9. Brady, J.M., *Artificial Intelligence*, Chapman and Hall, London, (1986), pp. 137-176

Chapter 11

Chromatographic Data Acquisition

D.W.Darkin

11.1 Introduction

This chapter presents an overview of the data acquisition systems which are available for chromatographic applications, and also considers how these systems can communicate with a LIMS. High Performance Liquid Chromatography (HPLC) and Gas Chromatography (GC) play such an important part in most analytical laboratories that they merit specific attention. This particularly involves the procedures for data acquisition and processing and their subsequent automation which leads to better efficiency.

The process of chromatographic data acquisition has developed significantly over the last decade with various degrees of complexity to suit most applications. The commercially available systems vary from low cost single-channel dedicated integrators to multi-channel systems based on minicomputers with intermediate systems available on a variety of PCs. Each system has its own area of application with specific advantages and disadvantages.

All data acquisition systems, regardless of size, require the detector analogue output to be converted or digitized to a form which can be handled and stored by computers. A variety of analogue to digital conversion techniques are available for this purpose and are discussed below.

11.2 Analogue to Digital Conversion (ADC)

Analogue to digital conversion is a process by which a continuously variable signal (e.g. voltage) from a detector or a sensor is converted to a binary number which can accurately represent the original data. These data can then be passed to a computer for processing and storage. It is necessary to convert the analogue signal to a digitized form as conventional computer systems only handle numerical information in the form of a binary code comprising of a series of zeros and ones. A typical ADC system is made up of several parts, shown schematically in Figure 11.1.

Analogue signals from a detector are fed into a multiplexor, then in turn via a filter and a gain amplifier to an ADC unit. Emerging from the ADC will be a series of numerical values, proportional to the original analogue signal, which is transferred to the computer. Each component in the ADC system is explained separately below.

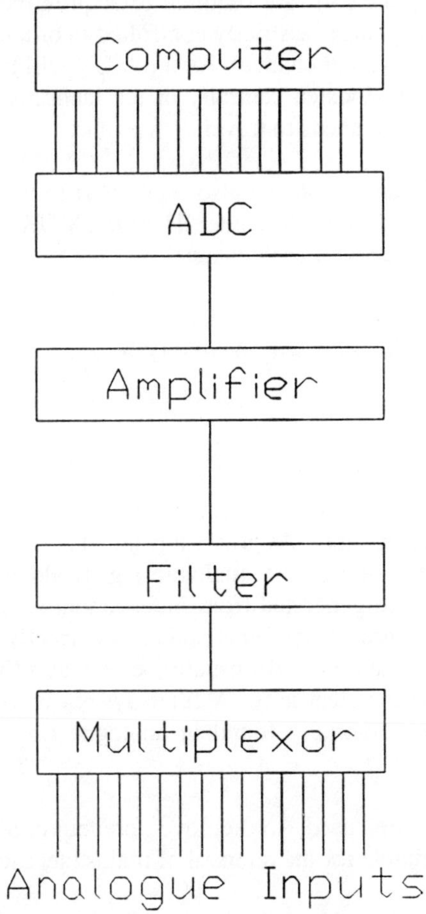

Figure 11.1 Components of an Analogue to Digital Conversion (ADC) System

11.2.1 Multiplexor

The multiplexor (MUX) is an optional part of an ADC system and is only used for multi-channel data acquisition. Its purpose is to sample each analogue input line in a fixed sequence, allowing sufficient time on each channel for the ADC to acquire a portion of the analogue signal, convert it and to pass the output to the computer. As each ADC takes several milliseconds to acquire a reading, several analogue inputs can be easily handled by a single multiplexor. However, the greater the number of channels connected, the less time is available for data acquisition.

The greatest advantage that a MUX data collection system offers is one of economy, as several chromatographs can be connected to a system. However, the cost differential is eroded by the software required to acquire the data; the sampling rate and sequence of the MUX must be strictly controlled in order to maintain the integrity of the data from each individual channel. As the MUX will be connected to numerous analogue inputs it must be located centrally in the laboratory in order to reduce the distance the analogue signal needs to travel.

The analogue to digital conversion can also occur next to the chromatograph, and the digital signal is transmitted to the computer. Here the MUX is still used to acquire the data and maintain the integrity of each channel.

Whichever approach is used the signal will have to travel a distance of several metres and background electronic noise will inevitably be generated which can degrade the quality of the signal.

11.2.2 Filters

Electronic interference is an inherent component of all analogue signals and it is impossible to remove completely. As noise, it originates from the detector and is also generated by the transmission of the analogue signal along the cable to the ADC. Further interference can be generated from independent sources such as thermostats, and are usually seen as large, sharp signal spikes. Incorrectly earthed instruments can also generate interference and this will usually be seen as a fluctuation of the signal at the frequency of the mains electricity. A filter system is used to remove or reduce much of the interference normally found in analogue signal sources and these are supplied in two forms.

1. *Notch filters* are used to remove predetermined frequency and are particularly suitable for the removal of mains electricity interference.

2. *Pass filters* are available as low pass or high pass filters and remove signal frequencies higher or lower respectively than the predetermined cut off point. Using a combination of both low and high pass filters will allow only frequencies of a certain band width to reach the ADC circuitry. This is particularly relevant to ADCs which have fast acquisition rates and are susceptible to noise interference.

Interference can be further reduced by ensuring the analogue transmission cable is properly screened. This prevents the generation of small currents in the cabling induced by external electromagnetic fields. Figure 11.2 shows how signal cables should be screened and earthed at one end to ensure induced currents do not flow along the screening.

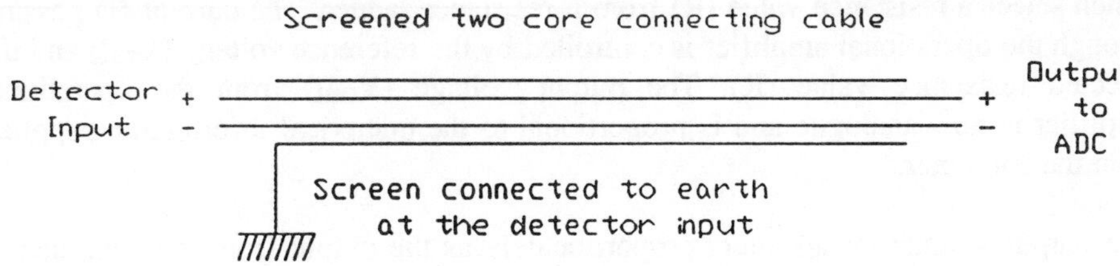

Figure 11.2 Screening of signal cables

11.2.3 Gain Amplifier

ADC circuits are optimized to operate at specific input voltage ranges, ie. 0 to 10 volts, -5 volts to +5 volts, and 0 to 5 volts. Modern chromatographic detectors have computer outputs which match the ADC inputs, but where these are not available the signal, usually in the order of a few millivolts, must be amplified to suit the ADC input. Gain amplifiers are used to achieve this; the gain on the amplifier, ie. the amplification factor, can be predetermined by switching resistors into the circuitry, or it can be controlled by the software of the computer.

11.2.4 Digital to Analogue Conversion

Several of the commonly used analogue to digital convertors use digital to analogue convertors (DAC) as part of their electronic circuitry and a simple explanation of a simple DAC is necessary before discussing ADCs. Figure 11.3 is a schematic description of a simple DAC.

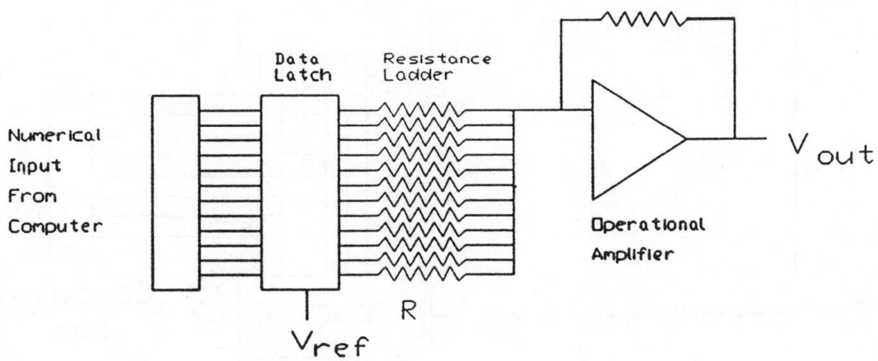

Figure 11.3 Simple digital to analogue convertor

The binary signal from the computer is compared via a series of solid state switches which select a resistance value (R) from a resistance ladder. The current (I) passing through the operational amplifier is controlled by the reference voltage (V_{ref}) and the selected resistance value (R). The output voltage (V_{out}) from the operational amplifier is now analogue and is proportional to the numerical information supplied from the computer.

The output voltage (V_{out}) alters proportionately as the output from the computer is altered. The analogue output is, however, not continually variable but is a sequence of steps determined by the summed current controlled by the reference voltage, and the selected resistance on the resistance ladder. If the number of resistors in the ladder is increased to match a greater number of digital lines or bits, then the proportional changes in current are smaller, as are the output analogue voltages. The greater the number of bits, the greater the resolution of the DAC. Most common DACs have 10 to 16 bit resolution, which corresponds to a discrimination of one part in 1024 (2^{10}) and one part in 65,536 (2^{16}) respectively. Therefore, a 16 bit DAC would provide a more precise and accurate representation of an analogue signal than one with a ten bit resolution.

Analogue to digital conversion is the reverse of digital to analogue conversion and can be achieved by different methods. Some of these methods utilize DACs as part of the electronic circuitry and are detailed below.

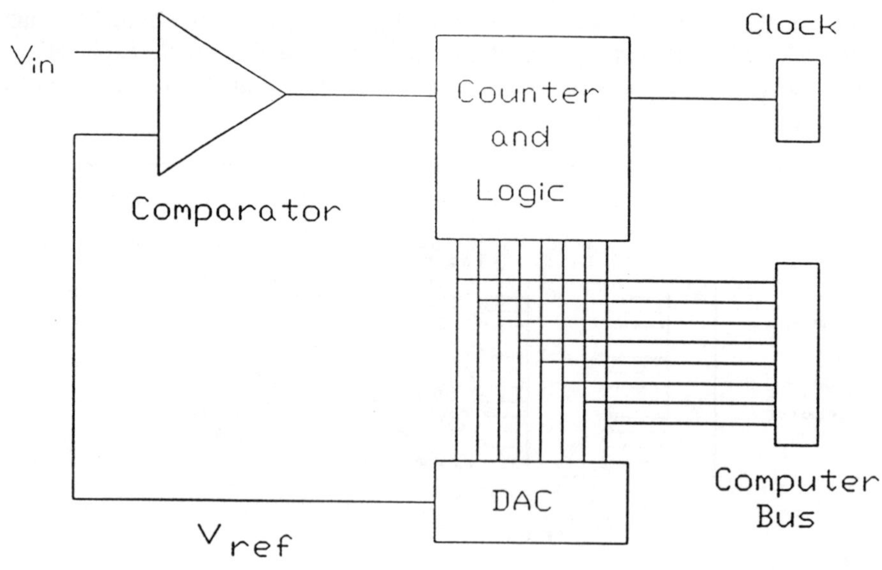

Figure 11.4 A counter type ADC

11.2.5 Counter Type ADC

The counter type ADC (Figure 11.4) is described in detail, not because of its applicability, but as an aid to understanding the principle of analogue to digital conversion.

The ADC begins to function when a start signal is supplied to the counter. This counter is also supplied with a continuous pulsed signal from a constant frequency clock and with each successive pulse the value in the counter is increased. The count value is also output to a DAC which converts the count value to a voltage, V_{ref}. This voltage (V_{ref}) is supplied to a comparator which compares V_{ref} to the incoming unknown voltage V_{in}. The function of the comparator is to output a change of logic state (high or low) when the voltage V_{ref} exceeds the value of the incoming voltage V_{in}. The change of comparator state stops the counter. The count value of the counter at this point is output to a computer and this numerical value is proportional to the value of the incoming voltage V_{in} on the comparator. Changes in V_{in} are reflected by proportional changes in the numerical output of the counter.

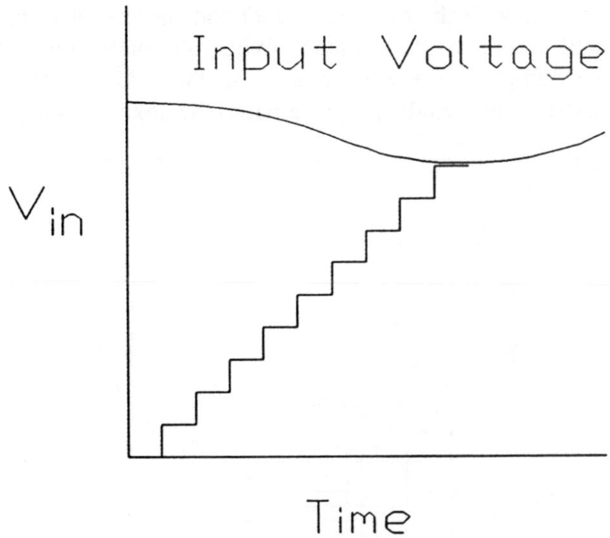

Figure 11.5 Output of a ADC unit with time

Figure 11.5 shows how the ADC samples an analogue signal over a period of time. This type of convertor takes a variable period of time to sample the voltage as the cycle limit is determined by the value of the unknown voltage, and the greater the voltage, the longer the time period required for measurement. This type of ADC would be unsuitable for applications which experience large changes in input signal, such as chromatographic detectors.

11.2.6 Successive-approximation ADC

The successive-approximation method of analogue to digital conversion (Figure 11.6) is a good general purpose method suitable for numerous applications. This ADC differs from the counter type primarily in the way it generates a reference voltage for input into the comparator.

The ADC begins to function when a start pulse is sent to the successive approximation logic module. This module outputs a numeric value to the DAC which represents one half of the maximum input range voltage on the comparator. The DAC supplies the reference leg of the comparator with this reference voltage, and the comparator returns the logic state of high or low if the input voltage is higher or lower than the reference voltage. The logic module then generates an output to the DAC which is one half of or one and one half the preceding value. The numeric value supplied to the DAC by the logic module is determined by the output state of the comparator, ie the comparator determines if the input voltage can be found in the top half or bottom half of the reference voltage. The sequence of adding or subtracting one half of the preceding reference voltage value is continued until all combinations are tested. This method of determining the input voltage is similar to weighing an unknown weight, starting with the heaviest weight and successively working through smaller weights until a comparison is made. When all combinations are tested the final numeric value output by the logic module to the DAC is also passed to the computer. This numeric value will be proportional to the unknown input voltage V_{in}.

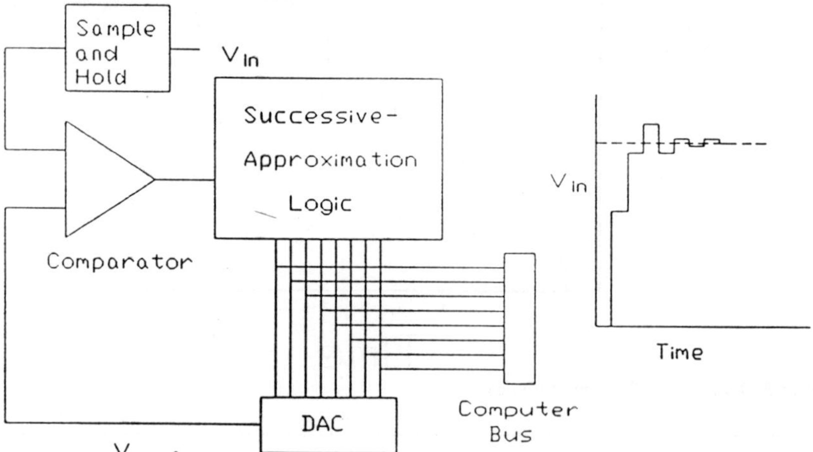

Figure 11.6 A successive-approximation ADC

Because this type of ADC takes a constant fixed time to measure the unknown input voltage, a sample and hold amplifier placed on the input line to the comparator holds the input voltage at a constant value. This provides the ADC with a relatively constant voltage to process. However, because the time frame in which these ADCs work is relatively narrow, approximately ten nanoseconds, they are generally very sensitive to noise spikes but useful for applications which require relatively fast data acquisition rates, typically, capillary gas chromatographs.

11.2.7 Dual Slope Integrating ADC

This type of ADC employs an integrator front end and is more likely to be used for chromatographic applications. Figure 11.7 is a schematic representation of such an ADC.

The input analogue signal is integrated over a relatively long time period of 20 milliseconds. The potential on the comparator ramps up from zero voltage to a value representing the incoming analogue signal. At the end of the cycle the input signal is switched to a reference voltage which causes the integrator to ramp down to zero volts. Also, at the end of the cycle time a clock counter is started and the time taken for the potential to ramp down to zero is measured. This time period is proportional to the integrated value of the analogue voltage and the digitized counter value is passed to the computer. The 20 millisecond cycle time is calculated from the mains AC cycle time and as a consequence interference at 50Hz is rejected. In the USA a cycle time of 16.7 milliseconds would be used to match the 60 Hz frequency of the mains voltage. Also, because the integration time is relatively long, the system is not so sensitive to noise spikes as are previously described ADCs. However, because of the relatively lengthy sampling period, this type of ADC is unsuitable for very fast data acquisition rates and is usually limited to a maximum of 20 Hz.

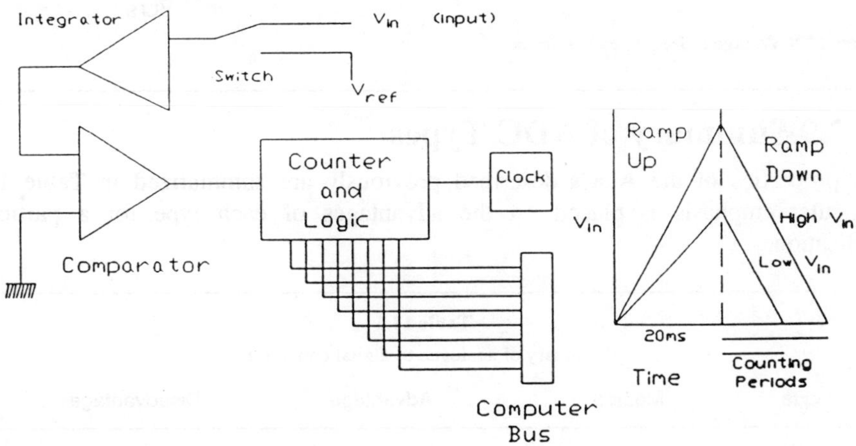

Figure 11.7 Dual slope integrating ADC

11.2.8 Voltage to Frequency Convertor (V/F C)

A V/F ADC converts the analogue signal to a frequency. This frequency is passed to a counter for a fixed time period and the value passed to the computer CPU. This process requires a relatively long time (25 ms for 12 bit conversion) and as a consequence an integrating amplifier is placed at the ADC input. A schematic representation of a V/F convertor is shown as Figure 11.8.

The integrating amplifier averages the input signal for a short time period and this smoothes the signal by averaging sharp noise peaks. The V/F type ADC is therefore particularly suitable for applications in noisy environments where electrical interference is a problem.

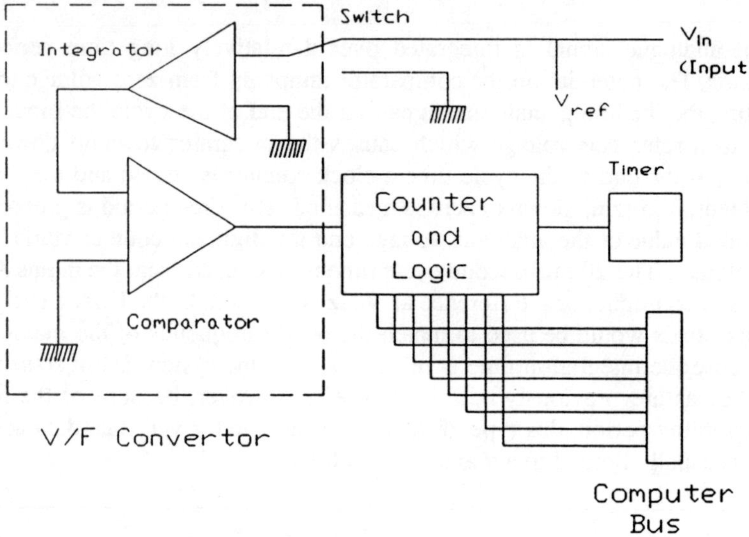

Figure 11.8 Voltage to frequency convertor

11.2.9 Summary of ADC Types

The properties of the ADCs described previously are summarized in Table 11.1; particular emphasis is placed on the advantages of each type for a particular application.

Table 11.1

Summary of analogue to digital convertors

Technique	Method	Advantage	Disadvantage
Counter Type	Voltage Comparison	Fast data rates 1ms to 10ns	Prone to interference Inaccurate
Successive Approximation	Voltage Comparison	Fast data rates 1ms to 10ns	Prone to interference
Dual Slope	Integrating	Insensitive to noise	Slow data rates 1ms to 1s
Voltage to Frequency	Integrating	Insensitive to noise	Slow data rates 1ms to 1s

11.3 Data Acquisition Systems

The data acquisition and processing systems that are available for the chromatographer are classified into dedicated, distributed or centralized systems, each is discussed in turn.

11.3.1 Dedicated Systems

Instruments falling into this category tend to be single channel integrators, devoid of the ability to re-analyse the digitized raw data. There are some exceptions which allow a single chromatographic run to be reprocessed and a new baseline fitted, but generally the data cannot be stored within the instrument. These instruments generally produce a report as shown in Figure 11.9 containing the peak retention time, peak area and peak integration method, ie. whether the peak terminated as a valley between another peak or whether the theoretical baseline had been detected.

They also have the ability to skim small peaks on the trailing edge of a large peak to provide a more accurate measurement of areas, as shown in Figure 11.10. This is particularly important for peaks which appear on the tail of a large solvent peak or a component on the tail of a large sample peak.

NO CAL => AREA %

RT	TYPE	AREA	
1.16		435	.037 83
1.60		3878	.337 3
2.91		218	.018 96
3.73		274	.023 83
5.84		7747	.673 7
6.37	M	408	.035 48
6.80		1685	.146 5
8.31		26060	2.266
9.09	M	402346	34.99
10.50		162762	14.15
10.85	T	3122	.271 5
12.77		300413	26.13
13.12	M	22709	1.975
14.13		37232	3.238
21.60		3096	.269 2
21.93	M	2769	.240 8
23.84		174722	15.19

DLY OFF STOP 30 REJECT OFF
MV/M .30 ATTN 16

Figure 11.9 Integrator Report

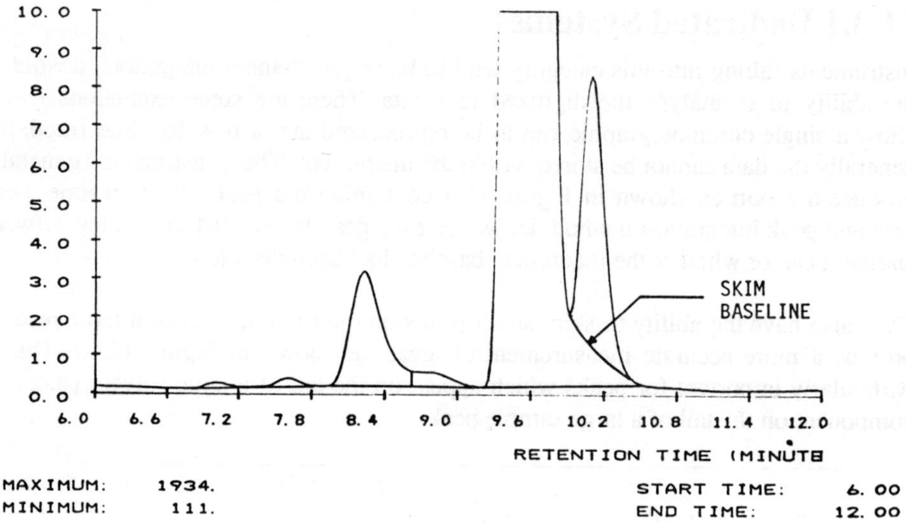

Figure 11.10 Peak skimming Options

Some integrators have the ability to communicate with computers via a serial (RS-232C) or parallel (IEEE-488) port and the peak report can be passed to a main computer for further processing or storage. This is particularly valuable if large numbers of similar samples are processed as the data can be captured automatically.

11.3.2 Distributed Systems

A distributed data acquisition system is the intermediate stage between the dedicated and the centralized approaches of data processing. A distributed system is based on small, single or multi-channel data acquisition units which collect and process data locally and transfer the processed data, in the form of peak area reports, to a large central computer for further analysis and storage. The functional requirement of a distributed system is that each of the local data processing units can store both raw and processed data and function independently of the central computer.

An approach to distributed data systems which is particularly cost effective is one based on a PC with resident data acquisition software. These microcomputers are single user systems and, depending on the software package, can acquire data from several chromatographs. Some of these packages use independent A/D convertors which can store raw data in resident memory and only pass data to the microcomputer for processing when requested. This can be considered an advantage if the PC is required to perform other tasks, such as access to the LIMS using terminal emulation software. Alternatively, other software packages can also be used whilst data acquisition is controlled by the independent ADCs. An important limitation of these ADCs is their storage capacity. The number of chromatographic runs which can be stored is a function of the sampling rate and the chromatographic run time. This limitation must be fully understood by the user and a rigid procedure of initializing the acquisition software must be followed in order to ensure data is not lost.

Other packages utilize A/D convertors on electronic component cards which slot into the communication bus of the microcomputer. This may be preferred by laboratories which want GLP compliance. These systems are also limited to two or three data acquisition channels, but from a practical point of view this is likely to be the ideal number of instruments accessed by one user. A useful feature of this type of data system is the ability to control HPLC equipment, particularly HPLC pumps. Mobile phase gradient profiles can be generated using different manufacturers' pumps and some systems can control several more HPLC instruments. This is a cost effective option as gradient controllers supplied by most manufacturers are expensive.

11.3.3 Centralized Systems

For the purpose of this chapter a centralized system is defined as a minicomputer running chromatography software either alone or concurrently with a LIMS. In the latter case, the LIMS is usually the main function and the chromatography software utilized because it is a cost effective option.

These systems range in complexity depending on the software packages which are used concurrently and the number of chromatographic instruments served. Some of the larger systems are minicomputer based instruments sporting upwards of 48 analogue channels.

Acquisition and digital conversion of analogue data can be achieved either by multiplexed ADC cards which slot into the computer or by remote ADC units which contain their own memory for data storage. These remote units are connected to the PC using either RS-232C or IEEE-488 communication protocols and a few can function as stand alone instruments. They acquire and process analogue data independently from the central computer but have to transfer the data after a run is completed. These data can then be stored, reprocessed or reported using software within the central computer environment. Each system has specific advantages and choice is based on the requirements of the application.

The distinct advantage of a system which holds both a LIMS and chromatography software is the way the two may be integrated - processed data can be automatically transferred to the LIMS database and reports generated automatically.

This type of system is particularly suited to a QC environment where similar repetitive analyses are performed. The progress of the samples can also be monitored carefully throughout the laboratory and data from other analytical techniques automatically incorporated into the report. The penalty, however, is paid when chromatographic data has to be reprocessed and interpreted. This is usually performed at terminals; when several users are working concurrently the performance of this system may be affected adversely if the processor size has been determined incorrectly.

Upgrading the system is very expensive and in most cases expansion of such a system has to be carefully controlled. Although additional ports can be added, performance decreases exponentially if the system is fully utilized. Expansion is best directed to systems which acquire, process and store data locally using PCs, as described in Section 11.3.2, and then transferring the completed reports to the LIMS database.

11.3.4 Summary

The preference for dedicated, distributed or centralized chromatography data acquisition systems is based on the circumstances, and also the scale, of the application. The choice between using a dedicated integrator or a computer based system is clear. But this is not the case when considering a distributed or a centralized data acquisition system. Table 11.2 summarizes the advantages and disadvantages of each approach.

11.4 Conclusion

There is a considerable number of chromatographic data acquisition systems available. It is important that a buyer addresses the analytical problem that requires solution, as a poor choice of instrumentation can have serious consequences. Bigger is not necessarily better and the type of work in a laboratory will narrow the choice of system to a certain extent. The following points are suggested for consideration:

1. Purchase the equipment which will provide the greatest degree of flexibility.

2. It is highly unlikely that any one system will provide all solutions to all the problems, and compromises will have to be made.

3. Consider the requirements of GLP compliance.

4. Consider the implication of compatibility with other scientific instrumentation and computer hardware - both present and future.

Communication between systems is vital for future development of data systems.

Table 11.2

Advantages and disadvantages of chromatographic data systems

Advantage	Disadvantage
Dedicated Integrator	
Always available for use, Little down time, User friendly, Simple, Minimum of training for use, Immediate local hardcopy report.	Unintelligent, Little scope for development, Few have reprocessing power, Little communication potential, Single user.
Distributed System (PC Based)	
Low cost per channel, Moderate initial expenditure, Small CPU failure liability, Availability of commercial software, Good communication potential, Good graphics support, Fast response for user, Local archiving, Potential for networking.	Single user, Single tasking, Little channel expandability, Requires established communication software for LIMS connection, Poor security, Poor GLP control, No immediate report.
Centralized System (Chromatography only)	
High computing power, Multi-channel , Multi-user, Good networking facilities, Good graphics support, Ease for standardization of methods, GLP compliance.	High initial cost, Significant CPU liability, No immediate hardcopy report.
Centralized System (Incorporating LIMS)	
High computing power, Multi-channel , Multi-user, Good networking facilities, Ease for standardization of methods, Customization of software, Good security facilities, GLP standards.	High initial cost, Highly significant CPU liability, No immediate hardcopy report, No local control, Single vendor software, Response problems with high CPU load.

Chapter 12

The Request for Proposal and the Functional Specification

M. Bertram and R.D. McDowall

12.1 Introduction

To derive full benefit from a LIMS, it is essential that proper consideration is given at the outset to specifying the analytical requirements in sufficient detail. This is to ensure that the equipment installed matches the analytical needs as closely as possible. To achieve this aim, there are four stages to specify a LIMS:

1. Review of the laboratory to decide if a system is needed.
2. Writing the request for proposal containing the laboratory's requirements (the *requirements specification*).
3. Choosing the supplier of the system.
4. Writing the functional and systems specifications from which the LIMS will be built.

The main purpose of this chapter is to discuss how a prospective LIMS should be specified via the requirements and functional specification documents. The material described here constitutes the first portion of the life-cycle of a LIMS project outline as shown in Figure 1.1. Depending how the system is implemented, stages three and four can be reversed.

A major stumbling block for any LIMS installation is the difficulty in defining the precise needs of an organization. Flexibility in the system is one way of overcoming changes in requirements, as described in Chapter 4. Properly specified, a LIMS will be a considerable bonus; a system wrongly configured will be disastrous (1). To remove as much uncertainty as possible from the planning of the LIMS two design documents called the requirements and functional specifications are written.

What are the requirements and functional specifications? These two documents are the fundamental building blocks for a LIMS (or any computer system). The aim of the requirements specification is to give an outline of what the laboratory does and how the LIMS fits into the organization, and to describe the tasks that the software should be capable of undertaking. It is incorporated into a document, called the 'request for proposal', which is sent to prospective suppliers to enable them to tender a quote for the system. The bulk of the document will be the requirements specification together with a description of the type and format of information requested from the prospective suppliers. This includes details of the supplier and the likelihood that it will remain in business throughout the lifetime of the LIMS project. A typical format for the response by a supplier is given in Section 12.5.

The functional specification takes the requirements of the laboratory further and describes the logic of how the system should function. This document is essential for programmers or system analysts to develop any custom LIMS software programs or to tailor an existing package. The logic linking the operations within the laboratory must be carefully thought out.

There are a variety of ways these documents can be written. Software consultants can be used to write both documents before submitting to potential suppliers (see Chapter 16). Alternatively, the requirements specification can be written by the users and computer staff for submission to potential suppliers as a basis on which to tender quotes. Whichever way is chosen, considerable effort needs to be expended by the users. The chosen supplier and user representatives can then write the functional specification jointly.

It should be stated from the outset that it is unlikely that a fully working and fully functional LIMS can be obtained from the specification documents at the first attempt, unless the software available matches the requirements exactly and no customization is necessary. The reasons for this are:

1. It is inevitable that some functions that looked attractive on paper will prove to be difficult to use practically.

2. There will be areas that have been underspecified or even omitted.

3. Sections will have been wrongly specified.

This is shown, quite dramatically, by the analysis of nine US Government software contracts (2). This showed that the main causes of failure were vague definition of system and user requirements coupled with an inability to upgrade the software for future requirements. Of software contracts worth \$6.3M only \$0.1M was used as delivered, a further \$0.2M was used after modification and \$3.2M was never used. This clearly demonstrates the need to define the system and all its functions carefully and exactly.

In general, depending on the complexity of the laboratory, between 75-90% of the system should work as specified at the first attempt. The remainder will require additional programming and/or enhancement for the reasons outlined above. It is essential that this eventuality is planned for and that monies are set aside.

12.2 Review of the Laboratory

The first stage in the implementation of a LIMS must be a review of the laboratory (3). This must critically review the operations, and their inter-relationships, carried out within the laboratory and highlight the areas where there are inefficiencies or bottlenecks. In theory, the process should be to step back and review the current working procedures within the laboratory with an open mind (3). A computer system can always be made to fit an existing laboratory environment. However, it may not be desirable to do so for a number of reasons such as the working practices being outmoded or inefficient. Other potential requirements have been outlined by Alan Braithwaite in Chapter 3.

An idea that can be useful whilst reviewing the laboratory is to identify the sample flow within the laboratory. How do samples enter the laboratory and what happens to them once they have arrived? For instance, if each test is applied sequentially to the sample then tracking and location of the sample at any time is important; or if the sample is divided into aliquots for testing and the results collated together, then labelling the sample and tracing which tests are overdue may become important considerations. In this respect laboratories tend to be individual, but the exercise helps the chemist to understand and review all the activites involved in receiving, assaying and reporting the results from any sample.

This review will yield information for a cost benefit analysis or net present value assessment to determine if a LIMS will be viable in the laboratory. It may be that alteration of working practices is all that is required to improve efficiency. The review will also provide the necessary information to write the requirements specification if a LIMS is deemed essential.

12.3 The Requirements Specification and Request for Proposal

If the review of the laboratory indicates the need for a LIMS, the managers of the laboratory then need to gain the support of senior management and the computer department (if not already involved) in the project. There is also a case for including systems analysts and programmers in the review process so that they can become familiar with the laboratory ready for the next stages of the process. Regardless of the approach, once approval or support is available the first step should be the formation of a LIMS Project Team.

12.3.1 The LIMS Project Team

As mentioned by Peter Johnson in Chapter 2, for a project of the size and scope of a full LIMS, a project team should be established at the outset. The membership must consist of representatives of the laboratory management, users, the potential system manager and computing professionals. The direction and leadership should come from the laboratory management. This will ensure that the needs of the laboratory are realized and also that the system is a natural development and it is not being foisted upon the users by the computing department.

The function of the project team is to manage the project in all its stages: to balance the users' requirements with the available budget and to consult with departments that will be affected by the installation of the system. Liscouski (4) recommends the use of pre-set criteria to measure the achievement of each requirement. This is very useful as it provides a measurement by which to test the finished product and demonstrate to management, the users and the supplier, that the LIMS has reached its goals.

The project team will be involved in developing:

- Plans and strategies for the project.
- Assessing alternatives.
- Setting goals for the system and tests to assess performance.
- Implementation.
- Drawing up the timetable for the project and monitoring progress.
- Obtaining resources.

Also, via liaison in the later stages of the project with the LIMS user group, the team should create a receptive environment for the success of the project by identifying users and involving them in systems development where appropriate.

If the group is to function effectively, it must work as a team, with regular meetings to review and ensure efficient progress. Positive backing by senior management, coupled with this inter-disciplinary approach will be major factors in the success of any LIMS project.

12.3.2 Planning the Requirements Specification

The requirements specification summarizes, from the laboratory's viewpoint, the overall aims of the LIMS. It does not include any detail of how it is to be achieved. The importance of this document cannot be over emphasized (5). Moreover, it is essential to formulate ideas before approaching a potential supplier or developer of a system. Substantial planning has to go into this area to evaluate the current, as well as the future, requirements of the laboratory. A rapidly changing laboratory environment can only be envisaged two or three years ahead, whilst one that is relatively static could be taken further.

The overall aim should be flexibility to accommodate present and future needs. At the forefront of all thinking should be the questions:

- What will be done with the data or information?
- How will the LIMS be integrated within the laboratory and the organization?

It is important to avoid using any jargon when writing the requirements specification because disciplinary boundaries will be crossed; if necessary define any key words (5).

There are two types of philosophy concerning the implementation of a LIMS. The first is a 'black box' approach where all data are acquired and processed automatically by the system with little intervention from the analyst producing final results for the supervisor or submitter. The alternative is to think of a LIMS as another analytical instrument to be used and controlled by an analyst whose responsibility is to accept or reject data or any calculations. The type of laboratory, staff available, workload and assays will determine the option to take.

The writing of the requirements specification may be the opportunity to redesign the operations and working practices of the laboratory. A LIMS can be tailored to fit almost any laboratory environment, but if the working is inefficient or outmoded then the introduction of the system can be used to bring about change. A note of warning, however, if this way is chosen then the introduction of the system into the laboratory should be carefully handled.

12.3.3 Content of the Requirements Specification

The requirements specification should contain five sections as a minimum, these are summarized in Table 12.1.

Section 1 is an overview of the laboratory. In this section the laboratory should be described along with the type of work undertaken. The number of staff, together with their qualifications, is a useful indicator as to the complexity of the work. The analytical techniques that are used in the laboratory are then described; a statement of the direction that has been taken with automation helps to complete the picture. The reason for this is that it is easier to introduce a LIMS into a modern automated laboratory than a manually operated one with old instruments.

The source of the samples entering the laboratory together with the ways in which the results are reported must be noted as this will give a prospective supplier an idea of how the LIMS should operate. The sample turnaround time should be stated and also how any further processing of the results is to be undertaken either by the laboratory itself or another department. (If any changes to this way of working are contemplated they will be outlined in the third section describing how the system will operate.) The past and present workload of the laboratory must be outlined to give an indication of the data volumes expected from the work. Finally, although this may be clear from the description of the laboratory, it should be stated if a specification or protocol driven LIMS is requested (see Section 12.3.5).

Section 2 concerns the future development of the laboratory. Here the workloads for the future should be estimated. This is required for assessing the processing power of the computer and volume of disk storage to prevent problems in running out of storage early in the life of the system.

The writer should also include any envisaged development of the laboratory over the same time frame: for instance, will new techniques be added or old ones discarded? This can be difficult but an attempt should be made; if a laboratory already has a development plan in existence then it should be appended here. Staffing levels should be estimated: are they going to rise or is it anticipated that the introduction of the LIMS will enable the laboratory to hold the staff numbers constant?.

Table 12.1 Contents of a requirements specification

Section 1: *The Laboratory Overview*

- Describes the laboratory and the work undertaken
- Staff numbers and their qualifications
- Analytical instrumentation
- Level of automation
- Sources of work
- Speed of turnaround of samples
- Further processing of results
- Past and present workload
- Specification or protocol driven LIMS required

Section 2: *The Future of the Laboratory*

- Estimated future workloads
- Future development of the laboratory
- New techniques or instruments to be added
- Older techniques discarded
- Estimated staffing levels and qualifications

Section 3: *Overview of the Proposed System*

- What instruments would be desirable to interface to the system?
- Should the LIMS implementation be phased?
- Description of the requirements
- Outputs from the system (Reports)

Section 4: *Integration with Existing Systems and Software Packages*

- Data input from other computers
- Data output to other computers
- Integration with existing communications systems
- Integration with word processing systems
- Processing of the data by specific software packages

Section 5: *Corporate Computing Policy*

- Hardware policies } at local and corporate levels
- Software policies }

Section 3 describes the proposed system and the requirements for data capture and dissemination. Included here is a description of the system in a logical order from when samples enter the laboratory through their analysis to the issuing of the analytical report. It must state which analytical instruments will be interfaced to the computer(s) and if they need A/D converters or already have digital output. Data input not only comes from analytical instruments but also from balances, pH meters, other computers and bar code readers. Data entered electronically will usually be free from transcription errors and every effort should be made to connect as many instruments and ancillary equipment as possible to the LIMS to ensure the highest levels of data integrity and to avoid transcription error checking.

The output from the system must be described. As reporting is a main benefit of a LIMS, every effort should be made to automate this as much as possible. For example, if a sample receipt is sent to a submitter, then this is easily automated.

Included in this section of the requirements specification should be a statement concerning the phased implementation of the LIMS. This is especially important when considering a large system where different parts of it are being introduced to the users at different times. Hong (5) suggests that this is not ideal as users will be faced with several different versions of the software before the final version is established. Whilst agreeing with this view, the authors maintain that a phased implementation is possible, providing that each new phase does not destroy the previous work nor change the overall operation of the system. Indeed, if the LIMS is highly customized, then a phased implementation may be most desirable.

Section 4 details how the LIMS will be integrated with the relevant computers within the organization, and describes any software packages. It is vital at this early planning stage to have a good idea of how the LIMS will be connected to the existing word processing systems within a company. How will reports be passed from one computer to another? What software packages are already used to process the data further or display it? It would seem a waste of money if a package such as RS/E were duplicated within a LIMS. It would be far more cost effective to transfer the required data to the computer running this software or provide an interface to the package if running on the LIMS computer, and then write a procedure to read the file into the package and analyse the data. Future developments in networking should allow this to occur without much user involvement.

Data input from other computers that will be connected into the LIMS should be considered at the same time as the data output requirements. Berthrong (6) has described how a supplies requisition computer can be interfaced to a LIMS and give the laboratory advance warning of samples arriving in a quality control laboratory. With the development of Open Systems Interconnect (OSI) standards, communication between different manufacturers' computers will be easier, therefore consideration should be given to the connection of existing systems.

Section 5 lists the computing policies at both the corporate and local levels with respect to both hardware and software. This gives prospective suppliers information regarding the site and company where the LIMS will be installed.

12.3.4 Production of the Requirements Specification

The production of a requirements specification can be achieved by senior analytical staff if the information flow in the laboratory is relatively straightforward.

Alternatively, computer consultants may be retained. This approach is described by John Knox in Chapter 16. Consultants may offer new ideas and fresh approaches combined with experiences from different installations. This may be attractive if there is lack of time or staff within the laboratory to write the requirements specification. However, much time must be spent in explaining the laboratory's requirements to the consultants before a specification is produced. Expressing a personal view, the authors consider it essential to involve the users in preparing the specification as they will be putting the system into operation and they know their own laboratory, regardless of which approach is taken.

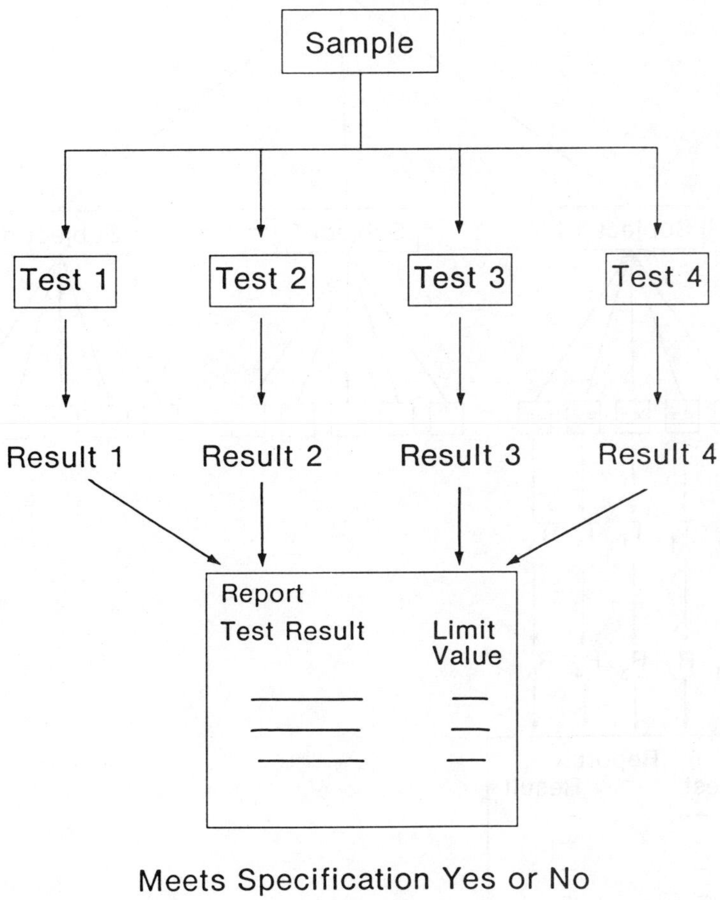

Figure 12.1 Specification driven LIMS

12.3.5 Specification or Protocol Driven LIMS

In choosing a LIMS the basis of operation must be borne in mind: most standard commercial systems are specification driven; that is, they are aimed towards a quality control environment in which small numbers of samples are put through relatively large numbers of tests. After splitting the original sample, ensuing results must be collated to see if they meet a predefined specification. This is shown in Figure 12.1.

However, there are applications which require a protocol driven system to cope with the reverse situation in that large numbers of similar, but unique, samples usually have a single test applied to them.

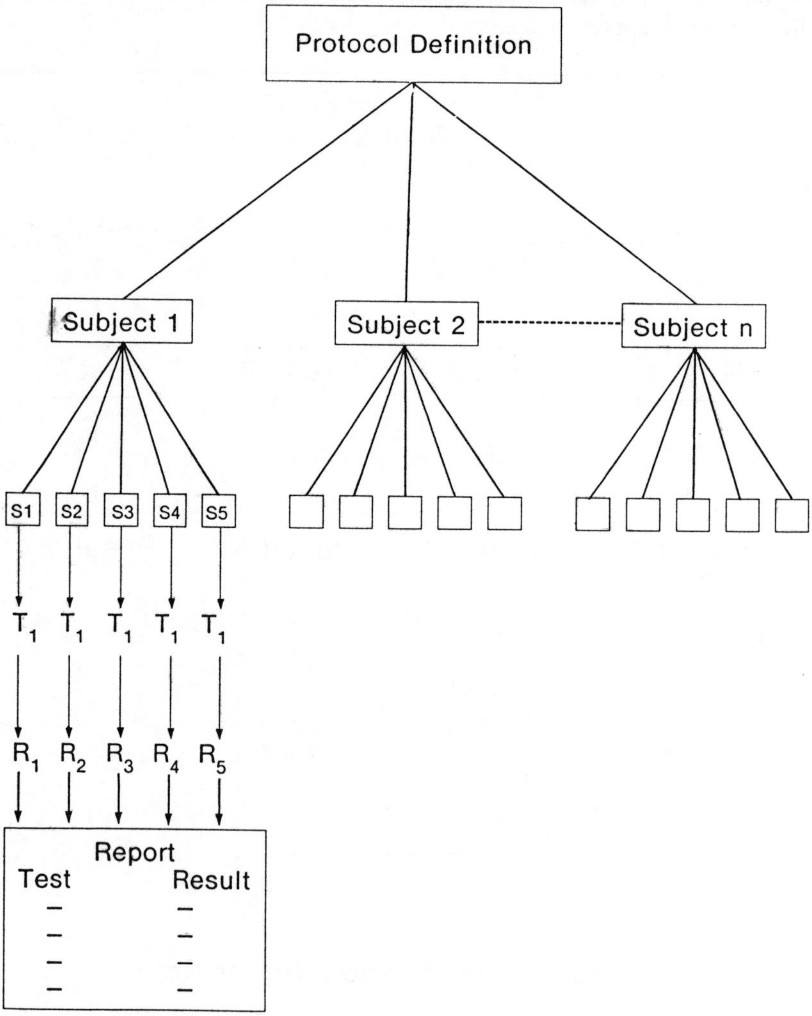

Figure 12.2 Protocol driven LIMS

This is shown in Figure 12.2. Frequently, individual samples are irreplaceable and of limited volume, being biological in origin. Although individual samples are important, it is the relationship and trends between groups of samples that is sought. The concentrations to be measured result from many pharmaceutical and biological factors and they cannot be compared to any meaningful standard value. This is a vital distinction for laboratories working in the analysis of biological fluids in the pharmaceutical industry.

12.3.6 Ease of Defining LIMS Requirements

How easy is it to define a LIMS for a laboratory? Figure 12.3 gives an indication. It represents both the nature of the workload and the type of operations carried out by a

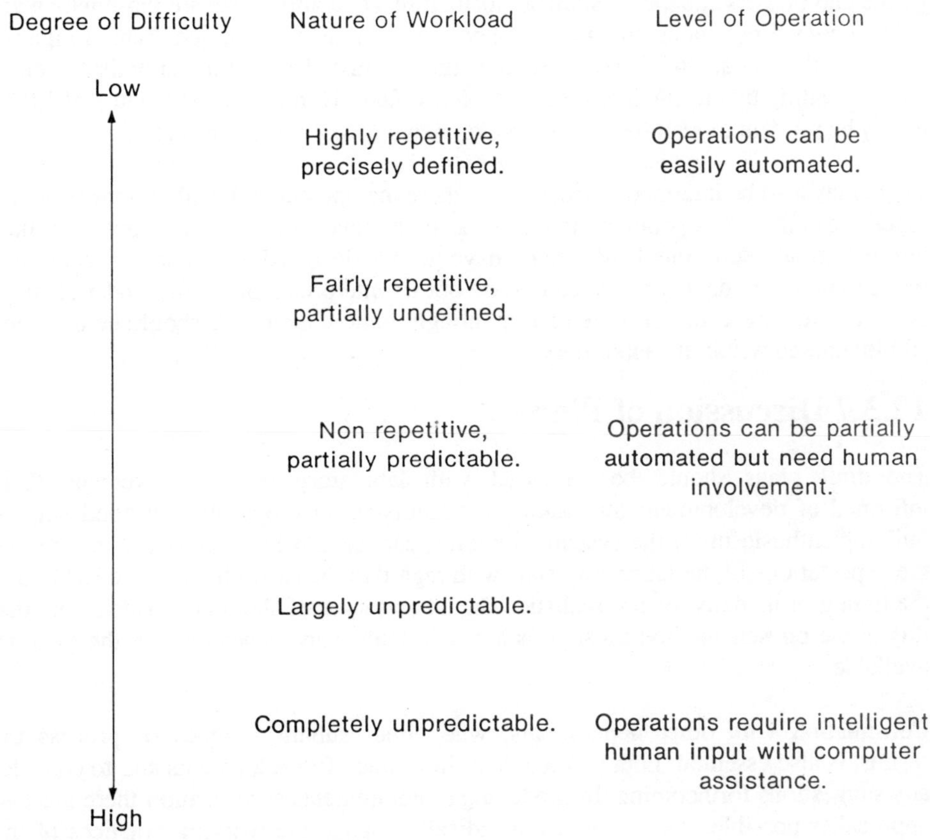

Figure 12.3 Defining LIMS requirements

The easiest situation to envisage is a highly repetitive laboratory where the needs can be precisely defined and the procedures can be easily automated. This is similar to a quality control laboratory and is where the majority of commercial LIMS are targeted. This situation is the easiest to specify as there is less need for flexibility because the laboratory is very regimented.

Moving down the scale of difficulty to situations which are more difficult to define means that flexibility of the software is very important to overcome the unpredictable nature of the work. Great care is essential from here down as computers cope with routine tasks well, it is the situations that have not been anticipated by the designers that will cause the most problems with software. All practices must be defined otherwise the users will have to work in a standard inflexible way (5). An alternative method to ensure a workable LIMS is to design for flexibility; this approach should ensure that if requirements change, these can be reflected by corresponding changes to the software, undertaken either by the users or programmers.

At the end of the scale the most difficult situations to computerize are those which are completely unpredictable and where the operations require intelligent human input to interpret the data. In these instances there must be a full appraisal before recommending that a LIMS will be suitable; in fact given the present status of LIMS it may be prudent to wait for future developments before installing such a system.

There may also be intermediate instances where the operation in a laboratory is highly repetitive but the operation requires a high level of human input for data interpretation. Here the LIMS may have to be designed like another analytical instrument where data is collected on-line but its interpretation is only undertaken by skilled staff. This calls for more careful thought in how the LIMS should be designed and integrated within the laboratory.

12.3.7 Discussion of Plans

The draft plans should be discussed with laboratory personnel; keeping them informed of developments and asking for feedback on proposals is a good way of building enthusiasm for the system. However, care should be exercised to ensure that the expectations of the laboratory staff with regard to the capabilities of the LIMS and the timing of its delivery, are realistic. The laboratory staff should be made aware that this is the chosen method of approach but it could vary depending on the systems available.

Consultation with other departments, who either submit samples or process the results, is also essential, both to keep them informed of developments and to consider any suggestions forthcoming. In producing a requirements specification there are two approaches possible: the first will not radically change the working practices of the laboratory that have been built up over the past years. In essence, this tailors the LIMS to the laboratory. The second method tailors the laboratory to a LIMS and could be used as a way of implementing changes in the working practices. In either case, it is better to aim high and incorporate all the wishes of the laboratory into the specification and then debate with the supplier each point where the product on offer does not match the specification requested. Otherwise, the resulting system may be a disappointment.

12.4 Source of the LIMS

There are three ways to acquire a LIMS:

1. In-house development.
2. From a software house.
3. From a commercial supplier.

The advantages and disadvantages of each source are discussed below.

It is at this stage that the project team assumes vital importance as the multi-disciplinary nature of a LIMS becomes apparent. Comments on the appearance of a commercial software package and how it would operate in a laboratory are relatively easy for the prospective users. However, in-depth knowledge of the computer operation and how the database stores and retrieves data come best from experienced computer personnel. (The advice is much better if they have laboratory experience to appreciate the problems of the scientists.) This experience and advice manifests itself in many areas: eg considerations of memory and disk size with respect to the anticipated workload, how powerful the database query module is, and where performance problems are likely to occur.

12.4.1 In-house Development

If a do-it-yourself approach is adopted, whereby a computer department or other individuals develop the software to the specification, this will mean a huge resource commitment by the site. The easiest method is to take an existing commercial database management system and use it as a basis for the software. The major disadvantage is the length of time needed to write the software, together with the resources required, eg. taking on, or re-deploying staff, and the on-going support needed.

The advantage is that laboratory staff have a system written specifically for their needs, although enhancement of the system can only come from within that laboratory. The huge demands of a LIMS project on computing resources make it doubtful whether any, except the largest, organizations would choose this route.

Examples of this approach are given in Chapters 13,14 and 15. Gareth Jones and Chris Triner in Chapter 13 describe a LIMS that has been developed by the Wessex Water Authority to automate the analysis of water and industrial water samples. Speed is vital in responding to problems and the LIMS has been specifically set up to do this. Automation, integration and communication are key components of the system.

Graham Martin, in Chapter 14, shows how development of a LIMS in the Chemicals and Polymers Division of ICI was undertaken to overcome the (then) limitations of commercial systems and to eliminate error in producing results from analyses of products. Commercial pressures to ensure that the quality control assays of products were correct and reliable were the driving forces for the development of this system.

In Chapter 15, Alan McLelland describes the application of LIMS within a clinical biochemistry department of a major teaching hospital. Clinical chemists have been using database management techniques for many years before LIMS were even conceived, in laboratories that have very high sample throughput. The turnaround time of samples is crucial and in some instances the results can help decide the treatment in life threatening situations.

12.4.2 Software Houses

The choice of a software house to provide the LIMS is a viable alternative. The end result should be exactly what the user requires but the development costs may be high. The time taken to write the software will depend on the expertise of the supplier and the programming staff employed. If the software company has been well chosen it should already have some expertise in this field and should be able to use or adapt existing software in this project and help keep costs down. As in the case of the in-house development, extensions to the system would come from within the laboratory and these would be chargeable if the system were not maintained in-house.

John Knox has detailed the benefits of using software houses for both consultancy for the introduction of a LIMS into a laboratory, and in the development of a system in Chapter 16. The advantage of this route is often the wider experience of the software house in installing LIMS for many different companies and industries.

12.3.3 Commercial Supplier

The final option is to use a commercial supplier. Over recent years analytical instrument companies have seen that information management is a logical extension of their product ranges, so much so that the major manufacturers now offer LIMS in one form or another. These commercially available systems are competitively priced, the development costs are spread over the whole customer base and the systems are sold in competition with other vendors. Competition means the product is under continuous development and is constantly evolving as ideas are incorporated for the benefit of all users (4). The overall development of the package can take many tens of man years and this will increase throughout the lifetime of the product.

Source code of the system software is useful if further in-house programming is considered. It must be remembered that commercial companies differ in their attitudes to this. Some suppliers will release the source code for the whole system, provided a non-disclosure agreement has been signed, whilst others will only give that relating to any custom software. In the instance where the LIMS supplier will not release the source code, they may be willing to sign an escrow agreement whereby a computer tape is deposited with a lawyer for distribution to the users if the supplier ceases trading. In all cases, care should be taken that any proposed changes to the software do not invalidate any maintenance agreements. Moreover, these should be discussed with the supplier before programming starts, to avoid any major problems.

Chapters 17 (Peter Berthrong), 18 (Mike Milano, Alan Williams and Lou Lesser), 19 (Steve Crisp and Johann Mollner) and 20 (Bob Hillhouse and Mike Head) describe the commercial systems of Beckman, Perkin-Elmer, Hewlett-Packard and VG Laboratory Systems respectively. These software packages need a minicomputer for their operation and are the offerings of some of the major analytical instrument companies. There are many other companies that offer LIMS, their omission from this book does not imply that the products offered are sub-standard or inferior but quite simply there is not sufficient space.

Often, if the laboratory is small, the size and cost of these large systems prevents their purchase; a LIMS that runs on a PC is described in Chapter 21 by Harmon Brown and Gloria Lawler of Nelson Analytical. This has taken a commercially available database and customized it for use in the laboratory.

12.5 Format of the Vendor's Responses

A 'request for proposal' document is required before buying a LIMS from a commercial supplier or a software house. This document should include a section describing the format of the response expected from the vendor. If this is not included, the responses are likely to be at different levels and some will not contain enough information to allow a choice of vendor to be made. This will, at the least, result in a delay in the decision making.

This section of the request for proposal, or a covering letter sent with it, should state a closing date for the receipt of proposals and the contact name to whom enquiries should be addressed, a name to whom the proposals should be sent, and the number of copies of the proposal required.

The type of information required from a prospective vendor would be as follows:

1. *Hardware* A detailed technical description of all hardware in the proposed configuration together with details of the systems software required.
2. *Software* A detailed functional description illustrating how it fulfils the requirements specification. Flow charts may also help to gain an overall perspective of the system.
3. *Documentation* The documentation required to be supplied with the system should be stated. It is also useful if copies can be sent for inspection to give an impression of their quality.

4. *Price* It should be clearly stated whether a fixed price is required for all modules in the system or if budgetary pricing is allowed. A decision should also be made before sending out the request for proposal whether one overall price for the system is required or that each module, whether hardware or software, needs to be priced separately. It is also useful to distinguish between the pricing of currently available software and that to be developed.

5. *Training* The amount of training included with the system should be specified, if that is not stated in the requirements specification.

6. *Existing installations* Discussions with existing users are a very useful way of evaluating proposals, so a number of reference contacts should be requested. The number of installations, both locally and in total, helps to give an idea of how viable the product is in conjunction with how long the product has been on the market.

7. *Maintenance* The proposal should state the terms, conditions and cost of both hardware and software maintenance agreements.

8. *Failure to meet specification* The requestor should state whether deviations from the specifications are allowed. If they are, the proposer should state what they are, and when they occur.

9. *Company background* The background to the company can give an indication whether it is likely to continue trading in both the long and short term. Requesting the company's business plan can give an indication of where the company and the product is planned to go in the future. The proposer may not accede to this request.

10. *Quality assurance* The QA standards to be followed can either be specified in the request for proposal or a statement of the proposer's standard procedures can be included in the proposal.

11. *Delivery* The delivery date, relative to an order date, should be requested. If the system is to be delivered in stages then the date for each stage of delivery should be requested.

12. *Software development* If software is to be developed as part of the proposal the requestor can ask for details of the elapsed time to develop each module, the number of staff involved in the development and the number of weeks' effort involved. This will show which modules need large amounts of resource and involve large amounts of development time.

This information will help the prospective vendor to format the proposal and give the user consistently stated proposals to ease the work of comparison and selection. A checklist for the format of the vendors' response is given in Table 12.2.

Finally, for information on how to evaluate the various LIMS proposals the reader is referred to Chapter 22 by Don Mattes.

Table 12.2

A suggested checklist for the format of the vendor's response

Hardware description
Software description
Documentation
Price
Training
Reference installations
Maintenance agreements
Failure to meet specification
Company background
Software quality assurance standards
Delivery dates

12.6 The Functional Specification

Once the choice of supplier has been made the hard work begins. The requirements specification produced earlier is used as a basis for producing a further document called the functional specification. This defines the functions of the system without detailing the methods by which those will be achieved. (The functional specification can be submitted to a supplier instead of the requirements specification as a basis for tender, this is especially true when computer consultants have been retained to write these documents.)

What is the difference between a requirement and a functional specification? A simple example of this is the need for a program to update records in a LIMS. The requirements specification would simply state that a program was needed to update records within the system. In contrast, the functional specification would give much more detail. This would state that the record update program would need to add, modify, list, copy and print records. It is also much easier to omit essentials such as a delete record option. As can be seen, the level of detail is far greater than that needed for the requirements specification.

The functional specification is drawn up by the user's representatives and the systems analysts of the supplier or software house. Essentially, it is a document that can be read by the users and therefore is written without reference to computer terminology. The functions to be performed by the system must be stated in a clear and concise manner. The problem is, if the wording in the document is ambiguous or does not adequately describe the function correctly, then the resulting program will not work as required by the users. Close liaison is essential as the initial problem to be overcome is the crossing of disciplinary boundaries. The systems analyst, who is drawing up the specification documents must become aware of the intimate workings of the laboratory. Great care is needed to include all working practices to ensure that flexible software is produced. Remember that one is dealing with another discipline where a word can have an entirely different meaning, for instance, sample and test can have different meanings to the personnel involved with writing this specification (5).

A note of caution: once the functional specification has been agreed between the two parties it becomes the basis for settling any problems in the future. It is therefore in the financial interest of the users organization that what is written is exactly what is required. If something is found to be missing and it is not included in the specification documents this will cost money. The designers of the system must ensure that operations are consistent from program to program otherwise the users will be frustrated. The functional specification should consist of a number of sections which will be discussed individually.

A frequent mistake is to concentrate on the analytical requirements of the system and to forget the management needs. It is important at this stage to consider these as well, because once the database has been set up it may be difficult or expensive to redesign it to provide the information required. As much thought should go into the management requirements of the LIMS as into the analytical needs.

At this stage it is also worthwhile to start thinking about what test data will be used to validate the system. This can lead to the recognition that certain functions have been left out of the specification.

12.6.1 Time, Thought and Feedback

Above all, the writing of the functional specification must not be rushed. Haste in specifying the user's requirements will inevitably result in a badly designed LIMS. It is tempting to think that because the requirements of the system are known there will be few problems in defining the functions.

Nothing can be further from the truth.

This stage is slow, complex, frustrating at times, but intellectually rewarding if the end result works well. It can come as a surprise to find that prior agreement between laboratory personnel at the requirements level can often result in violent disagreement at the functional stage. Again, the need to take time in writing the specification cannot be overemphasised.

As detailed by John Knox in Chapter 16, the requirements may change, and it is essential for all parties that the specification is agreed before programming is started. Mistakes made now or second thoughts later, are expensive to correct.

Once a section has been written it must be circulated for comment to the users. This is particularly important as the people most likely to be involved in writing the functional specification are the senior staff of the laboratory. This process will ensure that what is requested of the system bears a strong relationship to what actually happens at the bench.

12.6.2 Content of the Functional Specification

Similar to the requirements document, the functional specification is structured and consists of a number of sections, Table 12.3 summarizes these.

The introduction should set the scene for the whole system, outlining the objectives of the proposed LIMS and the tasks it will be undertaking. This section is important as senior management may only refer to this and a concise summary of the main features and the expected benefits of the system will be easy to understand.

The second part of the document will cover the functions required for the LIMS from when the sample enters the laboratory to when a report is sent to the requestor. Logically, it should describe the operations in the order that the system will be required to perform them.

Table 12.3

Contents of a functional specification.

Section 1: Introduction

- System objectives
- Tasks undertaken
- Overview of the system

Section 2: Detailed Functional Description

- *Modules and their operation*
  - Screen layouts
  - Operational logic of each module
- *Data inputs*
  - Format and volume expected
  - Files from intelligent instruments or other computers
  - Data input from instruments
- *Processes within LIMS*
  - Details of calculations used on data
  - Storage of results
  - Data transfer across modules
- *Output required*
  - Format for terminals, plotters and printers from LIMS
  - File format and volume for transfer to word processing systems and other computers
  - Files and volumes for downloading to analytical instruments
- *Operational limits*
  - Defined where appropriate
  - Ease of expansion if possible
- *Security*
  - Host computer functions
  - Database security
  - Audit trail, archive, retrieve and purge functions of database

---

Specifically, the following areas must be documented:

1. *All inputs to the system* This can be from other computers, analytical instruments or manually via terminals. It is important to provide detail of the format of each type of input and the volume of data expected. If any of the input data is to undergo any calculations these should be defined and documented in the specification. A chart describing the data inputs can be very useful for a summary. The users will be expected to supply the precise details of the screen layouts and the logic of the operation. It is here that the greatest problems in the implementation of a LIMS occur: if the logic is not right and consistent with what the users want, the system will not work.

2. *Details of the internal processes required must be described concisely* Any calculations used must be documented fully and the data to be transformed defined. Any linkages to other programs of data or results should also be stated. Results of calculations may be stored for recall or deleted from the database once reported; the users should define their requirements fully. If data are to be transferred between modules it will be pertinent to describe the procedures used.

3. *The output required must also be specified* Initially, this will be to peripherals of the computer system supporting the LIMS, other computers and word processing systems. However, increasingly in the future, output will be to the analytical instruments or the controlling computers containing instruction files for the method conditions and sample identities. As shown in Chapters 3 and 10 these intelligent instruments will set up and monitor the analytical conditions and then upload files containing the results for incorporation in the database. The interfaces to other computer systems must be defined along with what information or data is to be transferred. Details of the required output, for instance the formats of reports must be detailed. These examples must be realistic and contain data that the user is familiar with so that they can be commented on fully.

4. *The operational limits of the system must be defined in this document* Care must be taken to allow sufficient flexibility to ensure that any growth of the workload, or different working practices are accommodated by the system. Reiterating Section 4.3.3, the LIMS will be less flexible than the manual or semi-computerized system that it replaces. The users that specify the system must ensure that all work practices are stated at the outset and the that proposed design will accommodate them.

When all practices have been detailed, the resultant document must be very closely scrutinised by the users, as the writer will invariably be someone who is not totally familiar with the working of the laboratory, and may also not be a chemist. The wording of the functional specification is important and great care must be taken to ensure that the meaning of any sentences is not ambiguous and that words are carefully defined.

The type of development contract that has been entered into with a commercial supplier may affect or limit how much change to the specification may be undertaken. The amount of change, and whether it will cost money to do so, should be stated in the document.

It should now be clear that the functional specification forms the essential foundation for a workable and efficient LIMS. Failure to complete the task will result in much frustration and money fruitlessly expended. The last thing a user wants to hear is 'We didn't specify that' in answer to a request of how to use the system.

If the system is being developed from scratch then the security aspects must be considered. The first area should be the host computer with additional security for the database and any associated files. However, if the system is a commercial package, then the security features will already be in place and all that remains for the users to do is to implement them when the system is delivered.

If a protocol driven system is being considered the potential users must also consider complications to the other essential database utilities such as the audit trail, purge, archive and retrieve. Unless specified these utilities may operate at the sample level and not at the protocol level. This could mean that instead of one command to archive a protocol, many, possibly hundreds of individual commands may be necessary to completely remove a study to magnetic tape. If it takes much time and effort to archive data in this fashion, what would retrieving the data be like, especially if a regulatory agency inspector were waiting for the results?

Similarly, the audit trail is a record of everything that has occurred to every record in the database. It is a necessary system overhead for the regulated industries. However, if the audit file had a search program that reported only exceptional events, then this would be preferable to one that reported all events regardless. The user is not interested in audit reports that show routine progression of samples through the laboratory; it is only where data has been changed or an unusual event has occurred that there is concern. This again highlights the need for flexibility in defining systems and the need for commercial suppliers to provide it.

12.6.3 Systems Specification

Often a further document known as the 'systems specification' is produced. This defines the methods to be used to implement the required function of the LIMS. This is the specification which the programmers will use to produce the software for the system, and will include items such as the screen layout and the procedural logic for all the tasks that the LIMS will perform. This is discussed in more detail in Chapter 16.

12.7 Summary

Good planning and forethought is essential to lay the basis for overall success and provide for the growth of the system in the future. If an analysis of the laboratory's needs indicates a LIMS, then this book is designed to help. Some laboratory managers, after this analysis, may find that they do not require a LIMS, although the analysis in itself may make them more productive by highlighting inefficiencies. It should be pointed out that a LIMS will probably entail staff working with less flexibility compared with manual systems. However, the benefits of such a well designed system will outweigh this disadvantage. Mistakes will be expensive to correct and a badly designed system will be difficult to operate. Good planning will bring the full benefits of laboratory automation.

12.8 References

1. Stockwell, P.B., 'Commentary: The Changing Role of Computers in the Laboratory', *Journal of Automatic Chemistry,* 7 (1985) 57-59

2. Mahindru, A., 'Software Quality Assurance and Validation and Verification', *Proceedings of Third International Phoenix Conference on Computers and Communications,* IEEE Computer Society Press, Silver Spring, Maryland (1984)

3. Squirrel, D.C.M., 'The Why, When and Wherefore of Automatic Analysis', *Analytical Proceedings,* 22 (1985) 79-80

4. Liscouski, J.G., 'Planning an Approach to Laboratory Automation', in *Computers in the Laboratory Current Practice and Future Trends,* Ed Liscouski, J.G., American Chemical Society Symposium Series 265, American Chemical Society Washington D.C. (1984) pp 1-9

5. Hong, R.S., 'Evaluating Your LIMS Requirements', *Instruments and Computers,* 3 (1985) 32-36

6. Berthrong, P.G., 'Computerisation in a Pharmaceutical QC Laboratory', *American Laboratory,* 16 (1984) February 118-126

7. Dessy, R.E., 'Laboratory Information Management Systems Part 1', *Analytical Chemistry,* 55 (1983) 70A-79A

Chapter 13

Integrated Water Quality Data Management

J.G.Jones and C.J.Triner

13.1 Introduction

Wessex Water manages the entire water cycle within a 4000 square mile area which includes the UK counties of Avon, Dorset, Somerset, Wiltshire, and parts of Gloucestershire and Hampshire. The water cycle refers to the progress and processing of water from the time it falls as rain until it returns either to the sea, or by evaporation, to the sky. Wessex Water's main responsibilities include:

- Provision of water resources.
- Treatment and distribution of drinking water.
- Collection and treatment of dirty water.
- Disposal of dirty water and sewage.
- Maintenance and protection of bathing water standards.
- Monitoring and control of water pollution.
- Care and husbandry of fishing and fisheries.

Complementary to these duties are numerous other important responsibilities. For example, liaising with local authorities on development plans, licensing water abstractions, policing pollution and, when necessary, prosecuting offenders.

It is a multi-faceted role in which one responsibility dovetails with another. For instance, the disposal of dirty water has a direct effect on the provision of safe drinking water. Because each and every duty relates to another it is important they should be under single control to avoid conflicts of interest. For this reason, Water Authorities came into being in 1974.

Throughout the work of a Water Authority, the collection, dissemination and interpretation of scientific data is essential if operations concerned with water are to be carried out efficiently. In 1974, all data from the samples analysed at Wessex Water laboratories were handled and filed manually. The large quantity of data being dealt with in this manner soon brought about difficulties in data retrieval and, in 1975, a central computer Water Quality Archive was introduced. Laboratory report forms were sent weekly to the Wessex Water computing centre for punching and the data were stored on magnetic tape. Limited pre-formatted printed output options were available which took up to three days to be returned.

Increasing sample numbers due to legislation, in particular the introduction of EEC (European Economic Community) directives, and the need for data to be available earlier and at lower levels of detection, brought about dramatic changes in the Wessex approach to water quality data management in the early 1980s. Laboratory instrumentation became more automated and a mainframe based on-line Laboratory Data System was introduced to deal with sample registration, results input and the transfer of results to a new disk based Water Quality Archive. The Laboratory Data System was developed by Wessex's own computer services' personnel in close consultation with laboratory staff to ensure that the system matched the laboratories' requirements.

At the same time, increasing financial pressures from central government prompted Wessex Water to rethink its approach to its scientific activities. On the analytical front, the major outcome was the channelling of all chemical and microbiological analysis to one centralized chemical and microbiological laboratory at Saltford, between Bristol and Bath. From the 29 laboratories, 80 staff and 33,500 samples per year in 1974, this Regional Laboratory now employs 20 technical staff and handles in excess of 125,000 samples per year. Data handling practices within the laboratory have changed to cope with the large number of samples at one location. Many of the analytical tests undertaken are now controlled by local microcomputers linked to the laboratory's host minicomputer which is, in turn, linked directly to the Laboratory Data System.

Today, the Laboratory Data System is not just a vital part of the Regional Laboratory's activities. It is one of a number of complementary water quality data systems each of which is readily accessible from any of five hundred terminals situated throughout the Wessex region.

On-line computer time is a valuable resource. At any given moment a number of on-line systems could be competing for computer time, and for this reason each system has a priority rating. The commitment by Wessex Water to quality data management is such that the Laboratory Data System holds the highest possible rating alongside the direct billing facility.

13.2 The Integrated Water Quality Data Management System

13.2.1 Objectives of the System

The Regional Laboratory at Saltford receives up to 500 samples each day, seven days per week, ranging in type from clean drinking water samples to the dirtiest of industrial effluents. Certain laboratory instruments undertake several thousand tests per hour, playing their part in a total workload approaching two million results per year, each of which must be accurate and available quickly to operational staff.

Human beings do not possess the skill to make the necessary decisions on such a vast array of data within the required time-scales. Hence, Wessex Water designed its integrated system of water data management with the following aims:

1. To enable sample registration at the laboratory to be undertaken at a computer terminal.
2. To eradicate laboratory errors associated with manual data handling.
3. Enable the results of analysis to be available immediately following analysis at any of the network of terminals throughout the Wessex Water region.
4. To automatically validate analytical results against pre-defined limits prior to transfer to the Water Quality Archive.
5. Highlight exceptional results in order that operational or field staff are able to undertake any necessary corrective action without delay.
6. Minimize the time between sampling and the results being made available on the Water Quality Archive.
7. Enable the results of analysis held on the Water Quality Archive to be readily available at a computer terminal.
8. Provide, at computer terminals, a register of consented discharges and any associated analytical data, as laid down in the Control of Pollution Act 1974, Part 2.
9. Produce worksheets, exception reports, completed sample reports and laboratory management reports on request.
10. Produce printed reports and graphs from archived water quality data.

13.2.2 Components of the System

The inter-relationship of the independent systems and files which make up the Wessex Water data management system is indicated in Figure 13.1. Three separate on-line systems are utilized, each of which is available via the Wessex regional terminal network, namely the Laboratory Data System, the Water Quality Enquiry System and the Consents System.

Three disk based data files are essential to the operation of the overall system:

1. *Site Master File* This contains the following basic information for each of the 13,500 sampling locations used by Wessex Water: site name, site number (a unique eight digit number) and site type (a two character code to identify the type of sample, eg. consumer's tap, river, trade effluent). As an aid to

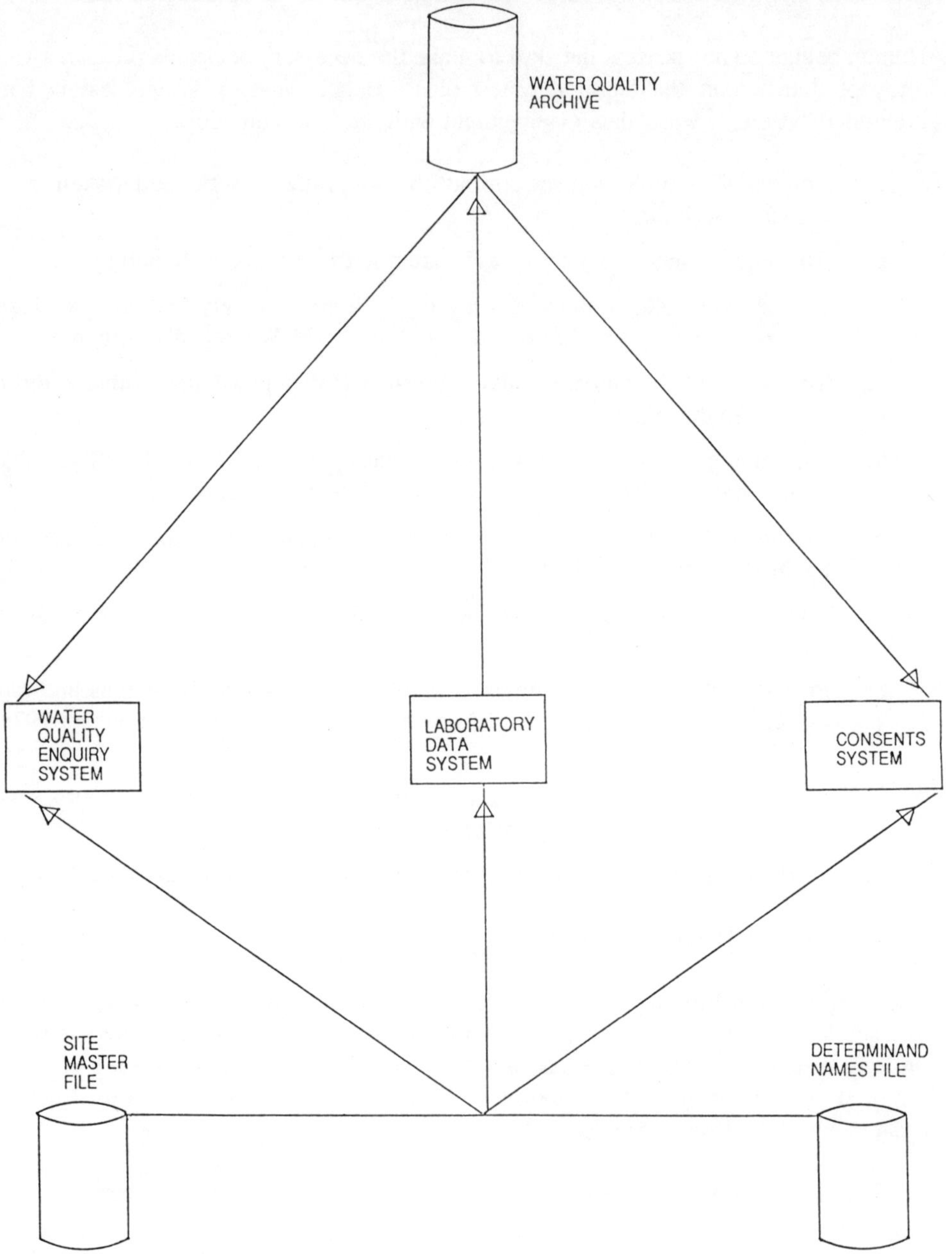

Figure 13.1 Components of the integrated water quality data management system

further identification, the National Grid (map) Reference and a hydrological reference are also included in this file.

2. *Determinand Names File* Contains a list of all determinands analysed at the Regional Laboratory, each being assigned a unique four digit reference.
3. *Water Quality Archive* A disk based store of the results for every sample analysed. Over 800,000 samples are currently stored, providing twelve years of data. Earlier data is held on magnetic tape.

The operation of each of the three systems is described in detail below.

13.3 Laboratory Data System

The operation of the Laboratory Data System is indicated by the flow chart in Figure 13.2. and the data handling equipment in the Regional Laboratory connected to it is shown in Figure 13.3.

The system utilizes the centrally-based Site Master and Determinand Names Files, but in addition, contains its own group of operational files. These are:

1. *Sampling Round File* This contains details, as lists of site numbers, of each sampling round undertaken as part of the routine sampling programme. Each list is allocated a unique reference number.
2. *Determinand Lists File* Contains details of the various analysis suites which may be assigned to a particular sample. Selection of a particular suite of determinands is available by site number, or site type.
3. *Worksheets File* This contains details of which samples are held on which worksheet for each determinand.
4. *Sample Readings File* Contains the up-to-the-minute validated analytical data for each sample currently resident in the Regional Laboratory. The file typically contains 2500 samples.
5. *Quality Control File* Containing both upper and lower consent/statutory and 'nonsense' limits for each sampling site.

13.3.1 Sample Registration

Sample registration at the laboratory requires certain basic information to be input from the bottle label via a computer terminal: the eight digit site number, the date and time of sampling, the sample purpose code which defines the reasons for taking the sample together with any on-site measurements and special analysis requests.

At registration, the sample is automatically assigned a unique laboratory number. The system then searches the Site Master File to ascertain the site name and type, and then using the site number and type and sample purpose code, searches its Determinand

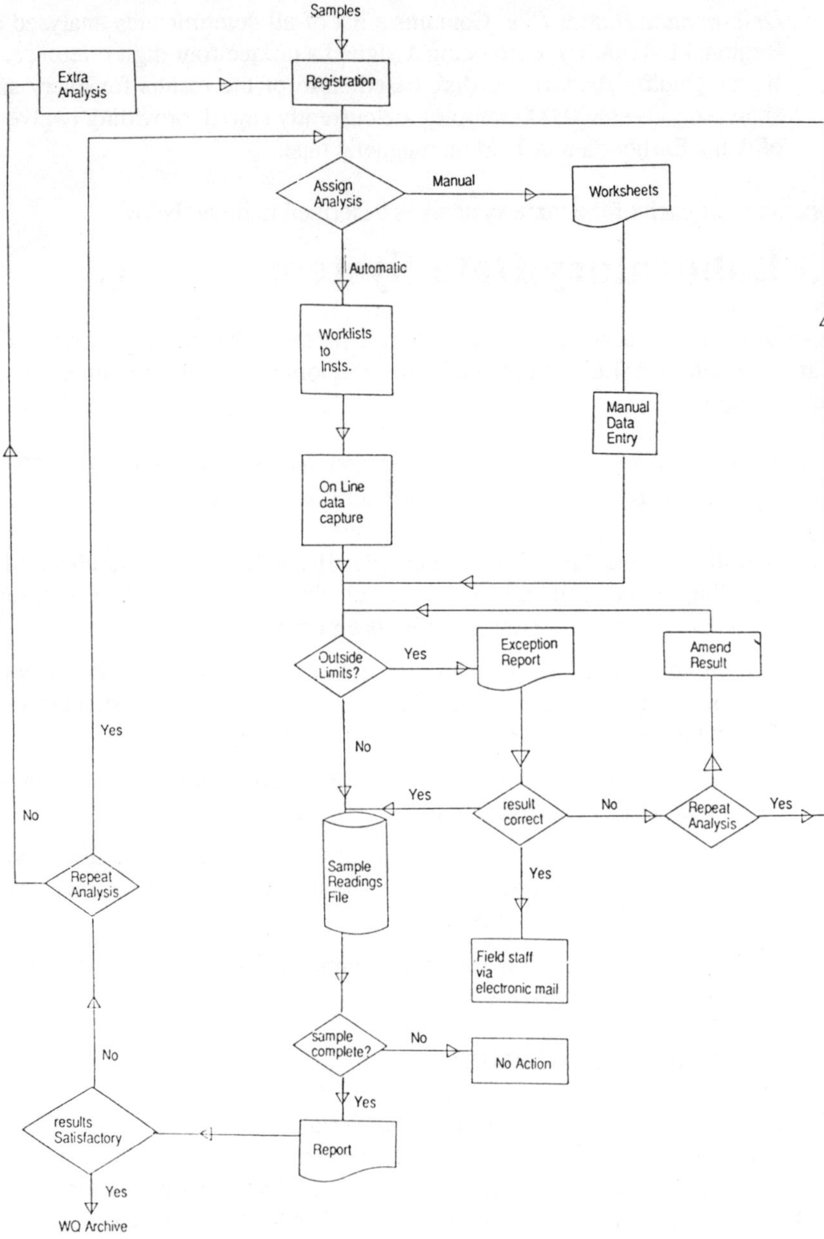

Figure 13.2 Operation of the laboratory data system

List File to dentify which analysis needs to be undertaken. On those occasions when special analysis is requested, for example during a special survey or for samples taken due to a pollution incident, it is a simple matter to add or delete determinands.

Following registration, the sample details and its analytical requirements are passed to the Sample Readings File.

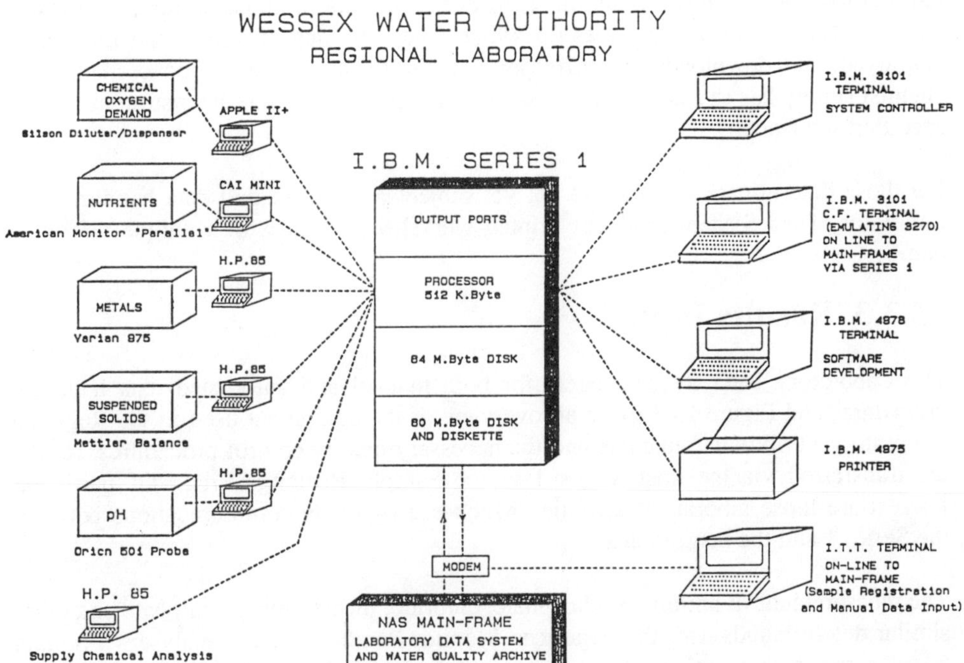

Figure 13.3 Equipment connected to the laboratory data system

Sample registration is currently being enhanced by the introduction of a facility whereby a whole sampling round can be registered on a single screen. The unique round reference will be input which will initiate a search of the Sampling Round File. The corresponding list of site numbers will be displayed, and the appropriate associated information, eg. date and time, entered in the fields provided. This method of registration is expected to reduce registration time by a factor of four.

Further enhancements are planned using bar coded sampling points. A light pen will be used to read the bar coded site label, and portable computer/printing facilities will provide a bar coded sample label containing all the necessary data for registration. This will not only ensure that all samples are taken at the required sampling point, but will also speed the laboratory registration procedure and lead to an earlier start on the analytical work.

13.3.2 Worksheet/Worklist Production

The initiation of the worksheet/worklist program causes the Sample Readings file to be scanned, searching for those determinands added since the last worksheet/worklist request. An essential aspect of this procedure is the involvement of the Regional Laboratory's host minicomputer, an IBM Series 1. This in-house computer sorts out worklists for the various analyses for which automatic data handling and transfer is employed, and downloads the corresponding list to the equipment concerned. The chemist simply has to ensure that the samples are introduced into the instrument in the specified sequence.

For those determinands which are not yet subjected to automated data handling, the Laboratory Data System produces printed worksheets, each with a unique reference number.

13.3.3 Results Input

The Laboratory Data System caters for both manual and automated data handling procedures and Figure 13.3 gives an overview of the equipment used in the Regional Laboratory. Following analysis and the necessary quality control procedures, results are transferred via the host Series 1 to the Sample Readings File. Of particular interest are those laboratory activities which use two way communications between the Series 1 and the instruments.

The measurement of ammonia, phosphate, chloride, nitrite, nitrate, sulphate and other similar determinands uses the American Monitor Parallel discrete analyser. For data transfer, the instrument's own computer is linked to the Laboratory Data System via the 'Series 1'. The laboratory number and the required analysis suite are contained in a bar code which is attached to the sample tube. The instrument reads the bar code and analyses only for the required determinands.

Chemical oxygen demand is measured with Gilson equipment via an Apple II microcomputer. The microcomputer is not only used as a data 'mover', but also controls various aspects of the test such as dispensing reagents, sample dilution, absorbance measurement and results calculation.

The determination of metal ion concentrations is achieved by a Varian 975 Atomic Absorption Spectrophotometer via an HP85B microcomputer. The microcomputer not only sends and receives data to and from the host Series 1, but also optimizes analytical conditions by adjusting the lamp turret, flame conditions and slit widths, and organizes instrument calibration.

Additional HP85B microcomputers are used in other areas of the laboratory for automated data transfer associated with pH measurements, suspended solids determinations, and certain physico-chemical tests on supply water samples.

For manual assays, following analysis and the necessary quality control procedures, the results are written on the appropriate worksheet. The results are then entered at a computer terminal against a screen-based copy of the worksheet and then transferred to the Sample Readings File. Typical analyses for which manual data entry is used are biochemical oxygen demand, detergents, microbiological analysis of supply and sea waters and miscellaneous organic analysis by GC and GC/MS, eg. oils and pesticides.

13.3.4 Exception Reporting

The laboratory has a commitment to identify and report any atypical results from amongst the thousands produced each day. Immediately following validation, exception reports are passed to operational staff via the Wessex electronic mail facility, thereby ensuring any problem is attended to without delay.

As results arrive at the Sample Readings File they are automatically subjected to validation against the upper and lower consent/statutory and 'nonsense' limits held on the Quality Control File. Any result falling outside a limit is flagged.

The exception reporting option initiates a search of the Sample Readings File for the flagged exceptions and then prints an Exception Report, which includes all results flagged since the last report. It is then the duty of a senior member of the laboratory staff to investigate each result in an exception state. This involves checking the original data within the laboratory, considering the viability of such a result relative to other results on the same sample, and possibly examining the sample itself. The chemist's decision of whether to accept or reject the result is entered via a terminal after calling-up a screen copy of the Sample Readings File for the sample. The chemist's decision may be one of the following:

1. If the result is valid, then the current sample results are automatically transmitted to the appropriate operational staff by the Wessex electronic mail facility.

2. If the result is incorrect, usually due to a keying error during manual data input, then the correct result is entered.

3. If the result is suspicious, then the particular determinand is identified for repeat analysis and is thereby included in the next worksheet/worklist request.

13.3.5 Transfer to Water Quality Archive

Only when all required analyses of a given sample are finalised and any result in an exception status has been cleared, can the sample be denoted as 'complete'.

Each day, a report of completed samples is printed at the laboratory which lists the analytical results for every sample completed since the previous day's report. A senior member of the laboratory staff will then scan the data for each sample, noting any which contain apparent anomalies and/or those which would benefit from further analysis. Samples which are satisfactory are passed to a file reserved for completed samples, and each night the entire contents of this file are passed to the Water Quality Archive, and any reference to the samples concerned is deleted from the Sample Readings File.

Any anomalous results are subjected to a procedure similar to that used for checking results in an exception status. Only when the senior member of staff is satisfied with the results will the sample be passed as 'complete'. Access to each of the above mentioned facilities within the Laboratory Data System, namely: registration, worksheet/worklist production, results input, exception reporting and transfer of completed samples to the Water Quality Archive is restricted to laboratory staff by the use of passwords. Within the laboratory, other passwords are used to ensure that certain facilities are accessible only to senior staff. However, a further facility is available which is not password protected as no data manipulation is involved; this is described below:

13.3.6 Sample Register/Results Display

The sample register is a list of all samples currently resident on the Laboratory Data System and may be interrogated for a particular sample or set of samples using site number, site number/date or site type as the key. The system then searches the Sample Readings File for the samples of interest and provides the facility to page through the samples, and their current results, on the screen.

This option, although used by laboratory staff, is used most frequently by operational and field staff to obtain the current analytical position on their samples, thus enabling any necessary follow-up to be undertaken at the earliest possible opportunity.

13.3.7 Laboratory Management Reports

Immediately following the nightly update of completed samples to the Water Quality Archive, a number of management reports are printed. The reports are central to the efficient management of the laboratory workload, for example, by highlighting backlogs. Typical reports are:

1. A print of the current sample register.

2. A list of laboratory numbers of all samples successfully updated to the Water Quality Archive.
3. A list of analyses, with their respective samples, which have not been undertaken within pre-defined target time-scales.

13.4 Water Quality Enquiry System

The Water Quality Enquiry System is an on-line means of recalling and displaying sample data from the Water Quality Archive. Only the central Archive and Determinand Names Files are utilized.

Data are available to the user in one of three formats via a simple menu screen:

1. *Results for one or more sites* Up to fifteen sites may be selected by entering the appropriate site numbers, together with the date range required. Data is presented as one sample per screen and any result falling outside a consent or statutory limit is highlighted.

2. *Results for one determinand on one or more sites* The operator enters the determinand number of interest, up to fifteen site numbers and the required date range. Data is presented as a chronological list of results with any result outside a consent or statutory limit being highlighted.

3. *Results for several determinands on one or more sites* Up to five determinand numbers and 15 site numbers may be entered together with the date range. Data is presented as a chronological list of samples with one sample per line.

Facilities are available within each option to page forwards or backwards through the data for each site, or to change sites. It was a deliberate decision not to provide a facility to list data from two or more sites for comparative purposes on the same screen. Such screens are difficult to organize, let alone read. If the user requires to compare data from various sites he has the facility to 'screen-dump' the required pages to a local printer.

13.5 Consents System

The UK Control of Pollution Act 1974, Part 2 requires that discharges to inland waters, estuaries, coastal waters and underground strata will conform to certain conditions regarding nature and composition. These conditions are included in a consent to discharge. One of the most important provisions of the Act is that a public register is available containing details of applications for consent to discharge, details of the consents granted, the analytical data on samples of waters and effluents taken for pollution control purposes, and any steps taken as a consequence of a failure to meet a discharge consent.

The on-line Wessex Consents System has been designed to encompass the requirements of such a register and contains details of 6000 consents to discharge. The system utilizes the central Site Master, Determinand Names and Water Quality Archive Files. The amount of information in a particular consent varies considerably, dependent upon the nature of the discharge and the number of discharges covered by the consent. To hold all the necessary information for all consents in one 'Consents File' would be wasteful of disk space and the system has been designed around a number of specific operational files, the most important of which are:

1. *Application File* containing all the information relevant to a particular consent application.
2. *Site File* containing site location data for each site, together with consent numbers.
3. *Name and Address File* containing the consent holder's name and address.
4. *Discharge File* containing specific details of each consented discharge such as consent number, national grid and hydrological references, and Water Quality system site numbers for both the discharge and upstream/downstream sampling sites in the receiving water.
5. *Condition File* containing the permitted limits for each determinand in each consent.
6. *Text Condition File* containing, for each consent, details of any condition in the consent which is expressed in words. A particular text condition may appear in many consents and to conserve file space the commonly used text conditions are allocated codes. The code is then held on this file and a separate reference file holds the complete wording.
7. *Action Taken File* contains information to indicate the steps taken by operational staff as a follow-up to the failure of a discharge to meets its consent.

The Consents System can be accessed at any terminal within the Wessex Water network. Access to a particular consent is obtained by entering the consent reference number or by using one of a number of search facilities, such as a district, site name, holder's name, receiving water, national grid area, etc.

Analytical data relating to consented discharges and associated sites on the receiving watercourse is also available. Any consent failure is highlighted along with the 'action taken' information from operational staff. The creation of a consent on the system is restricted to particular Wessex personnel, but the display of consents and associated analytical data is available to all staff. However, not all the information contained within the Consents System is available to the Public Register. This is the case when the holder of a consent has successfully applied to the Secretary of State for exemption.

13.6 Off-line Reporting Facilities

There are many occasions when the on-line Water Quality Enquiry System is unable to provide data in the required format. On these occasions the user has a host of off-line reporting options available, some printed, some plotted. The request is entered via a computer terminal but processing takes place as and when computer time is available, as on-line systems always have priority. The output is routed to one of a number of report centres around the Wessex region. Typical examples of such reports are:

1. The comparison of results from the various sites sampled during a survey.
2. The variability of a particular determinand at a given site(s) with time.
3. The calculation of maxima, minima and mean on the data from a given site over a period of time.
4. Calculation of half-yearly trade effluent charges for individual traders.

13.7 Future Developments in Laboratory Automation and Data Handling

The trend in the Regional Laboratory workload over the next few years is for more samples with more complex analysis suites at lower levels of detection. Embargoes on increased staff numbers have presented the challenge for further automation within the laboratory, especially for those tasks which are labour intensive.

Plans are well advanced to introduce further automation and a number of examples follow:

1. *Biochemical Oxygen Demand* Automated data handling with a microcomputer is to be introduced before the end of 1987. In addition, the analytical method is very labour intensive and, in view of the large number of samples being analysed, which is often in excess of 300 per day, is difficult to automate. Robotics is currently being considered as a possible solution.
2. *Microbiological Analysis* The Regional Laboratory currently uses the traditional labour-intensive membrane filtration method for microbiological analysis. Automated colony counting using image analysis technology is currently being appraised with a view to the introduction of data handling links to the Laboratory Data System before the end of 1987. The use of image analysis is intended as a stop-gap measure. The past three years have seen the initial development work, by the Authority's own microbiological staff, of a technique which is expected to replace membrane filtration and its associated procedures before the end of 1989. The equipment, the Malthus Microbiological Analyser, uses a sealed glass jar to which the water sample and the nutrient broth are added. If bacteria are present, the conductivity of the water changes as the organisms multiply, giving rise to a unique growth profile for each particular bacterium. The Malthus equipment continually monitors the conductivity of all the samples under test, looking for any increase caused by bacterial growth. The advantage of this technique, in addition to enabling automation to be introduced, is that the time necessary

to obtain an indication of the bacteriological quality of water is reduced from eighteen«hours to less than nine with no staff involvement other than with the initial sample introduction. The ultimate aim is to use telemetry links to the permanently manned Regional Operations Centre at Bristol to provide the earliest possible indication of unsatisfactory water quality, and to initiate immediate corrective action.

3. *Plasma Source Mass Spectrometry* Regional Laboratory staff are currently commissioning a VG Isotopes Plasma Source Mass Spectrometer, the eventual aim being to replace atomic absorption technology for metals analysis and provide simultaneous multi-element analysis. The instrument can be operated in three modes dependent upon the sensitivity and accuracy required:

    1. *Single ion mode* which provides typical detection limits of better than 0.1 ppb with confidence limits of 5%.
    2. *Simultaneous multi-element mode* which provides typical detection limits of 0.5 ppb with confidence limits of 10%. Sensitivity can be improved by reducing the number of elements analyzed for in a particular scan.
    3. *Rapid scan mode* which provides a semi-quantitative analysis of a given sample for over 70 elements, with typical detection limits of 2 ppb with confidence limits of 30%. Communication to the Laboratory Data System will be via the instrument's own IBM PC microcomputer.

4. *Bar Code Technology* The current use of bar code technology with the American Monitor Parallel discrete analyser has been indicated earlier. The potential for two other uses of bar code technology within the laboratory is currently under investigation, namely sample recognition, particularly for those areas where laboratory robotics are planned, and improved sample registration.

13.8 Summary

The last decade has seen increasing pressure upon water authorities to reduce costs. This, coupled with the introduction of various legislative quality standards, particularly in the form of EEC directives, has forced Wessex Water to reappraise its scientific activities. The challenge has been met by a firm commitment to the application of new technology in both water analysis and the management of resulting data.

Wessex Water has never viewed new technology as an end in itself, rather as an essential tool with which to keep pace with the ever-present need for greater operating efficiency.

Chapter 14

ACCOMPLIS

Graham Martin

14.1 Why ACCOMPLIS was Developed

The primary purpose of an analytical laboratory is to provide the organization it serves with information about the constitution of samples submitted to the laboratory for analysis. Such information has immediate impact on the organization when it is used as the basis for taking decisions. It follows that the value of the contribution made by an analytical laboratory depends upon the quality, correctness and promptness of the information it provides to the decision making process. Correctness is by far the most important of these criteria, since it is immaterial how efficiently laboratory information is produced and how quickly it is delivered if it is erroneous.

The maintenance of quality standards is particularly relevant when the output of the laboratory is information concerning the quality of manufactured articles. Clearly, the quality control procedures used to maintain the consistency of manufactured goods can be no more reliable than the analytical procedures employed to supply information to the quality control process. Furthermore, in order to facilitate effective quality assurance, the laboratory must do more than ensure the validity of its operational procedures; it must generate and save in a retrievable condition, reliable evidence that it has observed and maintained those procedures. Considerations of this kind are no longer confined to those laboratories which are obliged to comply with established regulatory standards such as Good Laboratory Practice. They are of increasing significance in a growing range of industrial sectors in which the achievement of proven and consistent quality is becoming a major aspect of product differentiation and an important route to securing competitive advantage.

It was this view of the fundamental role of the analytical laboratory in the industrial setting which formed the basis for ICI's requirements specification for a LIMS. It is a measure of ICI's commitment to maintaining the highest standards of quality in its manufactured products that the company was motivated to develop its own LIMS to meet the exacting specification it had produced. This was done after reviewing commercially available products.

The specification placed considerable emphasis upon issues concerned with eradication of error in the generation and interpretation of analytical information. It contained a number of practical proposals for combatting error sources and particularly sought to avoid the use of error-prone activities such as repetitive or lengthy operations at the keyboard. More than three years later, the system to have emerged is ACCOMPLIS.

The process applied to the development of ACCOMPLIS was a combination of classical system development life-cycle and prototyping. This process involved dividing the total development into discrete phases with evaluation and design refinement steps at completion of each phase. It has been a significant advantage for the development to be carried out alongside a fully operational laboratory environment and to harness, over a prolonged period, the practical experience of many analytical chemists and quality assurance experts throughout ICI.

14.2 What is ACCOMPLIS?

ACCOMPLIS was designed to be a comprehensive LIMS in any analytical department which would consist typically, although not necessarily, of several individual analytical laboratories which may be dispersed around a building or over a site. Each of these laboratories would usually specialize in a particular analytical technique or product area. To meet the requirements for error reduction, system reliability and performance, the distributed processor architecture shown in Figure 14.1 was adopted. For reasons of performance, flexibility and compatibility, VAX and MicroVAX computers, manufactured by Digital Equipment Corporation (DEC) were selected for this system. Each laboratory area is supported by a MicroVAX II which handles all the real-time data acquisition tasks, for which it is specifically programmed in the real-time language RTL/2. Laboratory instruments, balances, terminals, printers, bar code readers and the optional graphics digitizing tablet are all connected to the MicroVax II which serves their laboratory area. Each of these laboratory computers is networked to a host computer (VAX or MicroVAX II) responsible for the long-term data management, file-serving and reporting facilities of the system.

To ensure compatibility with other VAX systems it uses standard utilities and languages available under the VMS operating system, including RMS filing, Common Data Dictionary and Datatrieve (the data base query language). Communication between the host and laboratory computers is by means of a two way packet-oriented protocol (LIMPET) incorporating automatic error-checking and self recovery procedures. This protocol has been modelled upon KERMIT (Chapter 10) but works as a background task between dedicated computers. The software structure is shown diagrammatically in Figure 14.2.

For laboratories in the organization served by ACCOMPLIS which do not require real-time data acquisition from laboratory instrumentation and bar code readers, the remaining LIMS functions may be provided via terminals directly linked to the host computer.

An important aspect of the system design was to provide a high degree of flexibility, both for in-service modifications and for applicability in a wide range of analytical laboratory situations, while ensuring that long term support can be efficiently provided to all users. The approach taken to achieve these objectives was to implement system functions by means of highly generalized programs which are configured for an individual installation by means of extensive data files. The system code remains consistent across all installations and can be readily updated or expanded without disturbing existing data. Likewise, data may be modified to facilitate changing circumstances of use without disturbing programs or halting operation of the system.

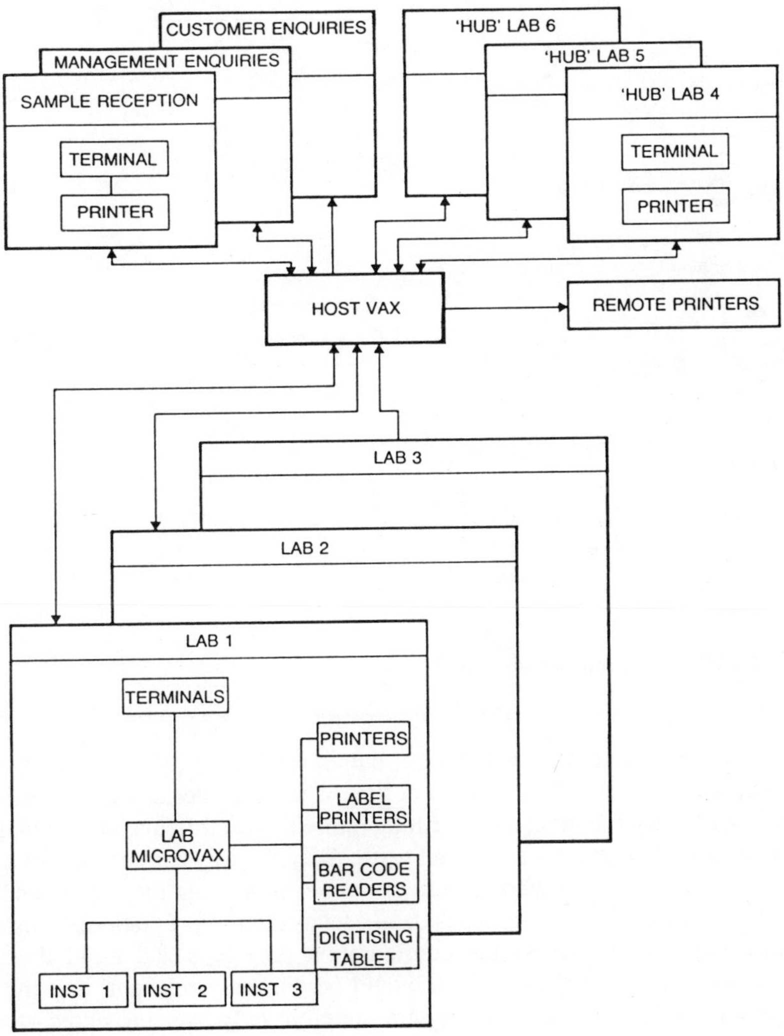

Fig. 14.1 ACCOMPLIS hardware configuration

A major advantage of this approach in the laboratory is that it enables a very wide range of RS-232C output instrumentation to be interfaced to the system. Instead of having to write a handling program for each individual instrument, a generalized program for instrument handling is configured by data for all of the RS-232C instruments which are on-line to the system. Coping with changes in laboratory instrumentation, which is quite a frequent occurrence, now becomes a simple matter of editing the relevant files. The application of data-driven and real-time programming techniques provided the combination of reliability, flexibility and consistency which was stipulated in the original specification.

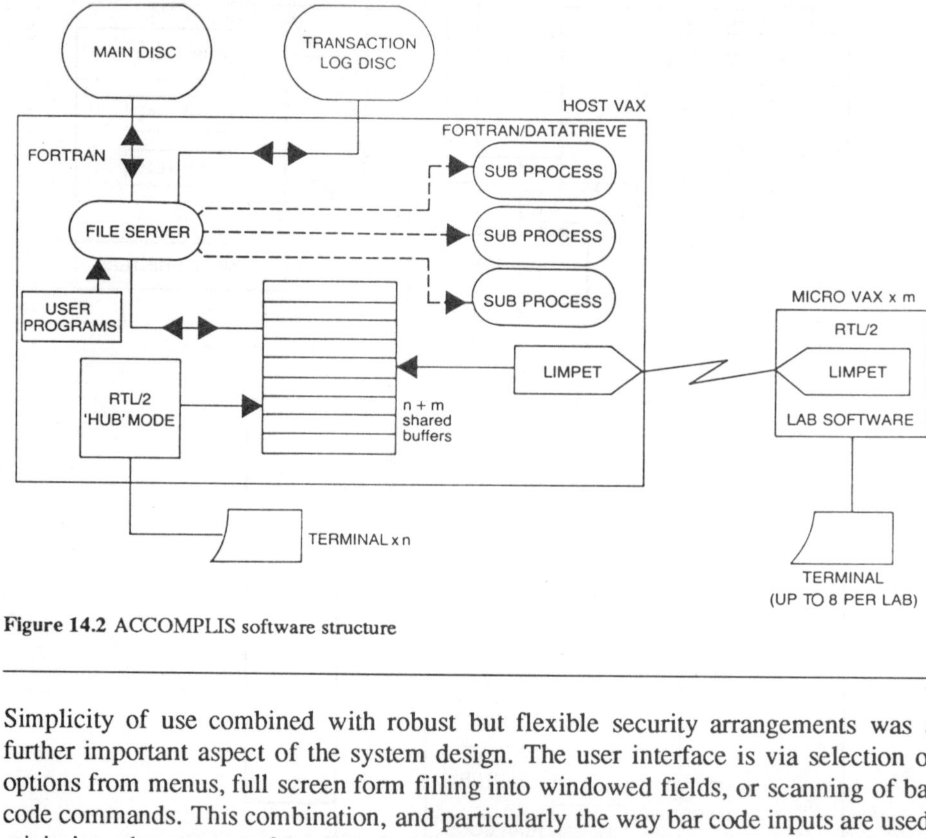

Figure 14.2 ACCOMPLIS software structure

Simplicity of use combined with robust but flexible security arrangements was a further important aspect of the system design. The user interface is via selection of options from menus, full screen form filling into windowed fields, or scanning of bar code commands. This combination, and particularly the way bar code inputs are used, minimizes the amount of keyboard activity involved in using the system and greatly reduces the risk of entering erroneous data. Menus are not programmed. Instead, the menu building routines access the configuration data files and build the relevant menus on the basis of options, which would be currently relevant, and the system privileges available to the user. Users are therefore only ever presented with menu choices which are currently meaningful and which they are authorized to use. Menu selections via single keystrokes enable experienced users to type ahead; the system automatically senses the typing speed and suppresses the unnecessary display of menus.

An ACCOMPLIS installation may have any number of privilege classes which are defined by the set of system facilities available to that class. Users may be assigned to different classes in each of the laboratories in which they work. Individual users are personally identified by selecting their name from the menu which lists the analysts in their laboratory and entering their own undisclosed password. Simple, menu-driven facilities are provided for users to change their passwords and for managers to modify the set of functions available to individual users by granting or revoking specific privileges. Even the system's colour scheme can be altered by simple menu selection.

An interesting problem at the design stage was to strike the right balance between freedom of action of individual laboratory workers and the need for data integrity and evidence of it in the system. To gain full user acceptance the system would have to respond to decisions made by its users rather than the other way about. Although it would have been much easier to have designed a system which called for a rigid work pattern requiring each analyst to do his work and enter data in a predetermined sequence, people do not like to work that way and it presents difficulties in dealing with unforeseen circumstances. The approach taken supports users in planning their own work and leaves them free to decide the sequence in which they complete their individual work plan. The system makes it simple for users to indicate the actions they are about to take, it checks for the validity of that action and then either executes and confirms the command or advises the user of the invalidity of what he is about to do.

The structure of the ACCOMPLIS database reflects the way in which samples are normally submitted to an organization of analytical laboratories and the manner in which results are reported to users of the analytical service. Indexed sequential files are used for the main files (DEC's Record Management Services). The three main files form a hierarchy, primarily storing work submissions to the laboratory complex, secondarily samples which are components of these submissions, and at the bottom level, analyses which are components of these samples. The primary key of each of these files is related to that of its parents in a straightforward numbering scheme. Fields within the record structures are of two kinds: either they are mandatory, being used by system code, or they are optional and can be configured to users' requirements at the outset. Textual material which forms qualifying additions to laboratory results is held in separate dependent files that are related to the main keyed files by file name, and small directory files are built to keep track of such texts. The structure of the ACCOMPLIS main database is represented in Figure 14.3.

14.3 How ACCOMPLIS Works

ACCOMPLIS provides a very extensive set of LIMS functions, many of which are optionally enabled for the whole system or for individual laboratories on the system, by means of configuration data files. The functional scope of the system is shown in Table 14.1. However, space does not permit a discussion of the entire range, so the remaining sections are devoted to the primary functions of the system which would be used in almost all analytical laboratories. These are discussed in the sequence in which work flows through an analytical laboratory with some practical examples included to illustrate how the system supports laboratory staff, their management and, most important of all, their customers.

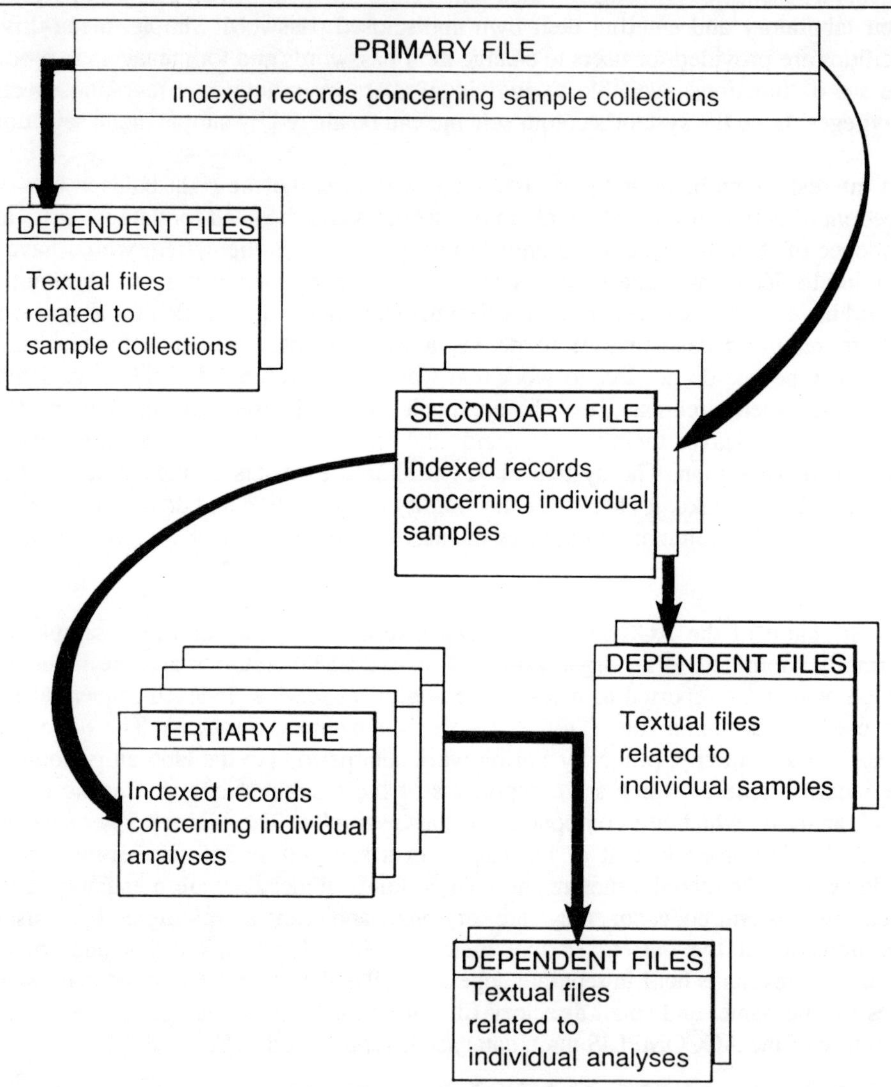

Figure 14.3 The ACCOMPLIS main database

14.3.1 Registration of Samples

Collections of samples, normally from a common origin, are logged into the system by entering the relevant details of the complete collection into named windows on a full-screen form. This is followed by entering details about individual samples in the collection on a second set of forms. In addition to the database fields required by the system, both types of form may also contain fields defined by users. If required,the positioning of windows on the forms can be user-configured to resemble existing paper-based record systems. The data entered to the forms are automatically checked for validity and whole or part screens may be copied and edited to speed and simplify the registration process. The system ascribes a reference number to all the samples which references each to the original collection and automatically prints a label for each sample. Each label bears the identity number, its bar code the sample name, and may optionally include other information such as hazard or storage conditions as shown in Figure 14.4. A customer receipt for the sample collection can be printed on demand.

Table 14.1

Functional scope of ACCOMPLIS

Assisted sample log-in on user-configured forms with automatic generation of bar coded sample labels and template options for routine work.
Flexible work scheduling with automatic batch building options.
Comprehensive sample tracking and work progression facilities driven by menu or bar code commands.
Menu-driven generation of command bar codes.
Real-time data acquisition from all RS232 output instruments and balances.
Automatic calculation of results taking account of calibration standards, blanks, quality control samples, test samples and their replicates. Methods are automatically validated as they are run.
Bi-directional protocol for communications with laboratory data systems.
Comprehensive method management utility.
Integral facilities for addition of text and diagrams.
Automatic and manual facilities for results validation.
Optional, multi-level approval schemes for analytical reports.
Automatic report construction using built-in or user-configured report generators.
Automatic delivery of reports by electronic mail or remote printing.
Menu-driven management information report generator, including a cost accounting package.
Inter-laboratory mail facility.
Automatic transaction logging and audit trailing.
Flexible, menu-driven archiving facility.
Integral diagnostics functions.
Menu-driven control over colour schemes of system screens.

Each sample log-in form contains a table in which the analysis to be performed is specified and which displays the final results of these analyses as they are generated. The table contains columns for the analytes or tests to be determined, the laboratory which will perform each test, the instrument which will be used and the method to be employed. It is only necessary for the person submitting the samples to enter the names of the tests to be performed. The system automatically completes the analytical route of the laboratory, instrument and method for each requested test. It can do this because it holds in a file which has the structure shown in Figure 14.5, a data model of the analytical repertoire of the entire organization it serves. ACCOMPLIS has automatic procedures for dealing with the situation where several different routes exist for the determination of a given analyte and the system ensures that only analysts will adjudicate in such cases. Frequently used groups of determinations may each be represented by a single key word which automatically expands to the full group upon entry.

ACCOMPLIS

[ICI] : 02–0468–01
Hazard: BIOLOGICAL
Whole Culture H1604/C
Refrigerate

Figure 14.4 A bar coded label

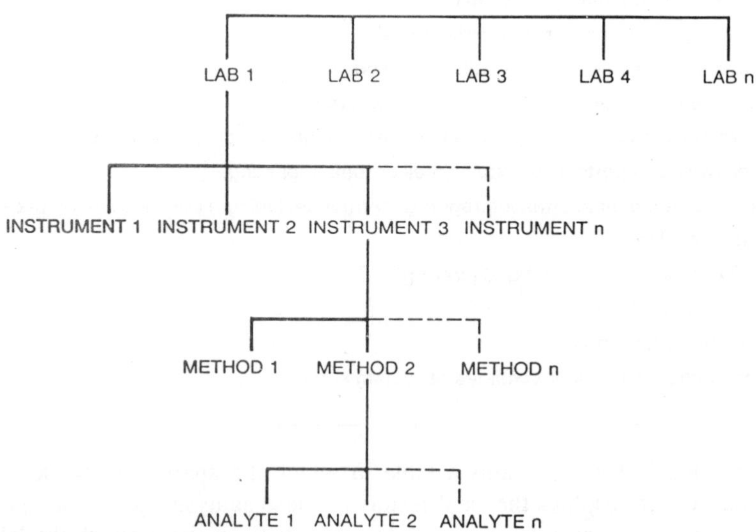

Figure 14.5 Data structure of analytical repertoire

Sample additions to any existing collection, or derivatives of any existing sample, may be registered at any time before the original collection has been completed, approved and reported. Complex parent-derivative relationships may be defined between samples and are maintained by the system and used in the generation of analytical reports. An option available with ACCOMPLIS allows a line diagram to be drawn to add further information about a sample at log-in. The diagram is entered to the system by means of 'drawing' with a special stylus on a digitizing graphics tablet. Once entered a diagram can be edited, displayed and printed at any time. This facility is particularly useful in laboratories elucidating molecular structures. In addition to the creation of a diagram qualifying a sample, users can, if they wish, create a further diagram for any or all of the individual determinations requested for a sample. All such diagrams are automatically reproduced in reports produced by one of the resident report generators.

14.3.2 Work Planning

The registered collections of samples may require work in any or all of the analytical laboratories served by ACCOMPLIS. All pending analyses in the database are accessed by a flexible, menu-driven work planning facility. This enables analysts in any laboratory to search for a particular type of analysis which is performed in their laboratory. This search of pending work may be as broadly or as narrowly defined as the analyst wishes. The work search report is presented to the analyst with the outstanding determinations listed in priority order. At this stage the analyst can select any sub-set of the work search items to form their next package of work or to top up an existing package. To aid this selection the full details of the sample and the collection to which it belongs can be brought immediately to the screen. This includes all results currently obtained and all the notes, diagrams and memoranda which may have been added by other analysts in the course of their work on the same samples. Then a worksheet is printed containing the accepted set of determinations, bearing the bar code identification of each sample and details of the tests to be performed.

Once determinations have been accepted onto a worksheet they cannot appear in another person's work search unless they are deleted from the original worksheet. At each stage of the work on the accepted set of determinations, right through to validation of the results, the system checks that each operation performed by the analyst is consistent with his worksheet and with the analytical method to be applied.

In practice, many analytical methods require samples to be subjected to a batch run on a particular instrument. A batch would typically consist of a particular set of aliquots of test samples and their replicates together with calibration standards, blanks and quality control samples. ACCOMPLIS supports the execution of batch methods of analysis by means of special batch method definition files which are created and maintained under the system's comprehensive method management facility. These methods are marked as 'batch' in the method database and they can optionally contain a specific run sequence for the method which defines the order in which standards, blanks, quality control, test samples and their replicates are to be loaded into an instrument run. The system will automatically build batches and provide the analyst with a full description of the correct loading sequence for all aliquots in the batch. Each batch may be represented on a printed batch sheet which is a special sub-set of the analyst's worksheet.

Samples requiring the same method of analysis may also be grouped into batches regardless of whether batch mode has been stipulated in that method. This type of batch working will frequently be more convenient when the method calls for lengthy operations between measurements, eg. drying or ashing cycles. It is clearly more efficient to perform such operations on a set rather than on individual samples. When work is planned as a batch, the system automatically enables batch entry of the data required to compute a final result for all samples. To give a simple example, if a gravimetric method of analysis is being used to determine moisture and ash from a set of four weighings for each sample, the analyst can call up the batch, select one of the four weighings and enter the values for this weighing for each sample in the batch, all on one screen.

The work planning facility in ACCOMPLIS enables analysts to select the work they will do in the light of full and up-to-date information. It ensures that the correct methodology and procedures are used to perform analysis while allowing the analysts to organize their work in the most convenient and efficient manner.

14.3.3 Sample Tracking

Having decided what work to do, the next step is to locate the samples. The whereabouts of the samples is printed on the bottom of the worksheet or batch sheet and can also be displayed at a terminal at any time. ACCOMPLIS provides a very simple means of tracking the physical location of all samples. The names of all the storage locations are held in a file and a list of all the samples currently stored in each location is maintained. Moving samples from one place to another is achieved by means of bar code commands. The use of command bar codes is a particular advantage of the ACCOMPLIS system. These codes change the function of the reader to that signified by the bar code, eg. storing samples in a particular location. The command bar code label would normally be fixed to the storage location concerned. The analyst scans the command bar code followed by the bar code labels for the samples which are being moved. The same simple, error-free technique is used for many other tasks within the system to avoid keyboard entries and as a means of rapidly entering or requesting information. Examples of these tasks include initiation of weighing from an on-line balance, loading samples to instruments and obtaining various printed reports. The command bar codes specific to each laboratory may be produced at any time via a menu-driven program.

Analysts 'sign-on' to a bar code reader with their own personal bar code before issuing bar code commands. This enables the system to check that the actions the analyst is taking are consistent with the worksheet and in the sequence which complies with the requirements of the method. A prompt error message is provided when this is not the case. This technique provides a simple to use, error-free means to track the correct progress of work throughout the standard analytical process.

14.3.4 Production of Analytical Results

In all analytical laboratories results are obtained by the application of analytical methods. Documentation of the latter normally includes a title, scope of the method, materials and equipment to be used, experimental procedure, checks to be employed, calculation of result(s) and the units of measurement. ACCOMPLIS mirrors this mode of working. A powerful method manager provides a secure environment where methods are created, copied, edited, authorised and regularly reviewed or revised. It maintains a log of method revisions and stamps results with the name and revision number of the method used. Methods are classified by status, eg. 'developmental', 'active' or 'lapsed'; only the latest revision of an 'active' method will be automatically called and executed. Previous revisions of methods are stored and can be examined or used as the basis for creating a new method.

The calculation section of the method enables variables to be defined, eg. weighings, volumes, dilutions, manual data, calibration coefficients, and instrument readings. The calculation of results can be defined in terms of an algebraic expression involving these variables and any substitutions required to produce results for other analytes determined by the same method. There is also provision for interim calculations used to check that the method is being properly conducted. If results of these interim calculations fall outside the predefined limits the system can be set, either to issue a warning, or to halt further execution of the method until the cause is rectified. As data are entered (either automatically or manually) for the variables in the method calculation, the system knows by means of the worksheet and dynamic instrument queues to which sample and method the data apply. As each data element is entered the system checks which calculations can be completed and automatically performs those calculations. When all the data are entered the final result calculations are automatically executed. The results may then be validated in the context of the calculation and, if satisfactory, forwarded to the database. Also stored in the database are the weighings, volumes, dilutions, instrument readings, etc, which were used to calculate the final result.

In the case of batch analytical methods, the data collected for every aliquot in the batch are stored in a unique batch file for further processing. ACCOMPLIS includes a batch file processing program which can, if necessary, be modified in order to create new programs to perform specific calculations on batch data files. This program accesses a library containing analyte concentration data for the calibration standards used in the batch method. It produces a polynomial curve-fit to the calibration data and uses this to calculate analyte concentrations in the test samples and quality control samples. These results are displayed and when approved they are sent to the database. The results for the quality control samples are stored in a special file logically associated with the current method file. The accumulated results for these samples are then always available for selective searching, display and plotting. This is particularly useful in providing evidence that a given method is under control and maintaining the required degree of accuracy. To aid this, the Method Manager includes a control chart display format for the accumulated results of quality control samples.

Some analytical methods do not involve the entry of numeric data or the execution of a calculation. These are methods which require some form of qualitative assessment, observation, or visual examination. These are classified in ACCOMPLIS as 'text' methods since the 'result' will be a block of text. The system contains an integral word processor to cater for the entry of this class of result. The word processor may also be used to append notes about analyses, whenever an analyst may feel this is useful. The full set of notes entered for a sample can be viewed at any time. A word processor was developed for ACCOMPLIS so that the creation and viewing of text would be an integral operation, and to ensure that all text files would be automatically named, stored and ultimately archived.

Part of the purpose of ACCOMPLIS is to support the work of analysts by provision of full and up-to-date information. Thus, an analyst working on a sample in one laboratory may easily obtain the results and notes produced on the same sample by other analysts in other laboratories. This can be very helpful because most analytical methods can be prone to interference effects due to the presence of other components in the sample matrix. The system provides a simple means to record and display warnings, advice or observations which have been made.

As a further aid to inter-laboratory communications ACCOMPLIS also incorporates a mailboard facility for broadcasting pages of information entered using the word processor. The system itself uses the mailboard to distribute routine reports regarding the completion status of work in hand and to notify analysts of the existence of batch data files which are awaiting interactive processing.

14.3.5 Validation and Approval of Results

Laboratories engaged in quality control work would normally be analyzing samples of products or intermediates for compliance to a given specification. ACCOMPLIS caters for the automatic checking of results against specifications by conspicuously flagging any out of limit values. The checking process is triggered by the presence of the product name or code on the sample registration form. The specification limits and their units are declared in product definition files. These files also cater for the definition of groups of products which share the same approval arrangements. Thus, a set of samples belonging to the same product group may be submitted for analysis as a single collection and will ultimately be approved and reported collectively.

Samples which contain completed tests are automatically brought to the attention of users with the authority to validate results. Results are validated with the full information pertaining to each result available for immediate display, and they may be set to the status of 'valid' or 'not valid'. The status 'not valid' is applied to results which are in some way defective but which cannot be repeated because the original sample has changed or is no longer available. Once all the requested tests on an original collection of samples have been completed and validated the collection is automatically advanced to the approval process.

ACCOMPLIS supports the creation and application of any number of multi-level approval structures. It caters simultaneously for approval schemes which apply to classified groups of products and to experimental studies. When any sample collection contains a reference to a specific experimental study or product group, then the system ensures strict adherence to the particular approval arrangements which have been specified. These arrangements may be as simple as a single approval step by any one of several authorized persons or as complex as several overlapping bands of approval authority with different people authorized for each band. No matter what mix of approval schemes is implemented, ACCOMPLIS automatically refers sample collections awaiting approval at a given level to the appropriate authorized personnel at that level. When an approver refuses approval to a sample collection, the system automatically moves into the word processor and prompts for the reason for refusing approval to be given. The approver identifies the test results which he or she believes to be suspect and ACCOMPLIS automatically refers these back to the laboratories in which the tests were performed. At this, or any earlier stage, laboratory staff with the requisite privileges can invoke a number of options to repeat or change results or to qualify them by adding further information.

'ACCOMPLIS' provides a controlled route via its menu system to enable users with appropriate privileges to amend its database records. Actions which change an existing record automatically trigger the audit trail. This is a formatted log containing for each record changed: the date and time it was changed, the identity of the person making the change, their reason for doing so, and details of the record being changed, including its original value. This information is collected before allowing the modification to take place. Any results which have been subject to entry in the audit log are flagged as such and this is indicated in printed analytical reports.

Simple menu-driven facilities are also provided to add new analytes to those already requested for a sample; to change the analytical route to be used for a determination; to cancel an analysis, a full sample or a complete collection of samples; to reschedule an analysis; and to add footnotes, memos and diagrams relating to any analyte. Additional sample labels and receipts can also be generated at any time, increasing the flexibility of the system.

14.3.6 Generation and Delivery of Reports

Once analytical work has been completed, validated and approved it is important that the required information is delivered to the point of use as quickly as possible and in an appropriate format. For this reason ACCOMPLIS supports a wide range of facilities for report generation and delivery. It will automatically produce and deliver reports for a sample collection as soon as the approval process of the collection is complete. Delivery can be by printing in the laboratory, any remote device on the system, or by electronic mail. The format and delivery mode of these reports is selected at registration of the sample collection. The report format may be selected from any of those included in the menu of available report styles.

Users with the appropriate privilege may print an interim report for an individual sample or a complete collection at any time. The desired report style is selected from a menu containing the names of the standard and any user-generated reports. The system includes a standard report (details of samples, methods, results and notes); a full report (as for standard but also including memos and diagrams); a current status report (showing the current progress on all tests and locations of all samples) and a Certificate of Analysis (which prints a user-defined header and footer, test results and specification limits and unique certificate identification). The system detects when a duplicate or interim certificate has been requested and marks the certificate as such.

ACCOMPLIS also allows users to develop their own report generators for analytical reporting and other tasks. These report generators may be written in Fortran, Datatrieve or any other VAX-supported language and would typically make use of the Datatrieve report writer and graphics functions. The reports can be named and saved in such a way that the name appears on an ACCOMPLIS menu of user-generated reports and can be selected to run from that menu at any subsequent time.

ACCOMPLIS also includes a very useful management information reporting facility. This is a menu-driven package which enables managers very easily to obtain statistics and trends relating to the performance of the analytical organization as a whole or any particular aspect of it. It provides numeric and graphic reports on backlogs, throughputs, turnround times, exceptions, and overdue or late reports. This part of ACCOMPLIS also contains a cost-accounting facility which will produce period accounts for customers, projects, production areas or account numbers. It also produces historic averages and trends which can be very helpful when formulating budgets. The accounting package can be driven, either on the basis of the allocation of different grades of analyst's time to particular samples, or on the basis of prices to be charged for each of the tests which can be performed in the organization of laboratories supported by ACCOMPLIS. An option is provided to switch off the accounting package if it is not required.

The pre-programmed, menu-driven and user-developed report generators of the types described above offer a convenient means of producing regularly required reports. In any large database system, however, the need will sometimes occur to make one-off enquiries of the results database, the analytical methods database or the audit log. As well as being a fourth generation programming language, Datatrieve can be used in interactive mode to construct *ad hoc* reports of this kind. The use of standard VAX utilities such as RMS, CDD and Datatrieve provides an economic and efficient means of supporting ACCOMPLIS users with advanced software tools. Furthermore, it means that data may be made readily available to other software packages running in the VAX/VMS environment.

14.4 Summary

ACCOMPLIS is a highly comprehensive LIMS whose development has been targeted at the primary areas of financial benefit which may be derived from the application of information technology in the analytical laboratory. It contributes directly to the control and assurance of quality of the sole product of the analytical laboratory, namely information. Its design was founded on the fundamental role of the analytical function in a business setting. The primary goal for the system was to reduce substantially the risk of error in the generation and interpretation of analytical information, because this is by far the most significant parameter in determining the value of such information.

Of the various measures employed in ACCOMPLIS, the application of real-time computing techniques and the substitution of many routine keyboard activities for bar code inputs are certainly the most important in reducing the risk of error. Of course, if a system is to make any contribution at all it must be used, and this means it must offer some tangible advantages and no significant disadvantages to all those who will use it.

How a function is performed, therefore, is just as important as the function itself. A paramount factor in getting these human interface aspects right in ACCOMPLIS has been the participation of users in the project team from its inception and the availability for trials of a typical target environment.

Acknowledgement

The ideas and experience of many analytical chemists throughout ICI have shaped the development of ACCOMPLIS and their contributions are gratefully acknowledged.

Chapter 15

Clinical LIMS

A S McLelland

15.1 Introduction

Development of LIMS in clinical chemistry departments has been unlike that in industry for two main reasons: first, many systems have their origins in work done in the mid 1960s, and second, the main objective relates to sample turnaround time rather than either cost effectiveness or accuracy.

The primary pressure on clinical chemistry laboratories for many years has been an escalating workload which has threatened, on occasion, to swamp available resources. It was clear that clinical laboratories across the UK faced serious problems in gearing their operations and logistics to a much increased demand during the 1960s and 1970s. Figure 15.1 shows the workload of the author's laboratory from 1950 to 1986, the numbers are patient analyses per annum; standards and quality control samples have been excluded from this graph.

Equipment manufacturers were quick to respond to the analytical needs of the laboratories, and clinical chemists themselves became involved in increasing the throughput of continuous flow equipment during this period (1,2). However, it was already clear that, with sustained improvements in analytical automation, the rate-limiting step on turnaround of analyses would be the handling and reporting of results (3). In effect, while one could purchase purpose-designed equipment to analyze specimens, the clerical activities associated with a workload of anything over half a million analyses annually required a similar degree of automation which was much less readily forthcoming from commercial suppliers.

15.2 History of Clinical Laboratory Data Processing

From the mid 1960s onwards, computer systems were developed in-house for clinical chemistry laboratories. Initially, the perceived need was for small computers to support the dual roles of process control and data acquisition, on the one hand, and record handling on the other (4,5). This total dependency on a single processor, and the implied assumption of 100% reliability gave many laboratories the occasional late night reporting session. Despite this, the single processor approach persisted into the seventies, and Raymond (6) noted in 1974 that such computers were often programmed on the naive assumption that arithmetic is the most critical problem facing the laboratory.

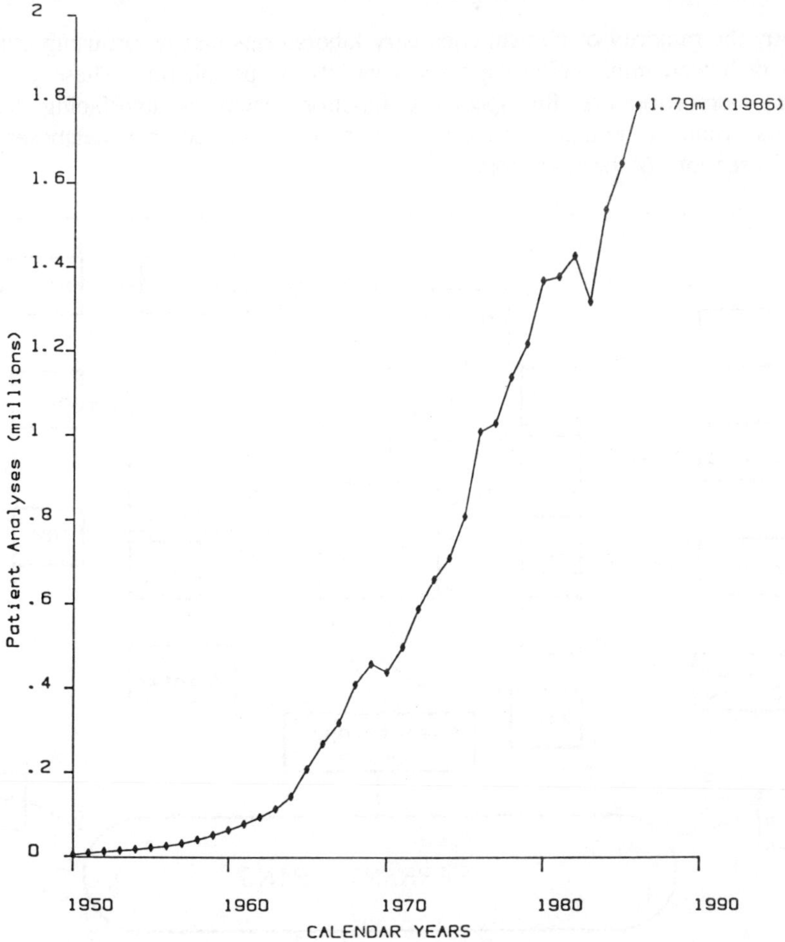

Figure 15.1 Workload of the Institute of Biochemistry, Glasgow Royal Infirmary (1950 - 1986)

With increasing complexity in analyser design and the increasing provision of on-board processors from 1970 onwards, the laboratory computer was freed from the analyser, and could concentrate on clerical activities.

During the 1970s commercial companies took an interest in this field, building on the work done before, during and after the development of the DHSS 3-Lab Project, and various suppliers have been licensed by the DHSS to use the 'Phoenix' trademark to describe their offerings in the field of clinical laboratory software. (The name 'Phoenix' originated with Wootton (7) who considered that this software had arisen from the ashes of the 3-Lab Project.)

15.3 Typical Current Computer Configurations

Currently the majority of clinical chemistry laboratories that report using computers have a dedicated mini which supports a variety of peripherals. These peripherals include microprocessors for specialist functions such as interfacing to major analysers, running graphics programs or maintaining private databases within specialist sections of the laboratory.

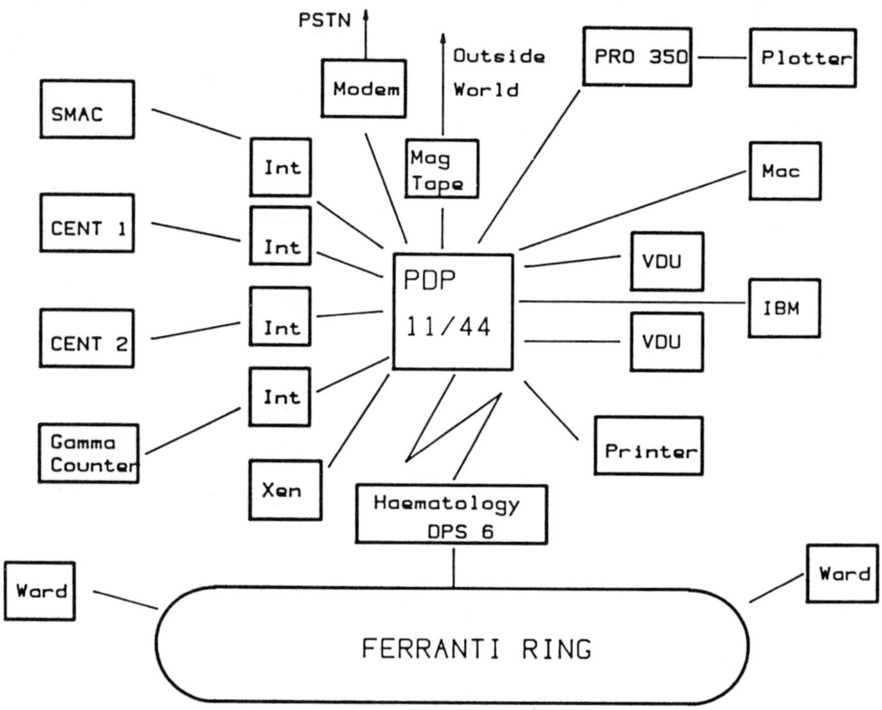

Figure 15.2 Computer hardware configuration and connections

Figure 15.2 shows a diagram of the author's LIMS hardware; showing connection from analysers (SMAC, Centrifichem 1 & 2, NE 1600) to the host PDP via microprocessors; access to the Public Switched Telephone Network (PSTN); connections to microcomputers for graphical presentation and other specialist functions, and networking to other laboratory computers within the hospital information system. The central processor in a laboratory with a workload of over one million analyses per annum would have at least one Winchester disk drive (at least 100 Mb) exchangeable disks (for security purposes and scratch space) of similar capacity; magnetic tape; at least one line printer and at least twenty terminals or microcomputers for system access. The access to the public telephone network is essential for communication with other sites, and for remote diagnostics in event of equipment failure.

There are exceptions of course. Some hospitals were planned with a strategy, often based on a central system which held the definitive patient record, and to which the laboratory analysers contributed data directly, rather than through a local dedicated reporting computer. Clinical laboratories have tended to find this approach unsatisfactory and, if given the choice, have taken the opportunity to maintain their own records.

Recently, it has been suggested that a network of microprocessors linked to a central file server, could achieve the same objectives as the departmental computer, but this concept remains unproven for the moment.

15.4 File Structure and Content

While it is difficult to generalize in this area, the majority of laboratory record computers, commercial or in-house, support four main types of file: patient identification, active requests, results and archive.

15.4.1 The Patient Identification File

This file contains the descriptive information about a patient undergoing treatment. It holds the patient's name and address, appropriate hospital data, and some means of linking this information with both active requests in progress for the patient, and any previous records. (These are no longer active in the sense that they have been analysed and reported some time ago, but are still required on the live system so that rates of change of analytes can be established.) Depending on laboratory, as many as 100,000 patient identification sets may be on file at one time.

15.4.2 The Active Request File

In addition to patient identification, this file contains fields representing the analyses requested on a particular specimen of blood, urine or other fluid. Clearly, one patient may have more than one request in progress at any one time, and these requests are differentiated by the unique laboratory accession number assigned to each specimen, and to the associated request form, as they are processed in the specimen reception area of the laboratory. The majority of searches of the system will involve the active request file as results are entered into the files from the various work stations throughout the laboratory. Most telephone calls for results will also relate to work currently in progress, where a full printed report has not yet been received by the clinician in charge of the patient. One of the primary objectives of a clinical laboratory is to make these records more accessible, both to laboratory and clinical staff.

As results are entered, requests are cleared from this active file, so that while the contents of this file vary dynamically during the working day, they are unlikely to exceed 10,000 records.

15.4.3 The Results File

This file is a temporary resting place for reported analyses which, it is felt, are still needed in order for the laboratory and clinical staff to interpret changes between current and previous results. The length of time for which past results are valid varies enormously with the analyses: serum electrolytes more than one month old are virtually useless for patient care purposes, but if an infant is diagnosed as having a thyroid disorder, a record of the infant thyroid function tests 20 years later may prove extremely useful. Thus the decision for deletion of results from the live files requires careful consideration. On the other hand, while it is tempting to store as much as possible, the longer data is allowed to accumulate, the larger the result file becomes and the more disk space it consumes. In the author's laboratory, the result file is over 25 Mb in size and contains over a quarter of a million result records, each linked to the appropriate record in the patient identification file.

15.4.4 Archive

The archive file is, typically, a sequential file, without keyed access. Each reported record, in addition to being written to the results file, is also appended to the archive file. Each archive record is self-contained in that it carries all available information on the original request form: full patient identification plus analytical results.

Access to the archive file is slow, as one would expect from a sequential file numbering over 300,000 records annually. Retrieval from this file is frequently the subject of an overnight search where each record is compared with a set of selection criteria, accepting or rejecting each one in sequence, so that, when the end of the file has been reached, an extract file containing records which conform to the specifications is available for further processing.

Note that there is no detriment to patient care associated with this rather leisurely archive extract. All records relevant to immediate patient care are available on the three live, keyed access files described above. Archive enquiry tends to relate to laboratory management information or clinical research, where an overnight delay is not critical.

15.5 Objectives of Computer Implementation in Clinical Chemistry

Sections 15.3 and 15.4 briefly sketched the typical computer hardware and file structure in a clinical chemistry laboratory in the 1980s. It is now appropriate to consider the aims and objectives of installing such a system in a laboratory and to what extent they are met by a laboratory records computer.

Despite changes in the pattern of workload during the last two decades, the objectives of the laboratory manager have remained fairly constant and are summarized in Table 15.1.

Table 15.1

Objectives of a clinical laboratory computer system.

1. Improve turnaround
2. Improve access to results
3. Reduce transcription errors
4. Permit *ad hoc* enquiries of the database
5. Count and monitor demand
6. Provide graphical display of results
7. Integrate with the hospital information system
8. Improve productivity

15.5.1 Improve Turnaround

Disease processes alter the dynamic equilibrium within the body. Treatment of these processes requires that the results of analyses which reflect the changes aree made available as soon as possible after sampling. This time pressure on laboratories is one of which clinical chemists are extremely conscious, and is one which has led to some radical re-thinking of the role of the analyst in health care: for example, the move to home glucose monitoring for diabetes and the use of desk top analysers applying thin film chemistries in acute situations.

Nevertheless, most chemical analyses are carried out in traditional laboratories with all of the associated time delays of specimen transportation to the laboratory; queueing samples at the analyser; batching less frequently requested or more labour-intensive tests into weekly runs; batching of reports; printing and validation, and delivery of validated reports back to the source of request (8).

Computers have a part to play at several points in the request, result, report cycle. They are central to many of the large analysers which expedite the analytical process. Analysers such as American Monitor Parallel or the Technician SMAC, are capable of delivering 7,200 and over 3,000 analyses/hr respectively. Both rely heavily on internal process control and data acquisition computers coupled with real-time quality control techniques such as the Westgard Algorithms (9) to ensure fast production of valid results.

Of equal importance, however, is the ability to deliver the results back to the requester, and the system design must ensure that this need is accommodated. It should ensure that requests from the less frequently performed analyses do not hold back results from the high workload, heavily-automated, fast turnaround sections, and it should incorporate some means of delivering data back to source.

Examples of this are illustrated in Figures 15.3 and 15.4. Figure 15.3 represents a standard cumulative report form used in the author's laboratory. The majority of the analyses on the form shown are processed on automated analysers, and are normally available within one to two hours of the specimen,s arrival in the laboratory. Some tests on the same blood specimen, however, because of cost, lack of appropriate instrumentation, or labour intensiveness of analysis, may be batched and may consequently be delayed. The report issued at 16.04 relates to a specimen received at 12.45 that day. 'F' codes indicate tests, such as magnesium, which were requested but were not available at the time of printing and will appear on a subsequent report. Cumulative reports like this help assess changes in the patient's biochemistry with time, and allow the laboratory to report all available results each day. Cumulation assists both clinical chemist and requesting clinician. To the chemist it is a final quality control check, in that various disease states and treatments can be expected to have a predictable effect on the rate of change of results across a time period: a result outside of predicted limits may imply an error in analysis. To the clinician, the previous results represent the context within which the current analyses must be considered.

BIOCHEMISTRY DEPARTMENT GLASGOW ROYAL INFIRMARY B

Name Sex M D.o.b. 31/ 3/28 Hospital GRI Ward 13

Address GLASGOW Consultant SOO Hospital Number 786655 B

Received 16/ 3/87 1245

Accession Number 320456 K

Clinical Information CRF

Withdrawal Date	Time	Adjusted Calcium 2.2-2.6 mmol/l	Calcium 2.2-2.6 mmol/l	Phosphate 0.7-1.4 mmol/l	Total Protein 62-82 g/l	Albumin 36-56 g/l	Globulins 22-33 g/l	Bilirubin (Total) 3-22 µmol/l	Alk Phos 80-280 U/l	AsT 12-48 U/l	AlT 3-56 U/l	LDH 230-525 U/l	CK <150 U/l	Comments (see overleaf)
24/ 2	1000				48	17	31							
26/ 2	0900	2.45	1.90	0.90	50	20	30	5	390	34	28			
4/ 3	1400				52	21	31							
9/ 3	0900	2.55	2.10	1.55	57	24	33	5	375	25	23			
14/ 3	0900	2.60	2.20	1.50	61	25	36	7	400	30	18	320	45	
16/ 3	0800	2.55	2.10	1.55	60	26	34	6	375	23	18	390	45	

Withdrawal Date	Time	Magnesium 0.7-1.0 mmol/l	IgA 0.8-4.5 g/l	IgG 6.0-15.0 g/l	IgM 0.4-3.0 g/l	Bilirubin (Direct) < 8 µmol/l	Acid Phos (Prostatic) <4.3 µg/l	Gamma GT <36 U/l	Urate 150-520 µmol/l	C Reactive Protein <10 mg/l	Comments (see overleaf)
6/ 2	0731							16			
23/ 2	0900	0.65						28		49	
9/ 3	0900	0.87						62		25	
16/ 3	0800	F	F	F	F			57	F	19	

COMMENT ISSUED AT 16:04:47 ON 16/ 3/87

Fig. 15.3 Cumulative report form

To minimize the effects of delay incurred by manual delivery of reports to wards, terminals for enquiry access to the laboratory database are located in wards. Figure

15.4 represents the frequency distribution of access times at three terminals strategically positioned within Glasgow's Royal Infirmary. The distribution shows minimal activity in the morning, with a heavy surge around 1pm as the ward staff enquire for results on in-patient specimens sent to the laboratory that morning, and sustained activity throughout the afternoon. The fall-off in enquiries in the early evening is partly due to the nine-to-five working pattern of the non-emergency side of most UK laboratories, but it is also due to the remote printing of a list of all high-volume analyses in ward order at the terminals from 17.15 hours onwards, thus obviating the need for an enquiry in many cases (Figure 15.5).

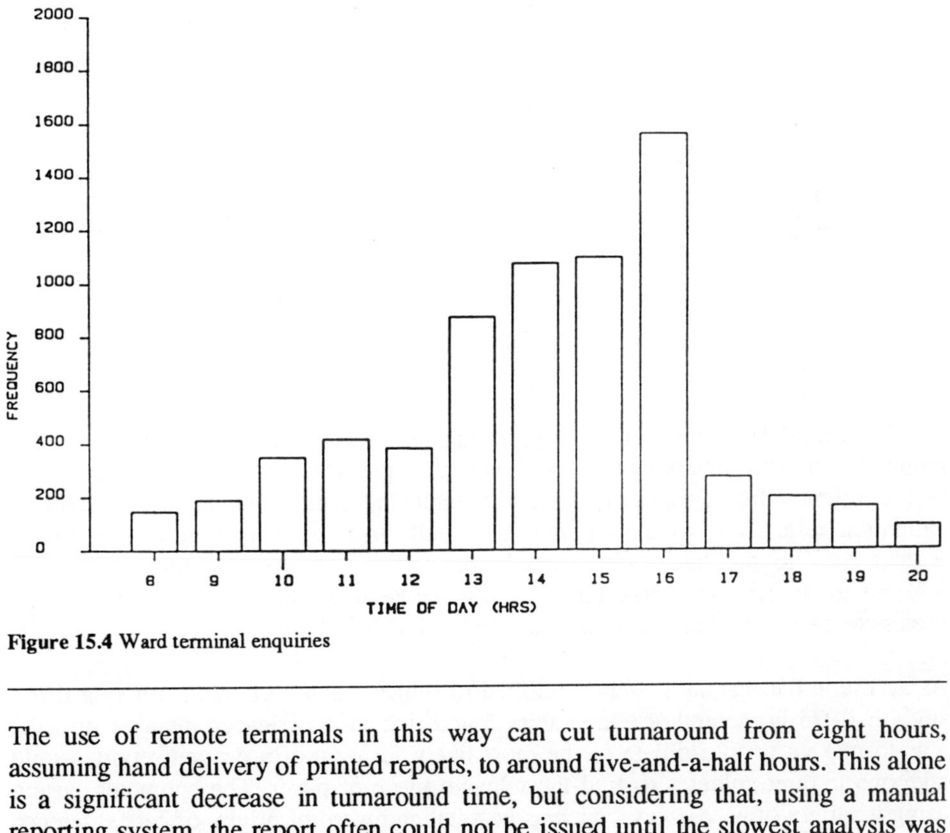

Figure 15.4 Ward terminal enquiries

The use of remote terminals in this way can cut turnaround from eight hours, assuming hand delivery of printed reports, to around five-and-a-half hours. This alone is a significant decrease in turnaround time, but considering that, using a manual reporting system, the report often could not be issued until the slowest analysis was completed and that this time delay could extend to several days, the true benefits of computerization in reducing turnaround in clinical chemistry departments can be more properly gauged.

15.5.2 Improve Access to Results

Remote terminals allow laboratory information to be returned to the clinician more quickly. Despite this, because a laboratory processor cannot readily support a terminal in every area from which it receives specimens, telephone enquiries for results and their interpretation are a significant drain on resources. Clerical activity in retrieving data may require the equivalent of a full-time person in the larger laboratories.

W A R D 3

FINAL VALIDATION OF THE FOLLOWING RESULTS MAY NOT YET BE COMPLETE.

IF IN DOUBT PLEASE CONTACT BIOCHEMISTRY DEPT.

PATIENT'S	NAME	DATE + TIME W/D		NA	K	CL	CO2	UREA	CREA
HA	JAMES	21/ 7/87	0900	133	3.4	95	29	2.9	60
RO	JAMES	21/ 7/87	0900	142	3.7	102	28	4.4	85
BE	JOHN	21/ 7/87	0919	138	5.7	100	26	21.4	1090

W A R D 4

FINAL VALIDATION OF THE FOLLOWING RESULTS MAY NOT YET BE COMPLETE.

IF IN DOUBT PLEASE CONTACT BIOCHEMISTRY DEPT.

PATIENT'S	NAME	DATE + TIME W/D		NA	K	CL	CO2	UREA	CREA
MC	JULIA	21/ 7/87	0900	129	4.7	96	23	5.7	75
TH	GERALD	21/ 7/87	0900	142	4.0	105	26	3.4	85

Figure 15.5 Listing of results by ward

Result retrieval from manual filing systems is something of an art form, since the required results physically change location as they progress through the various stages of transcription from worksheet to report, matching with previous records, typing, validation, re-analysis where required, and finally transfer to the outgoing mail desk with a copy to the 'previous records' file. Manual reporting systems have been referred to as crowded repositories of mis-sorted data and described as virtually useless to the point where they may have to be abandoned (10).

As access in the manual system is limited to subject name, it is not surprising that a study in 1975 in the author's laboratory found that an experienced member of staff, able to make a good estimate of the most likely location of a desired report, would take two to three minutes to find the relevant piece of paper. The computer system using a variety of access keys, such as hospital and ward of origin, or case reference number, has improved access to the extent that results for all patients in a given ward can be retrieved in under ten seconds.

15.5.3 Reduce Transcription Errors

Studies on wholly manual laboratories have produced a generally accepted figure of 3% error rate (11): with the best will in the world, 3% of everything one does, be it analytical or clerical, is in error. With a workload close to two million analyses annually, the implications for patients of that 3% figure are frightening.

Fortunately, if a laboratory interfaces its computer controlled analysers with a computerized reporting system it will avoid transcription errors in transferring the data; indeed a recent study in the author's laboratory showed a reduction of the 3% figure by more than an order of magnitude (12).

The current trend is to use a microcomputer to link the analyser and the central computer. This microcomputer, referred to as a data manager, is often required to perform the dual function of instructing the analyser which tests have been selected on a particular sample, and of functioning as data buffer between the analyser and laboratory computer. Where this is the case, it is not uncommon to find that the manufacturer is able to supply and interface such a microcomputer with the analyser for a small additional cost. Alternatively, if sufficient detail about the communication protocol is supplied then laboratory staff can write software to read or send data to the central system. Linkage between this micro and the host laboratory computer has traditionally involved terminal emulation software, but one can go a step further and integrate data transfer to the host simply as an option on the micro, using the powerful background routines available on many of the current microcomputers.

Figure 15.6 Microprocessors as buffer/interfaces.

Figure 15.6 shows an example of the use of such a micro. The Apple IIe is linked to the microprocessor of a Baker Centrifichem centrifugal analyser in the author's department. The Apple buffers data from each batch of 28 results on a single chemistry, and has enough capacity to store over a day's output (around 1500 analyses). It communicates with the central PDP 11/44 using a VT100 emulation package. The commands which initiate transfer of data to the host were subsequently written in an assembly language subroutine within a BASIC program which also incorporated various editing and printing options.

15.5.4 Permit *Ad Hoc* Enquiries

If a computer system is to be effective, it has to offer to the laboratory staff the prospect of their being able to do work which was previously considered too difficult or too time consuming to be contemplated.

Given that records in manual reporting systems are held in name order, answers to questions like 'What percentage of Alanine Transaminase results exceed 100 U/1 and 250 U/1?' or 'Can you identify results for all diabetic patients seen within the last year?' require extremely tedious and labour intensive measures.

Archive analysis programs which can examine, sequentially, each field of each archive record to determine whether or not a particular set of criteria are met may be no less tedious, but, so long as the computer is doing the work it will be much faster, much more accurate and considerably less labour intensive. Nor does this type of research activity require commercial database software. Questions like those above are answered on the author's system by a BASIC program which can identify and save to file all records meeting the search criteria. For instance, diabetic patients can be identified through the existence of key phrases like 'Diab', 'Keto', 'D.M.' and 'Glyc' in the clinical diagnosis field within the record. (See Figure 15.7.) Again, the extracted records can simply be saved to file for statistical or graphical processing.

```
SEARCH FOR PATIENTS WITH DIAGNOSIS SUGGESTIVE OF DIABETES
LAST AMENDED 06-MAR-87
DATA FILE = DT:[200,210]ARCHIVE.DAT
FILE CONNECTED AT 12:25:25
NAME    HOSP   WARD          HOSPNO    DIAGNOSIS              DATE W/D          LABN
M.A.    GRI    2             004388    DIABETIC               31/ 3/87        A 32512
R.C.    GRI    6             009094    DIABETIC VOMITIN        1/ 4/87        A 50412
R.C.    GRI    6             009094    NIDDM                   1/ 4/87        A 50412
R.C.    GRI    6             009094    DIABETIC                1/ 4/87        A 50412
M.D.    LIG    1             010204    MODM                    1/ 4/87        A 32525
J.E.    GP     THC                     HYPERGLYCAEMIA         31/ 3/87        A 10909
M.F.    LIG    3             010211    DEHYDRATED DIABE        1/ 4/87        A 32525
W.G.    GRI    20            323576    VASC RASH DIABET        1/ 4/87        A 32522
C.K.    LIG    DH            008219    NIDDM                  31/ 3/87        A 32511
J.K.    GP     GP                      DM                     31/ 3/87        A 32510
H.M.    GRI    4             936791    IDDM WT LOSS            1/ 4/87        A 32534
M.P.    GP     EHC                     MODM                   31/ 3/87        A 10914
E.R.    GRI    64            048856    DIABETIC                1/ 1/87        A 10000
J.T.    GRI    5             939026    ? DIABETIC              1/ 4/87        A 10918
M.D.    LIG    1             010204    MODM                    1/ 4/87        B 32525
M.F.    LIG    3             010211    DEHYDRATED DIABE        1/ 4/87        B 32525
W.G.    GRI    20            323576    VASC RASH DIABET        1/ 4/87        B 32522
```

Figure 15.7 Retrieving patients by diagnosis.

Most clinical chemistry departments that have a records computer can set their own priorities for the system, and archive analysis programs run overnight both provide a convenient way of using available data processing capacity, and extend the utility of the LIMS into areas of clinical research.

15.5.5 Count and Monitor Demand

A special case of the *ad hoc* enquiry relates to laboratory management. Like most laboratories, clinical chemistry is demand led, and reacts to changes in work patterns such as the escalation in workload shown in Figure 15.1. Increasingly, questions are being asked about the source of the demand and the cost of various analyses. Clinical chemists therefore require detailed information about their workload on a continuing basis. Again, the contrast between a computerized reporting system and a manual system is evident since, in the latter case, data stored only by patient name is not susceptible to this kind of analysis.

REQUESTING CHARACTERISTICS OF 175 WARDS FOR FEB'87 FOLLOW:

-WARD-	-TOTAL-				-EMERGENCIES-						-ROUTINE-		
	A	B	E	E(O)	ELU	GLU	AMY	OSM	PH	PCV	ELU	GLU	AMY
GRI 1	242	127	23		20						170	27	
GRI 1 OP	9	8	1								8		
GRI 1 OPD	1	1	0								1		
GRI 10	211	165	31		16	4	2	1	14		163	23	2
GRI 10 ONC	1	1	0								1		
GRI 10 OP	4	2	0								4		
GRI 10 OPD	1	1	0								1		
GRI 10/11	21	26	0								21		
GRI 11	197	158	31		10	1			20		156	10	2
GRI 11 OP	7	4	0								6	1	
GRI 12	51	40	9		7	1					37		
GRI 12.13	1	1	0								1		
GRI 12/13	26	23	1		1						22	18	
GRI 13	191	115	70		40	5	2	3	31		116	10	1
GRI 14	7	6	2					1	1		5	1	
GRI 14/15	10	8	1		1						8		
GRI 15	134	98	7		7			2	1		117	6	
GRI 16	61	58	1		1						54	26	
GRI 18	1	0	0								1		
GRI 18.19	3	0	1						1		2		
GRI 18/19	46	19	7		6				1		39	2	1
GRI 19	39	17	8		6	5			1		30	3	
GRI 2	141	117	18		14	3	1		6		114	8	
GRI 20	4	3	0								4	2	

Figure 15.8 Ward requesting patterns.

Figure 15.8 shows part of the monthly printout used to monitor the distribution of requests to the Royal Infirmary, Glasgow, and affords an instance of how, and to what extent, laboratories can influence their level of demand. In 1985, lack of capacity on the Centrifichem analysers (Figure 15.6) on which were processed the 'Cardiac' panel of enzyme requests led to a review of workload, and it was clear that a disproportionate increase in requesting from certain units in the Royal Infirmary had contributed to the problem. During 1982-85, requests for cardiac Enzyme (ENZ) profiles, at 200% increase, rose much more steeply than other high volume profiles like Urea and Electrolytes (U & E) or Liver Function Tests (LFT). Presentation to each unit of the results of the analysis, coupled with details of the steadily increasing number of requests from that unit over a three year period elicited a sympathetic response from the clinicians, followed by a reduction in the number of requests (Figure 15.9). That this voluntary reduction in demand can be obtained, and has now been sustained over a two year period without detriment to patient care, is in great measure due to the reasonable attitude of both clinical and laboratory staff, but also illustrates the potential benefit of good laboratory management data and the use to which it can be put in monitoring demand.

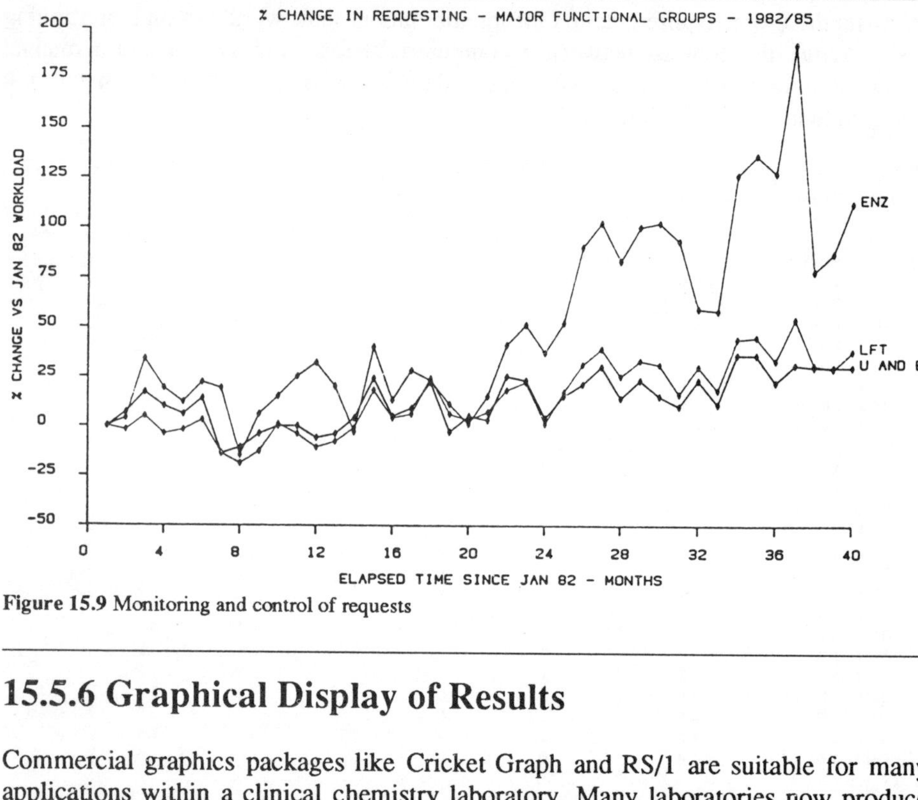

Figure 15.9 Monitoring and control of requests

15.5.6 Graphical Display of Results

Commercial graphics packages like Cricket Graph and RS/1 are suitable for many applications within a clinical chemistry laboratory. Many laboratories now produce graphical output for selected patients by transferring extracted data from tne main results files to an associated microprocessor for plotting. The only significant problem encountered, which may well be peculiar to health care, is the frequent need to plot data against a time-base. Most graphics software expects the X axis to be a linear scale, and are unable to accept dates and times of sampling, or to convert these to 'time elapsed' in order to create the kind of graph shown in Figure 15.10. Note the two requests for urea and creatinine on the 6th and 12th, separated by only a few hours, and the absence of results on the 14th. This is a graph which is designed to give clinicians a better understanding of the type of data which clinical laboratories present in their cumulative reports (Figure 15.3).

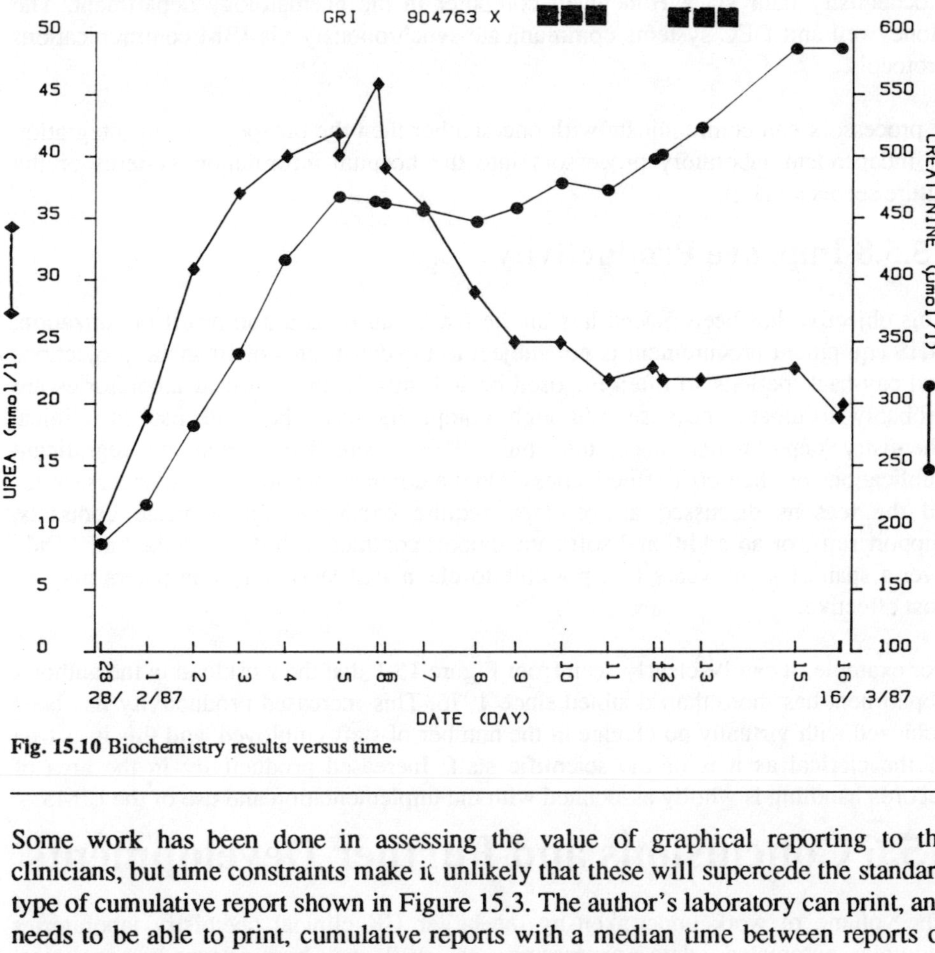

Fig. 15.10 Biochemistry results versus time.

Some work has been done in assessing the value of graphical reporting to the clinicians, but time constraints make it unlikely that these will supercede the standard type of cumulative report shown in Figure 15.3. The author's laboratory can print, and needs to be able to print, cumulative reports with a median time between reports of seven seconds: graphical reports take at least ten times longer.

15.5.7 Integrate into the Hospital Information System

As implied in the introduction, clinical chemistry laboratories were one of the first areas of medicine to use computers as a potential solution to the problems of record-keeping. This has resulted in a typical situation where clinical laboratories have dedicated records computers in place prior to the specification of a central hospital records system which will handle all patient data and which may extend to over a million patient records. Logically, the records system ought to be the lynch-pin of the hospital information system, but there is a risk that a system will be specified which is incompatible with the existing laboratory computers. It is essential that communications between existing and proposed systems are fully investigated. The lower portion of Figure 15.2 illustrates networking of computers within the Royal Infirmary into a Ferranti ring system, where, rather than have the ward enquiry terminals connected directly into the clinical chemistry computer, they will access the biochemistry data via a Honeywell computer in the haematology department. The Honeywell and DEC systems communicate synchronously via IBM communications protocols.

If processors can communicate with one another then the prospect of full integration of independent laboratory processors into the hospital information systems of the future seems assured.

15.5.8 Improve Productivity

This objective has been placed last on the list because, as a non-profit organization, NHS equipment procurement is not subject to the constraints of financial projections and pay-back periods so often imposed on industry. In this, clinical laboratories are probably fortunate, because, although computers have been in use in clinical chemistry departments since the mid 1960s, there have been no significant publications on their cost-effectiveness. On the contrary, while they are necessary for all the reasons discussed above, they require expensive maintenance contracts, support staff, or an additional software support contract with the manufacturer. Only over a span of many years is it possible to claim that laboratory computers may be cost effective.

For example, it can be clearly seen from Figure 15.1 that the workload in the author's department has more than doubled since 1976. This increased productivity has been achieved with virtually no change in the number of staff employed, and this is as true of the clerical as it is of the scientific staff. Increased productivity in the area of records handling is wholly associated with the implementation and use of the LIMS.

15.6 Conclusions and Further Developments

The volume of work undertaken by the larger UK clinical chemistry laboratories requires automated data processing as well as high capacity analytical instrumentation.

Computers help reduce turnaround time of reports to the extent that the more efficient laboratories can process and report the bulk of the workload in the same half-day in which it arrives. Computers extend enormously the accessibility of the biochemical database for research and management purposes; they minimize transcription and so reduce errors, and it can be argued that they may even be cost effective in the long term.

Yet the future remains challenging: integration with the hospital information system is not yet a reality in many sites. Little has been done to explore the various methods of making results more easily understood; if distributed analysis at the patient's bedside becomes a reality, how does the laboratory respond to the challenge of updating its database from widely scattered analytical sites?

Above all, artificial intelligence and expert systems hold the promise of computerized interpretation of clinical data, and it is clear that this will be the next major opportunity for laboratory information management systems in clinical chemistry.

15.7 References

1. Fleck. A., Begg, R.D., Williams, P.I. and Racionzer. A., 'Some Fundamental Aspects of Continuous Flow Analyses', *Annals of Clinical Biochemistry*, 8 (1971) 13-15

2. Walker, W.H.C., 'Curve Regeneration in Continuous Flow Analyses', *Clinica Chimica Acta*, 32 (1971) 305-306

3. Gaddie, R., 'The Contribution of Biochemistry to Clinical Practice', *Journal of the Royal College of Physicans, London*, 1 No. 2 (1967) 148-156

4. Blaivas, M.A. and Mencz, A.H., 'Improved Standardisation and Peak Analysis Utilising a Process Control Computer and Auto Analyser Systems', in *Automation in Analytical Chemistry*, Vol. 1, (1968) Mediad Inc., White Plains, NY. pp133-138

5. Whitby, L.G. and Simpson, D., 'Experience with On-line Computing in Clinical Chemistry', *Journal of Clinical Pathology*, 22 (1969) Suppl. 107-124

6. Raymond, S., 'Criteria in the Choice of a Computer System I. The Computer in Theory', *Journal of the American Medical Association*, 228 (1974) 591-594

7. Abson, J., Prall, A. and Wootton, I.D.P., 'Data Processing in Pathological Laboratories: The Phoenix System: 1 The Overall Design and the Main Facilities', *Annals of Clinical Biochemistry*, 14 (1977) 307-314

8. McLelland, A.S., 'Laboratory Computing' in Computer Techniques in Clinical Medicine, P.W. Macfarlane (ed), Butterworths, London (1985) pp. 121-133

9. Westgard, J.O., Barry, P.L. and Hunt, M.R., 'A Multi-Rule Shewhart Chart for Quality Control in Clinical Chemistry', *Clinical Chemistry*, 27 (1981) 493-501

10. Scottish Home & Health Department, 'The Use of Computers in Hospital Laboratories', Report of the Hospital Laboratory Computer Committee of the Information & Computer Services Advisory Group, Scottish Health Service Planning Council (1978)

11. McSwiney, R.R. and Woodrow, D.A., 'Types of Errors Within a Clinical Laboratory', *Journal of Medical and Clinical Technology*, 26 (1969) 340-346

12. Chambers, A.M., Elder, J. and O'Reilly, D. StJ., 'The Blunder-rate in a Clinical Biochemistry Service', *Annals of Clinical Biochemistry*, 23 (1986) 470-473

Chapter 16

Software Houses - Consultancy and Bespoke Development

J. Knox

16.1 What is a Software House?

'Software House' is a generic term used to describe a wide range of computer services companies specializing in software. Some of these may specialize in software packages, bespoke or custom software developments, or a mixture. Most will offer consultancy services, providing a strategy report or a requirements document. There may also be specialization in a particular manufacturer's hardware, or in a particular applications area.

If the company has a large computer department, then the laboratory may turn to them for help. However, there may be a number of reasons why in-house development may be unsuitable or impractical. For instance, the computing department may be geared towards commercial computing, with little or no experience of the problems involved in laboratory applications. Also the staff may not be available for LIMS development, or it may have low priority, resulting in long projected lead times.

The specialist software house, however, can offer the following advantages:

1. Specialist knowledge of laboratory applications, for example instrument interfacing, work station networking, or chromatography.
2. Experience in the design and implementation of laboratory systems and an understanding of how the laboratory works.
3. The resources to put a large team on a project, reducing development lead times.
4. Development standards and project management capabilities appropriate to the laboratory areas.
5. The ability to budget and control costs through fixed price development.

16.2 Choosing a Software House

Having decided that consultancy is required to solve the LIMS problem, the user has to find a software house which fits the bill. Companies in the field can be found by asking the hardware manufacturers for advice, by reference to the *Computer Users Yearbook* or by recommendation. If possible, the user should evaluate the merits of two or three companies before deciding to deal with one. However, at present there are not many companies which specialize in the laboratory area, so this may be difficult. The other approach which is often used is to send out invitations to tender, this is discussed later. The type of services offered by a software house in the laboratory area would include the following:

1. *Consultancy* Specialist consultants can be brought in to plan a computing strategy for the laboratory; to prepare a requirements specification; to undertake a feasibility study for a proposed system; or to prepare a full system design.
2. *Project-based development* A project team may undertake the development of a complete system, possibly a full LIMS, or an add-on to a LIMS package.
3. *Turnkey systems* Some software houses will offer a total system, hardware and software, covering all aspects of development, testing, installation, training and maintenance.
4. *Applications packages* A software house may offer its own application package as a system solution.
5. *Customising packages* Here a proprietary package may be tailored to the user's exact requirements, using the software house's experience of the package in several environments to provide the best solution.
6. *Time-based assignments* These are undertaken for clients who have an overload of development work, or to provide particular experience or skills.

Some of the questions to ask when choosing a software house include the following: Is the company well established and liable still to be in business in five or ten years time? Remember that any software bought has to be supported for the life of the project. Does the company have experience of laboratory systems? What are the consultancy rates for systems design and programming, to give some ideas of the likely costs? Does the software house have its own development standards and facilities?

It is important to choose consultants which the users can work with, as the LIMS development will be a partnership between them and the consultants.

16.2.1 Costs

The cost of systems development by consultants is based on the charge rates of programmers, analysts and other staff involved in a project. These rates, in the UK, may vary from a base of around £700 up to £2000 or more per week for specialized staff. Rates will vary considerably from company to company and from area to area.

These rates may seem high when compared to the salary of someone working in a computer department. However, if all the costs of employment are taken into consideration, then the cost may be seen to be comparable, the difference being the profit element for the consultants. The advantages of employing consultants are that the staff are used only for the duration of the project, and may do the job more quickly because of the skills and experience they can apply.

On top of the basic fee rates, the consultants may charge for machine time if using their own computer for development, whilst travelling and hotel expenses will be passed on directly to the company.

16.2.2 Types of Contract

The simplest type of contract for software development is known as 'Time and Materials'. This means that the consultants charge for staff time as it is used, and for expenses as incurred. Usually, a budget will be decided for each phase of the work, to provide the element of cost control.

A time and materials contract has the advantage for the user of complete flexibility in when and how the consultants are employed; and has the obvious advantage for the consultants of ensuring a profit on the work. But it has the disadvantage for the user of not placing a limit on expenditure, and therefore may be undesirable for the development of a complete system where costs could get out of control. It may be more appropriate to the development of enhancements and improvements to an existing system, where the work is allocated in smaller units.

For a full LIMS development, a fixed price contract will give the user the tight budgetary control required for what is equivalent to a large item of capital expenditure. However, it is unlikely that the consultant will agree to a fixed price on a full system before it has been specified in detail. It is very difficult to produce accurate programming estimates even with a full specification, especially for a LIMS, where the requirements vary considerably from laboratory to laboratory. Therefore, the consultants are likely to propose a fixed price for the system design phase, with estimates for the remainder of the development. Following the system specification, the consultants will then undertake the programming, systems testing and documentation, as a separate contract.

Competitive tendering is often used to try to obtain the best price for a full system. A requirements specification is sent out to prospective consultants, who then provide a detailed system proposal and a fixed price quote. This is fine for package based systems, where the marketing effort is part of the cost of the package. However, for completely bespoke solutions, the software house may have to put a large amount of effort into a proposal for possibly no return. For the user, this means that the best software houses may not tender, because they are too busy to put the required effort into a proposal. This factor should be considered carefully if it seems likely that only a bespoke solution will be acceptable. In some cases, a company will pay two or three software houses to cover their costs and produce alternative proposals for evaluation.

16.2.3 The Strategy Report

A strategy report is desirable where the laboratory or department has a few isolated computer applications, and needs some guidance to progress towards an integrated laboratory system over a specified period. The consultant will examine and document the current and future requirements of the laboratory, and propose one or more hardware and software solutions in an integrated plan.

The advantage of using consultants for a strategy report is that they have detailed knowledge of all the options available, for example the benefits of local area networking as against more traditional communication methods, or the important features of the different LIMS packages available. They can also take a more detached and objective view of the laboratory than perhaps someone involved in day to day operations.

16.3 The System Development Cycle

Full or partial development of a LIMS is a partnership between the users and the software developers. The user must contribute at each stage of the development, from defining the requirements, through designing screens and reports, providing test data for programmers, and testing the system before release.

The more time and effort the user puts into the development, the more suited to the user's actual requirements the final system will be. In certain environments, particularly in the pharmaceutical industry, this contribution is a requirement for GLP validation of the system.

Before embarking on a development project, the system requirements must be well defined in either a requirements or a functional specification. In general, the larger the system, the more formal the specification required at this stage. After a functional specification or a detailed requirements specification of the proposed system has been produced, then the stages of development begin. Each of these stages will be discussed in the remainder of this section. Ideally, every stage should be complete before the next is started. If there is to be any overlap, then the system should be formally partitioned so that the separation of stages applies to each part. In this way, the requirements for each stage will be fully defined and approved before work is started on it, and the amount of effort required for each stage can be effectively estimated.

16.3.1 Selecting the Development Tools

The final decision on the major development tools to be used: the hardware, the programming language(s) and the database management system, must rest with the user. The software house may advise on the advantages and disadvantages of particular tools, but in the end the user must be responsible for choosing them. If necessary, the client's own computer department should be brought in to advise (if it is not already involved), so that the correct decision can be made with regard to compatibility and communications with existing hardware and software in the company.

Very often, the computer hardware will have been selected in advance, because of company policy. The exception may be where the user has put the complete system out to tender with no hardware specified. It should be remembered that if a bespoke development path has been chosen, then the choice of hardware may affect the cost of developing the software. A program which takes, say, five days to write on one computer may take ten days on another, because of differences in the operating system and the development tools available on each computer.

Selecting a database management system and programming language should involve the same considerations. Users in the laboratory may be familiar with Fortran and consider it the best language for scientific purposes. However, a LIMS consists of more data processing (ie. entry, manipulation and reporting of data) than scientific work (ie. complex calculations), and other languages such as Pascal or Basic may be more appropriate depending on their implementation on the chosen computer.

The trend in systems development is towards using fourth generation languages (4GLs), which combine a database management system with a set of tools geared towards rapid system development, for example a screen and report program generators. However, the danger is that the requirements of the laboratory for data manipulation may be too complex for a 4GL to handle. It is therefore important that the 4GL should have a flexible and reliable host language interface, to allow a language like Fortran to be used where the 4GL utilities cannot cope.

16.3.2 Establishing Development Standards

Before system design and specification begins, the user and the software house should agree a set of development standards for the project. The user's computer department may have established standards to be followed, or the software house may use or adapt its own standards. The development standards should define the following items:

- Content and layout of system and program specifications.
- Design methodology.
- Program coding and documentation standards.
- Systems testing and validation procedures.
- Project management procedures.
- Project change control procedures.
- Responsibilities of the software house and the users.

Many of these standards may be established as part of a project plan. This document should define clearly the objectives of the project, the responsibilities of all involved (including user representatives), a timetable for the different stages of the project, the number of staff involved at each stage, and the significant milestones or target dates to be met. The plan should also establish a regular series of project review meetings.

16.3.3 System Design and Specification

Designing the LIMS is a process of translating the requirements as laid out in the functional or requirements specification, into a set of program descriptions, screen designs, report layouts and file structures, with a detailed description of the processing required. The design may go through several interactions as users review the proposals and problem areas are resolved. It is important that sufficient time is allowed for the review process to work properly, and that users have the commitment to make the required effort at this stage. It is all too easy for the user to opt out of the review process because of what are perceived as more pressing short-term problems.

The consultants should present the system design to the users in a form which can be easily assimilated, such as overhead slides in a presentation, with example screens and reports if possible. The design should be discussed in a forum with the consultants, so that the users can hear each others views as well.

One particular advantage of using a software house for the system design is that they will have had prior experience of designing a LIMS, and can apply the experience, techniques and ideas to the new system. LIMS are generally more complex than standard commercial systems. For example, a financial system may only have to deal in pounds and pence, whereas a LIMS has to cope with a large variety of measurement units, to be expressed with varying degrees of accuracy.

The end result of this process will be the system or technical specification. The system specification should contain enough information for both the user to confirm that the design meets requirements, and for the development team to proceed to a specification for each program in the system. The system specification should therefore include:

- Overall system context and description.
- List of programs in the system, with a brief description of the function of each program.
- File layouts or database design, with details of each data item in each file.
- Screen and report layouts, with examples as necessary.
- Description of the functioning of screens.
- Detailed description of any calculations required.
- Estimates and plan for the remainder of the system development.

It is necessary that the development tools are known before the system specification, as they will affect screen layouts, operation of screens and file design in particular.

The development plan should allow ample time for revision of the system specification, as inevitably the users' ideas of what they require will evolve as they begin to see what can be achieved.

16.3.4 Prototyping

Prototyping is a technique for specifying a system which has become popular with the advent of 4GLs. The development tools within the 4GL are used to produce a skeleton system: screens and reports without the detailed processing and error handling behind them. They can be easily modified following user review until an agreed set of screens and reports is reached. These can then form the basis of the system specification.

Prototyping is a concept which suits the user/developer relationship in a bespoke development. However, it should be used with caution, and the temptation to use the prototype system as the foundation of the real system should be resisted. What works on a small scale with limited data sets may be slow and cumbersome on a large scale. This is especially true of laboratory systems, which generally involve complex data structures and relationships.

16.3.5 Program Specifications

Once the system specification has been agreed by the users, work can begin on the program specifications. There may also be some detailed data design required: for example, final file structures and pointers, and data names. This data design should ideally come first, but it is likely that the fine detail in the program specifications will require a change to the data design, so the process is iterative to some extent.

Each program specification should allow the programmer to code and test a working program with a minimum of reference back to the system designer.No decisions on program functions should be left to the programmer, whose job is to structure, code and test the program so that it operates reliably and efficiently to specification.

The program specifications will raise further questions on the details of design, so the user must be prepared to spend time at this stage responding to these, and reviewing each specification. Design flaws at this stage will be costly to change later on. Because the program specifications will be technical, the user should appoint a representative with some knowledge of computing to review them.

16.3.6 Programming and Unit Testing

The programming and unit testing phase of a project is generally the most costly in terms of total effort, although with 4GLs the ratio of programming to specification is being reduced. Therefore, before this phase begins, each program must be as tightly specified as possible, to minimize time lost through mis-coded programs.

The work during this phase can take place either on the user's site, or at the software company's office. The former has the advantage of closer liaison between developer and user, but may be constrained by machine availability or lack of work space. Where development is not on-site, the user should be prepared to monitor the development closely, with regular meetings between the user representative and the project manager.

16.3.7 Change Control Procedures

Inevitably, some changes to the requirement or the specification will occur between the approval of the program specifications and the installation of the system. All such changes should be recorded and agreed under a change control procedure, as an important part of project control. The exact procedure will depend on whether the development is fixed price (in which case a cost must be agreed before implementing the change) or time and materials. In general, the initiator of the change will complete a change control form, which must then be agreed by the user representative and the project manager before the change is implemented.

Although a change control procedure is also desirable with a completely in-house development, it becomes essential with a contract development, because of the relationship between the user and the consultants.

16.3.8 Systems Testing

Systems testing involves testing all the programs as part of the complete system, rather than as individual units. It is important that the systems testing phase of a project is planned well in advance, so that the user has time to prepare the required test data. The project manager will prepare a System Test Plan and agree it with the client. The plan should include the following as appropriate:

1. Schedule of events and target completion dates, indicating the persons responsible.
2. Assess the impact on user staff and operations.
3. Impact on computer operations.
4. System test file set-up requirements.
5. System test data and expected results requirements.
6. Details and checklist of system test runs to be performed, including the type and quantity of data to be processed and reports to be produced.
7. Details and checklist of acceptance criteria.

The system testing itself should include the types of test listed in Table 16.1, where appropriate. Testing should be performed with the aim of proving that the system does not work. Where possible, user staff should be involved in system testing, as well as the user representative.

It is important in system testing to organize and retain the documentation produced. For each run there should be a checklist of test conditions, a list of data entered, a list of expected results, and a print of actual results, with any errors highlighted.

Table 16.1

Types of System Testing

Requirements Testing	This involves checking that each program unit conforms to its specification, and also running several representative sets of test data through the complete system and checking againstexpected results. All errors should be logged, rectified and re-tested. It is important here to differentiate between errors and changes to the specification.
Negative Testing	This involves trying to 'break' the system, for example by attempting to make the system accept impossible or incompatible data, or by putting an untrained user on the system.
Stability Testing	This involves testing the system to its limits of throughput, data volumes, terminals logged in, etc., and also testing that programs will run simultaneously without conflict.
Security Testing	If required, this involves trying to gain unauthorised access to the LIMS, or from the LIMS to unauthorised areas in the computer system.
Recovery Testing	This involves testing the ability to recover the LIMS database after a system crash.
Performance Testing	If performance criteria have been defined for the system, they should be tested on the major transactions and reports.
Regression Testing	When all system test runs and consequent changes are complete, all runs which produced errors should be repeated. This process should be repeated until no errors are found.

16.3.9 Documentation

The documentation to accompany a bespoke system should include at least a user manual, system manager's manual and programming documentation.

It is very difficult for the software developers to write a good user manual: first because of the tendency to use computer terminology, second because they do not have first-hand experience of the environment where the system is to be used, and finally, because they know the system so well that it is not clear which parts are obvious and which need in-depth explanation.

One approach to producing a good user manual, therefore, is for the software house to write the skeleton of how the screens are operated, what facilities are available, and generally how the system works. The users then take over the manual and expand the detail where necessary, putting the system into context. The end result should be a document which describes the complete system and which the user can understand.

The system manager's manual should cover all the detail of running the system, which ordinary users do not need to know. This will include loading or generating the system, maintaining system and configuration files, archiving and back-ups, and other details which vary from system to system.

The programming documentation is essential for a bespoke system so that the user can develop the system further without having to go back to the software house. In certain industries it will also be a regulatory requirement. The documentation should include the final program specification with detailed file descriptions, fully documented source code, and the program building instructions.

16.3.10 Validation

Software validation is the process of approving the system in the environment in which it is to be used. This involves checking not only that the system functions according to specification, but also that the specification meets the actual requirements of the laboratory in practice. Often, this will involve parallel running the existing procedures with the new system and comparing the results.

Validation must be the responsibility of the users, not the software house. The software house can advise on a program of testing, and procedures to be followed, but the management and execution of the validation must be performed by the users. This is because only the users have the first-hand knowledge and experience of the environment that is required to validate a system, and also because it is difficult for the software developers to step back and take a detached view of the system.

16.4 Summary

If all the procedures described above are included in the system development, then the end result should be a system which performs well and meets the user requirements. However, two things are essential:

1. The development must be a partnership between the software house and the users.

2. The development plan must allow time for that partnership to work effectively.

Chapter 17

The Beckman CALS Lab Manager System

Peter G. Berthrong

17.1 Introduction

In 1975, a New Jersey (USA) based company called Computer Inquiry Systems (CIS) developed a new chromatography data acquisition and analysis system called CALS (Computer Automated Laboratory System). In 1980, the CALS product expanded with the introduction of Lab Manager, one of the first commercially available LIMS. The original chromatography portion of CALS was called GC (General Chromatography) by the users and has been supplanted by PeakPro, the current CALS chromatography product.

Both major elements of CALS (Chromatography and LIMS) were originally specified by pharmaceutical companies and hence compliance with both GLP and GMP regulations was incorporated at the design stage.

In 1983, Computer Inquiry Systems was acquired by Beckman Instruments (itself part of the SmithKline Beckman Corporation) and became a wholly owned subsidiary. However, even now CALS and Lab Manager are referred to as the 'CIS System'.

17.2 Software Configuration

One of the keys to the success of Beckman's Lab Manager is the heavy use of configuration rather than custom code. A set of utilities within the system software enables changes to be made without altering or compiling programs and, in most cases, without any interruption of system operation. For instance, if a screen needs to be changed or a new report is required, these are accomplished with a utility and a report writer respectively, not by writing more programs. If the structure of the database itself needs changing there are utilities to accomplish these operations.

Of what advantage is this to the user? A system built with a combination of both standard and customized programs may give that elusive 100% solution to the user laboratory. However, a system of this type lacks the flexibility to accommodate changes either when the users start to actively accomplish their work or if new functions are added. Changes, however small, will require reprogramming, recompiling and linking. Re-configuration is a quicker and more effective alternative.

The real crux of this argument is support. Beckman provides a complete new version yearly. Sometimes these revisions completely change the way sections of the system work, both internally (performance improvements) or externally (increases in user functions). Will revisions work with custom code? Maybe. Beckman understands this problem, because early systems were written with both configuration and custom code. The result was very favourable at first but a support nightmare later. As the number of installed systems grows more rapidly each year, support for the entire base of systems becomes an ever more pressing and urgent issue.

Configuration also puts the ability to change into the hands of the user laboratory. The laboratory is a dynamic place: instruments, methods and people all change. The laboratory needs to alter its LIMS to reflect these changes and does not necessarily want to call in the vendor every time.

17.3 Global Features

Any good LIMS can do the normal routine of logging samples, testing, review and reporting results; it would indeed be a short-lived system that did not perform these essential routines of laboratory life. What differentiates the various systems are larger issues much more vital than the presence or absence of a specific feature. Particular topics that require expansion are: security and safety, hardware independence, group structure, GMP/GLP compliance, and interfaces to other systems.

17.3.1 Security and Safety

Lab Manager has sixteen levels of class-based security. Thus, if a user is assigned to classes two and four, he may do only the functions of those classes. It is not a hierarchy, this user has no access to class three functions. If clearance was amended to two to four (inclusive) then he would have class three functions. Every function, report, menu and soft-key label on the system has one or more classes of people who may execute them. This combination of class level and functions provides for a different pattern for each person in the laboratory. This becomes an ease-of-use issue as well because people (even high-security level managers) need see only the functions they require and do not have to wade through all the menus searching for the item needed.

Lab Manager addresses the data safety issue directly with 'Safegard'. This program allows data shadowing (See Chapter 7 and Figure 7.1) and duplicate copies of 'active' data are maintained on different disk drives, in addition, there is another large disk area devoted to 'completed' data.

The completed database is backed-up with magnetic tape; if required, multiple copies can be made and one stored off-site as a contingency against disaster. There is normally no need to back-up the 'active' database as one copy is automatically an instantaneous back-up of the other. The division of the data into active and completed allows 24 hour access to the active data, even when the completed database is inactivated because of back-up activities.

17.3.2 Hardware Independence

Lab Manager is hardware independent. In 1987 it runs on Hewlett-Packard A, E and F 1000-series computers, on DEC VAX computers, and on IBM 30xx and 40xx mainframe computers. The goal is to provide a software solution to the laboratory: a solution independent of computer hardware.

Where the laboratory has control over its computer destiny, appropriate recommendations are made commensurate with performance requirements and real-time data acquisition needs. In many cases, laboratories are advised by Computer or Management Information Systems (MIS) departments as to what hardware is acceptable from a corporate viewpoint. Working with MIS is essential to success for LIMS because of the need to interface to other systems, both before and after the LIMS process. The goal is to provide the CALS product line on computers from the three major computer companies: IBM, DEC and HP.

17.3.3 Multiple Copies

Lab Manager includes a unique group system which allows multiple laboratories or groups within laboratories to utilize the same software/hardware installation, yet maintain their individual identities and security systems. Group one could be a quality control laboratory with a well-defined set of rules and procedures controlled by the system. This laboratory would have a certain database design, custom screens and reports, and a SOP (Standard Operating Procedure) mandated flow of data through controlled stages of laboratory activity.

At the same time, group two could be a research laboratory utilizing the same computer and licensed software. This group has its own security, database, screens and reports which are completely different from those of the Quality Control Laboratory. Importantly, neither group has to be aware of the other; their data and configurations are completely separate. Of course, some users such as high-level managers or support personnel responsible for computer operations may have limited or full access to both groups. This group structure allows for the existence of eight completely different groups on each Lab Manager system.

17.3.4 Regulatory Agency Compliance

There is a need for systems that are compliant with Good Manufacturing Practice (GMP) and Good Laboratory Practice (GLP). The need for such compliance for the pharmaceutical industry and the emerging biotechnology industries is obvious, but other industries have decided that to comply with these rules puts them in a better position when future regulation occurs in their own industry.

A compliant system does certain things quite well. It provides a secure environment where people only perform the functions for which they are trained and qualified. It time/date and user stamps every action taken by the system and if a supervisor decides that a re-test is required, he must enter a reason (document) why he is making the request. All data, whether correct or incorrect, must remain on the system; data can never be destroyed or over-written; it should only be amended with a reason, by a qualified person, with the original result still visible.

Interestingly, while much concern centres on the sanctity of sample data, it is wise to remember that concern must extend to all the tables that a sample uses such as tests, calculations or specification limits. These data elements must themselves be protected and audited.

The maintenance of these tables (called dictionaries) is in itself configurable, for different rules may apply to different dictionaries with some or all features turned on or off on a per dictionary basis. In the most compliant case, each record has an individual password which protects it against change even by people who are otherwise authorized to maintain that dictionary. Again, in the compliant case each record requires approval. A change made in a product specification, for example, where the assay lower inside limit changes from 98% to 97%, will not take effect until approved by a supervisor or reviewer. Thus samples logged in will continue to use the old specification even though the new one exists; once approved the new specification takes effect.

Each revision has its own revision number along with the identification of the modifier and the approver, with appropriate dates and times for each. Utilities exist to allow qualified individuals to review the history of the revisions since a variable number of these are kept on-line with the current revision. The review can be accomplished by comparison between a revision and its successor, or by asking for the revision in effect on a particular date.

17.3.5 Computer Communications

The interface to other computer systems is another global concern. Many manufacturing operations utilize Management Resource Planning (MRP) systems. These automate the ordering of raw materials and components, the scheduling of production steps, and the tracking of all materials from receipt as raw materials to final disposition and movement of the final product through distribution channels. In these cases the sample log-in operation is done electronically between the MRP and LIMS and final laboratory disposition is posted from LIMS back to MRP. In the most sophisticated cases the LIMS may post intermediate status information or approved intermediate results back to MRP.

After results are collected by Lab Manager they must be available to other packages such as SAS or RS/E. The system allows export of data into ASCII files, along with imbedded control language, to access the other packages. These may exist on the same computer as the Lab Manager or on another, either at the same site or in a different location. While Lab Manager is capable of reviews of trends including averages, standard deviations and confidence intervals, for more complicated statistical manipulations use of these other, more directed packages is advised.

A powerful example of computer to computer interaction meeting the needs of various areas within a corporation is Executive LIMS. Samples are logged in to the Executive Lab Manager (usually a large mainframe computer such as an IBM) and the Executive distributes the sample to any number of departmental CALS systems, which are usually HP or VAX based systems. Thus, one sample in the central computer may generate multiple samples on departmental computers with results flowing back to the Executive as they are approved by the laboratory. Intermediate sample status information is periodically transmitted from the laboratory to the Executive making relevant information available to authorized users in Marketing, Production and Scheduling areas. In this case the master tests, calculations and specifications are kept on the mainframe system and downloaded to the appropriate departmental computer upon approval.

17.4 Lab Manager Functions

The next nine sections deal with an expanded view of what happens to a sample when it flows through the Lab Manager system. All of the steps apply whether this is a single computer Lab Manager or the Executive scheme, even though their execution may differ slightly.

17.4.1 Sample Log-in

Samples are logged to the system either by a trigger from an MRP or by an authorized person at a sample check-in station. Alternatively, some samples may be logged in by development personnel on special projects and may require non-standard testing patterns. All of these mechanisms may exist simultaneously on one system. Log-in screens can be different for each category of samples with specialized screens for raw materials, in-process samples, packaging components and finished products. Another mechanism is an optional module called Autolog which is covered later in this Chapter.

In some installations, multiple labels are printed at the log-in step and affixed to sample containers; these labels may be in human readable format, in one of several bar code formats, or a mixture of the two. Indeed, bar code technology may be heavily used in the installed system as each terminal and instrument interface coupler is capable of supporting bar code wands or scanners.

17.4.2 Sampling

There is next an optional sampling or check-in step. If the sample was logged in remotely, the laboratory may choose to invoke the sampling step to provide laboratory management with a time-date and identification stamp for the actual arrival of the sample in the laboratory. In some facilities, samples may immediately go to an intermediate retention area such as a freezer, in which case the sampling step may track the new location or other biographical information about the sample. Sample labels may be issued if this was not done at the log-in step, or they may be re-issued for sample splits and/or container changes.

17.4.3 Reports - 1

Although reports are the expected result of the laboratory process, it is important to remember that reports are available for a variety of purposes once samples are logged into the system. Thus, where samples need collection from a remote area a Sample Collection Report is issued for a messenger, directing him to the proper place. Some laboratories require reports for acknowledging sample receipt to the submitter.

Status reports tell the testers of priority samples for their laboratory, testing station or instrument, or tell reviewers about samples where testing is ready for their review. Back-log reports can inform the laboratory manager about samples that are overdue or about the number of priority samples processed during the last week.

Reports and reporting needs are as individualized as the many laboratories using the system. The report generator allows a new format to be prepared and then inserted into the menu and security scheme for use by authorized people. Reports are one of the key portions of a system which give it an individual personality.

17.4.4 Testing

After log-in and sampling the samples move to the testing area of the laboratory. Lab Manager allows testing from on-line instruments, from terminals or the keyboards of Mk5 Digimetry units. Within these two broad classifications of on and off-line testing are two mechanisms which are called test-by-sample and test-by-test. Essentially, in the former method all the results for a single sample are entered. This is a common mode of testing for a high priority or a unique sample. In the latter instance, all results for a given test are entered. The system looks for all samples in the laboratory which require that test, and lines them up in a worksheet by sample I.D. In many cases this worksheet also exists as a report on the system and hardcopy is produced as desired.

If this test is performed on an automated on-line instrument, the test-by-test facility becomes another utility called Runsheet. This allows the ordering of samples within a run and the insertion of calibration standards, blanks or pool samples, at any point, as appropriate to the testing technique used. Runsheet is commonly used to set up trays of samples for chromatography, continuous-flow analysis or atomic absorption techniques. Then the run or tray is submitted to the instrument as a whole. Each run has a unique sequential identifying number which allows later questioning of the database. Since the element number (or tray position number) within the run is also stored in the final test result, questions can later be posed, such as, 'what other samples were run with sample Z and what was in the two vials on each side of sample Z?'

To be truly effective, a LIMS cannot be only an off-line operation; facilities must exist to obtain data from as many connected instruments as is economically and electronically feasible. Most results from on-line instruments arrive as already calculated values from the analysis package appropriate to the technique. As examples, the mechanism for a chromatographic run and a spectroscopic assay determination are discussed below. These illustrate the primary ways that CALS gets instrumental data into the LBMGR database.

For the chromatography run the Lab Manager system sends instructions to the PeakPro chromatography package concerning the presence of calibration samples and any standard and sample weights required. Lab Manager also informs the chromatography system which instrument to use and what chromatography method is to be executed. In the case of Runsheet, the entire run is submitted to PeakPro and the run proceeds, PeakPro will assign baselines and peak names according to its methods and then the processed results are posted to the Lab Manager database.

For the spectroscopic analysis a choice exists for calculations on the host device or at the local interface. If the instrument can calculate and manipulate results, there is no need for a separate interface. The spectrophotometer has RS-232C output and is also capable of responding to data sent to it by an external device. There is a Mk5 Digimetry unit between the computer and a number (up to eight) of RS-232C output instruments on a lab bench (including the spectrophotometer). The Digimetry runs a unique program written in the Laboratory Interface Language (LIL) and prompts the user for his or her 'User ID'. After correct entry of this, the device prompts for a sample ID and the reply is given. Note that in many cases the user and/or sample IDs can be entered using bar codes. The interface knows from the test list attached to that sample what task it is expected to perform. From a table within the program, it may provide the spectrophotometer with a wavelength to use and then prompt the user to insert a blank. After reading the blank it requests a standard followed by the sample. (In some cases the internal calibration scheme of the spectrophotometer is used.)

The calculations are performed and a result is mailed back to the LBMGR database. This can include the raw data that was used for the calculation. (In many cases, this raw data is important enough to require posting.) Alternatively, the raw data could be transferred back to LBMGR and a calculation performed subsequently, especially if input is required from another test.

Using the IDAS (Instrument Data Acquisition System) module a laboratory can make many intelligent choices for instrument automation. The communication between IDAS and an instrument is two-way, so that if the Mk5 or the instrument itself requires other information from the system, such as another result for inclusion in a calculation, this can be accomplished. If the instrument is capable of responding to external control it can be set up by a computer-based method so that critical parameters such as flowrates, temperatures or wavelengths will not be overlooked. Data from other instrument automation systems has successfully arrived through the IDAS facility into the Lab Manager database, including many of the best known chromatography systems.

There is also data entry from other, less automated instruments and techniques as well as text written in a laboratory notebook or on a report form. One type of test handles Pass/Fail data in that an attribute, or set of attributes, is output to the screen and the tester replies 'Pass' or 'Fail'. He might also say 'Yes', or 'No', or 'Maybe' but the system is equipped to enforce compliance with a set of rules. The rules in this case might say that pass, fail and warn are the only acceptable values for this test. Additionally, each test type, whether numeric or alphanumeric, allows the scientist to append a free-form comment where he can amplify a given response.

There is a test type where the observations are entirely free-form and the reponses may vary from 'OK' to comments on the unknown peaks in an IR run. Another test type allows selection of one or more test attributes from a menu, an excellent device for inspections of incoming components where certain 'standard' types of faults are detected. Another test type allows input of numeric data and the subsequent calculation of results such as the entry of a titration value, normality and sample weight to produce a percentage assay value. All of these secondary calculations are stored in one of the tables of the system and thus adherence to the SOPs of the laboratory is assured. Importantly for data review purposes, once this calculation has been properly validated there is no reason for a data reviewer to calculate all results a second time: a productivity improvement available to all users of the system.

Facilities exist within the Lab Manager to pass data from one test to another or from one sample to another. This is important as often the result of one test is simply the input for calculation of another. Thus, an anhydrous assay result may be desired and the moisture value must be passed from a moisture test into the assay calculation. In some cases, one standard factor calculated at the beginning of a run will serve all the members of the run, and so the system collates results from a calibration standard and passes these results into tests on each of the samples of the run. A user site has reported use of this facility to keep standard factors for volumetric solutions, to ensure that only the latest approved factor is used.

17.4.5 Other Test-related Activities

Before passing on to the review of test information, there are other activities related to test information which should be discussed. Some tests are declared optional, meaning that they can be deleted by someone with the appropriate security. The test so deleted is time/date stamped along with the identification of the person performing the action. A reviewer must subsequently approve that this test was deleted. Other tests are declared to be 'For Information Only' and they will not appear in official Certificate of Analysis Reports or in reports, unless specified. There are also tests designated to be performed by supervisory or review personnel, which require no further review.

There are facilities to allow authorized users to display test results, and others to allow users to add a test, or tests, to a sample. This is quite important in the Research Laboratory where the unexpected results of a particular analysis may lead to pursuit of other directions of inquiry rather than simple confirmation of the original result. There is also the Retest command which allows authorized users to elect repeats of tests along with selection of a number of replicates for these retests. Although this facility is commonly given to supervisory or review personnel, this retest capability can separately be given to trusted analysts so that the work of the laboratory can proceed independently of decision making by supervisors.

17.4.6 Test Review

Once results are available they can be reviewed. This review, like many of the other LBMGR functions, is configurable and may not exist on systems where it is not desired. Also, results can be reviewed while other testing proceeds on a sample or group of samples; there is never a need for all work at a stage to be complete before review begins. The reviewer has the same overall methods at his disposal as the tester. Results may be reviewed on a sample basis, where all results for all tests on a given sample are displayed for review, or the reviewer may elect to review all results for a particular test across a range of samples.

When reviewing a particular test result by either of the two broad methods, the choices are varied. The result in a general sense is either valid or invalid: a terminology that has nothing to do with any limits that may apply. The reviewer is signifying that he believes the test was performed properly and that this result, is a 'true' result. The supervisor does not question the methodology or veracity of the result. On the other hand, some results are declared 'invalid', a status reserved for test results of questionable content. In either case, the reviewer will specify whether a retest is desired, the number of times the retest is to be performed, and a reason for doing the retest.

In some cases the reviewer is responsible for all test results on a particular screen whether they are multiple tests on a given sample or the same test on a number of samples. There is a 'Validate All' key which allows the reviewer to quickly review and accept large amounts of data. This mechanism is conveniently merged with the normal review procedure so that a reviewer encountering a screen with many values could signify a retest for one test of a given sample, and then with 'Validate All' accept all the other test results on the screen.

Many of these test-related activities are combined together in screens and can be performed together. For example, a reviewer may select a retest for a particular test and go on to provide the test information on the same screen. Alternatively, a reviewer could invalidate a particular result and instead of retesting, could select another approach through the add test facility also available on the same screen.

17.4.7 Approval

Once all the test results are reviewed the sample is ready for approval, usually by another level of reviewer or management. In some laboratories, approval is the only review mechanism because they have elected to by-pass the validation step. The primary difference between validation and approval is that validation is a test-related activity, normally applied to one test at a time while approval is a sample-related activity examining test results on that sample. The differences blur with time because the 'Validate All' key gives an approval-like flavour to validation and the re-test ability given to approvers gives them a test-review functionality.

Approval ranges from rejection or acceptance of the sample together with other categories, such as restricted or noted, which allow the approver to categorize and/or place cautions on his acceptance. Many systems will prompt the approver for other data such as comments on the sample or expiration dates for the sample. In many raw material laboratories, this type of date is known as a re-test date and causes the approval of that material to expire on that date unless re-testing is performed. The approver is initially presented with a convenient chart showing the test status of all tests on the sample, whether they are within specification, and have been reviewed by authorized persons. At any time the approver can go into a full display mode and see all of the actual results down to the finest detail. The latest revision of the system allows multiple approvals in that many samples can be approved from the same screen. The ability to view complete results for any given sample is, of course, retained and is accomplished at the touch of a key.

17.4.8 Reports - 2

Once a sample is approved many laboratories wish to generate some sort of certificate of analysis or official report, in many cases physically signed by a manager or certifying chemist. Lab Manager facilitates such reporting by allowing only one official report to be issued, at the proper time (after approval). Any attempt before approval will cause issuance of a report labelled 'preliminary'; any re-issue of the report following the one 'official' copy will be labelled 'duplicate'.

17.4.9 Completed Database

Once samples are reported they remain in the active database awaiting transfer to the completed database. This activity may occur nightly, weekly or even monthly, at the discretion of local system management. To the users of the system this is not a significant event. It is a tool to provide a mechanism for availability of the active database to laboratory work, even while the much larger completed database may be unavailable while it is being copied to magnetic tape.

17.4.10 Reports - 3

Samples in the completed database tend to lose their individuality in a practical, although not a real sense, as the system is always capable of printing out a report for a given sample. Most reports on completed data are of an aggregate variety, not single sample reports but reports of ranges of samples that meet certain criteria, eg. all raw materials purchased from Vendor Z, all products approved in December, or all samples where the pH test failed. These types of reports are often reported out in a matrix fashion with the individual sample IDs going over the columns or down the rows and the tests performed going in the complementary direction. Optionally, this data can be plotted in trend charts which show long-term trends either in line, bar or point formats, with or without specification limits or confidence intervals. Importantly, data from the report generator can be organized into input files for many of the favourite statistical and data manipulation tools such as Lotus 1-2-3, STAT-80, SAS or RS/1.

17.5 Optional Packages for Lab Manager

There are a number of optional packages designed for specific applications which directly interface with the primary Lab Manager system; in fact, all except PeakPro cannot exist without Lab Manager.

17.5.1 Autolog

For the refinery, chemical processing plant or environmental analysis laboratory where a fixed sample schedule exists once a process or study is underway, Lab Manager provides the Autolog option. Within the specification dictionary, a sample schedule is kept that governs how often a sample is to be logged automatically to the system for this item.

For example, in the continuous process for production of a chemical, once the process is started it is desired to remove samples for complete analysis in the middle of each shift, Monday through Friday and at the end of each shift on Saturday and Sunday. During the week, Autolog would automatically log samples at 0400, 1200 and 2000; on Saturday and Sunday the samples would be taken at 0759, 1559 and 2359 hours corresponding to the end of each of the three shifts.

Autolog offers a wide variety of scheduling patterns that work up to a yearly basis in a cyclical fashion. Thus samples could be specified for July 14th, Oct. 14th, and Jan. 14th at eight am., or other schedules could specify the taking of samples on the 15th and 30th of each month with provision made for the 28th of February. A sequence could begin on a certain date or time and then a frequency pattern could be established, eg. beginning on the 15th at 0800 beginning sampling every two hours until 2000. Again, all of the patterns specified above apply to all tests on a sample's product specification.

Autolog includes another type of scheduling because a given test within a specification may have a test schedule that is different from the schedule of the sample, or from the schedule of any other test. In the original example of samples at the mid-point of each shift (0400, 1200, 2000) where all tests are performed, it is quite easy to have a supplemental schedule for specific gravity that occurs at the beginning of each shift (0000, 0800, 1600 hours). For the sample as a whole, all tests (including specific gravity) are run at 0400, 1200 and 2000 hours), while the specific gravity test *only* is performed at 0000, 0800 and 1600 hours. Thus the specific gravity is actually tested on six samples each 24 hours.

Autolog includes all the utilities necessary for stopping downed processes and clean-up of sample schedules where expected samples did not arrive. In fact, many types of management productivity reports can deal with the absence of samples and their failure to arrive at the laboratory; this, in effect, is information of importance. Samples logged in using the Autolog option are then subject to all of the features of the Lab Manager system decribed earlier with trend reports and plots being the favoured method of report.

17.5.2 Batch Data Downloading (BTLAB)

Working with the transaction processor that is included with standard Lab Manager, BTLAB provides a mechanism for downloading data from an existing database into that of Lab Manager. The user site must format the file from the foreign database, BTLAB will then process the file and the data will be entered in Lab Manager's sample file. This is a great benefit for users who must switch systems because of obsolescence of either the hardware or software.

17.5.3 Environmental Waste Database System (EWDBS)

This Lab Manager option is primarily a tracking and reporting system for environmental data reported to the Environmental Protection Agency (EPA) and state environmental agencies in the USA. The software is able to determine the amount of various materials entering the environment and determine whether the amounts allowed by various permits have been exceeded. Violations or excursions are tracked on monthly or yearly cycles. For many sites the software automatically fills out the appropriate forms (NPDR - National Pollution Discharge Report) for submission to the governmental agencies.

17.5.4 Instrument Calibration and Maintenance Scheduling (ICMS)

For laboratories in the regulated industries the availability of data concerning instrument calibration and maintenance is critical. Data for submission to governmental agencies could conceivably be recalled if it was not produced with instruments properly calibrated, maintained and documented. This option to Lab Manager keeps a dictionary of instruments which includes the calibration and maintenance records. (The instrument does not have to be interfaced to the system to participate in the ICMS system.)

When a tester attempts to use or enter data collected from an instrument in this dictionary, the system checks that the instrument is within its allowed calibration and maintenance periods. If not, the data is not allowed. In some cases, the system will suggest alternative instruments that meet the ICMS criteria. In most cases the appropriate instrument must be properly calibrated or maintained and returned to service by an authorized person. Many different schemes are supported, such as by time period (each three months, monthly, weekly, etc.) by number of uses, or indeed, calibration may be required before each use of the instrument. If allowed by local rules, grace periods for each activity can be implemented.

Statistics about the use of the instrument such as number of runs or hours of use, are also collected and stored. This information is invaluable in the management of laboratory resources because the local manager can now graphically demonstrate the need for new equipment when the loads on existing equipment reach maximum levels. But the primary benefit of ICMS remains the documentation. In many laboratories the actual calibration and maintenance work is done but cannot be proven because the documentation is not available. ICMS centralizes and enforces the necessary documentation procedures.

17.5.5 Chromatography Automation System (PeakPro)

In the past 20 years the growth of chromatographic techniques has changed the analytical laboratory. Automated data collection and processing from chromatographs historically shows a great return on investment for the laboratory in terms of time and staff saving. A good LIMS will provide either a chromatography package itself, interfaces with chromatography systems from other vendors, or in the ideal case, both.

The chromatography system affiliated and integrated with Lab Manager is the Beckman PeakPro System. PeakPro is the latest version of the original system delivered in 1975 and successfully used since.

PeakPro schedules and supervises collection and analysis of data from multiple instruments; it is completely multi-tasking and multi-user. (One installation on an older model computer had 100 defined chromatographs on the system, two thirds of which would be acquiring data at one time.) The system then uses a unique, multi-pass algorithm to assign component names, draw baselines, integrate and calculate. The algorithm is experienced; it understands what a chromatographer wants to do in most cases and makes quite reasonable choices. If necessary, the chromatographer always has the option of over-riding the automatic system and graphically adjusting baselines/results; these manipulations may be saved as a method for later samples.

The PeakPro system retains all data, both raw and processed along with the method used to generate the result. The results are posted to the Lab Manager database and are available for all subsequent operations. In fact, the Lab Manager system will, in one scheme, completely direct PeakPro by supplying all methods, file names and orders for the run, or will accept results retrospectively from PeakPro after the chromatographer is satisfied with the run. Working in concert with Lab Manager, PeakPro has a very tight security scheme which controls allowed use of methods, files and instruments by different people, or groups within the laboratory.

17.5.6 Protocol Studies - Drug Stability

Until now all samples described in this chapter have been discrete: they existed temporarily in higher orders, such as appearing in a runsheet or on a work list, but the sample was logged in as a single entity. A protocol-driven study defines a pattern of related samples to be received over a defined period of time. The protocol may govern large numbers of samples and tests over periods of time up to five years or more.

A drug stability study answers questions about the stability of therapeutic compounds over extended periods of time. Many of the studies performed are defensive in nature; the purpose is simply to continually prove that a previously determined expiration date on a product is still viable. Therefore, stability studies are done on all products, not just the new ones. Of course, whenever a new variable is introduced, such as a new raw material or packaging component, a new study might prove the validity of the original date, or by using accelerated storage conditions over shorter periods of time, try to predict a shelf life for the product.

The Lab Manager stability option (DSTAB) provides a protocol-driven system that tracks all the storage conditions (room temperature, 37°C, 40°C, high relative humidity, different light conditions, or different combinations of conditions) and the times for sample removal (3 months through 60 months, for example). The system will remind appropriate personnel to take the samples from storage and assigns a testing pattern to those samples consistent with the stability protocol. If unforeseen patterns develop during testing, the system allows changes in the later stages of the protocol so that additional data collection may be specified.

17.5.7 Protocol Studies - Drug Metabolism

A more complex protocol study is represented by drug metabolism (see also Chapter 12) where different subjects are treated with compounds over periods of time and are then tested at various time intervals, as well as at study end (sacrifice) for some subjects. Here, the time resolution for the study is much different from stability studies since the action of therapeutic agents and their distribution through the tissues can in some cases be measured in minutes.

Furthermore, drug metabolism studies are complicated by the multiplicity of sample types, the need to allow for different treatment groups, requirements for cross-over studies, and the need to relate information from several types of test through calculations to come up with the results required.

The protocols for these studies are divided into structured and unstructured areas. The structured elements are usually the ones common to all studies with a fixed format such as a study number or other study identification. This includes fields managed by the system such as revisor's ID, revision date and the password allowing access to this protocol record. Unstructured protocol areas allow for liberal use of comments to effectively document the study as well as the use of 'keywords' to trigger action by the system. This area of the protocol governs study design and scheduling of data elements including division into treatment groups, identification and sex of the subjects for each group, the schedule of testing for each group, the specimen types for testing, and the names of tests to be performed.

In addition to the major work done on protocol definition, the drug metabolism option includes a number of specially designed screens and reports to facilitate the flow of samples through the Lab Manager system. One major enhancement is 'review in context' which allows related data sets to appear on the screen when the reviewer must ponder the acceptability of different results. Review in context fits exactly the requirement of a protocol-driven system that it is not so much the individual data point, but its appearance and value within a particular context, which is important.

17.6 Conclusion

Lab Manager today presents an advanced, primarily user-driven set of software and options that run on the computer hardware of the some of world leaders: IBM, DEC and HP. It is a standard system utilizing a table-driven, rather than custom program implementation, and is amenable to periodic enhancement through the requests of user groups that exist in the USA and Europe. The system follows a 'toolbox' approach in that it is customer supportable to the extent that the customer wishes to be involved; the same tools available to Beckman for installation and support are also available to the customer site. The system has its roots in GMP and GLP compliance and intends to carry this forward in future revisions. Primarily, it is a laboratory system designed to laboratory specifications, installed and maintained by a staff that came from a laboratory background. It is this attention to the laboratory world of LIMS that has and will differentiate the Lab Manager system.

Chapter 18

Perkin-Elmer LIMS 2000

Michael J. Milano, A.D.Williams, and Louis A. Lesser

18.1 General Description

LIMS 2000 is designed to facilitate and to automate the labour intensive data handling process in laboratories that analyse 1,000 or more samples a year. The system is built around a high-performance database management system designed specifically for the laboratory.

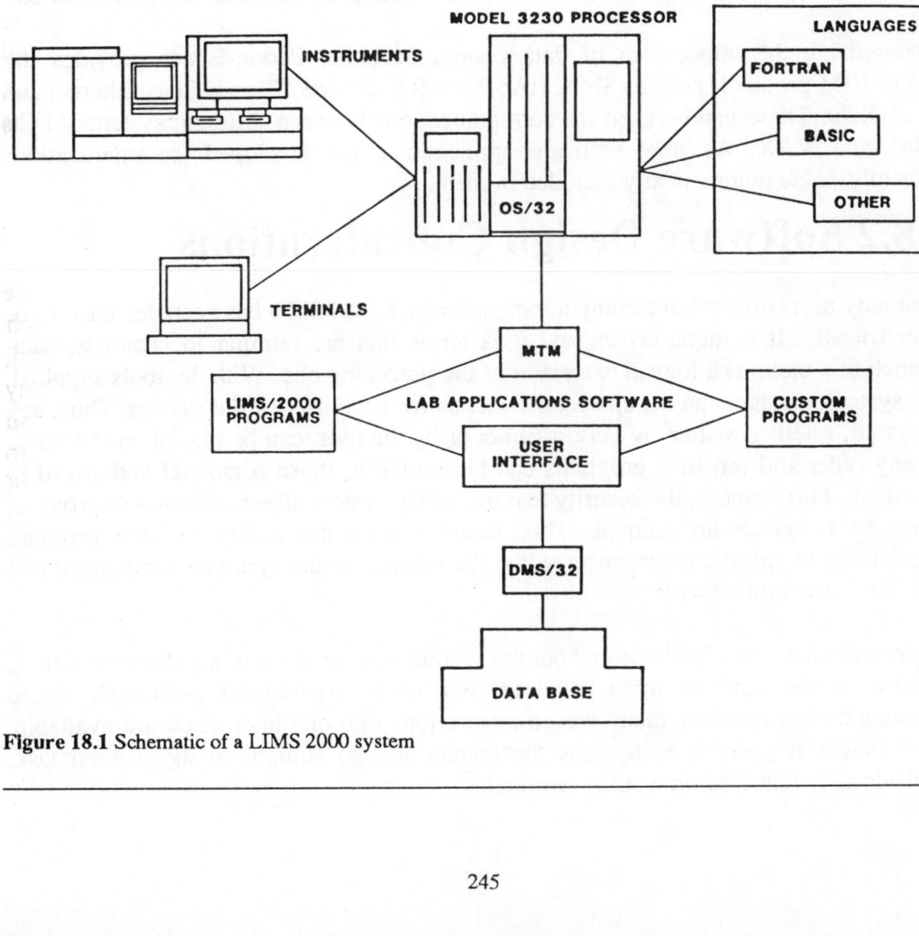

Figure 18.1 Schematic of a LIMS 2000 system

In addition to the database, it consists of a computer, and the LIMS 2000 program modules, as shown in Figure 18.1. Despite its apparent size and complexity, the system is very flexible in that it can interface with a wide variety of instruments and peripherals from most manufacturers as well as with spreadsheets and word processing programs. To make data entry quicker and more accurate, a variety of approaches can be used, including automatic electronic transfer from laboratory robots, analytical instruments, and bar code readers. Later, the system can be upgraded with both hardware and software to encompass additional functions.

The hardware is configured around one or more centralized minicomputers which can range in size from systems capable of supporting five to eight simultaneous users to ones with many tens of users.

Since the configuration of the system is very flexible, the hardware can be combined in a variety of ways. For example, it is possible to have one or more processors acquiring chromatographic data and communicating the results to a central system which holds the database. Facilities also exist to allow certain parts of the processing to be distributed, which allows the database to be searched, the relevant information extracted and passed to a PC running, for example, a statistics program. Thus, the user can manipulate and interpret data without the central processor becoming overloaded.

Recognizing the importance of data communications, Perkin-Elmer provides the major IBM protocols such as SNA, HASP and RJE 2780/3270 as well as Ethernet and X.25/X.29. These can be used for communications between different systems - PCs, IBM and VAX. As new software applications are developed, communications capabilities are automatically included in them.

18.2 Software Design Considerations

As many users do not have computer experience, LIMS 2000 has been designed to be user-friendly. It is menu-driven and uses terms that are familiar to chemists; each branch of a menu is a logical extension of the preceding one. With the tools supplied, the system manager can set up specific menus for each user of the system. Thus, any program, whether written by Perkin-Elmer or by the user, can be placed on the menu in any order and sensitive programs can be limited to those personnel authorized to use them. Furthermore, the security features of the system allow each user or group of users to be given an account. This, coupled with the ability to limit program availability to specific users, ensures that the security of the system is maintained and that GLP is complied with.

Help screens are available throughout the system with prompts to aid the user if he is unsure of the kind of input required. For more experienced personnel, macro commands that combine many screen interactions into one instruction are available. This makes it possible to by-pass the menus and go straight to the desired task, considerably reducing time at the terminal.

Included in the system are several tools that facilitate the tailoring of the system to meet special user requirements. These tools cover a spectrum of capabilities from menu-driven system set-up modules to full-feature, high-performance programming languages. Here are a few examples:

1. *Extended sample entry* Users can easily design screens and procedures for logging in different types of samples.
2. *File templates* Users can define how the system should handle data files transmitted from instruments or personal computers. Once the templates are defined, LIMS 2000 can automatically receive and process data files without user interaction.
3. *LIMS toolkit* This is a high-level, interpretive language that contains many special functions that are designed specifically for use with the LIMS 2000 database.
4. *LIMS FORTRAN library* Over 300 Fortran subroutines for database access, screen handling, device input/output, and graphics, are available. This makes Fortran programming much easier and allows user-written programs to be consistent with standard LIMS 2000 software with respect to peripheral device independence, macro capability, and database definition.

If special applications are called for, LIMS 2000 will accept custom software written, either by the user, or by Perkin-Elmer's LIMS Implementation Department. Examples of special software written include unique sampling protocols, provisions for customer billing, and special facilities required to interface with the corporate computer.

18.3 The Database

The foundation of LIMS 2000 is the Perkin-Elmer database management system, designed for real-time processing and data security. The database itself is a large, contiguous disk file, the size of which is limited only by the physical size of the disks used to store it. Multiple independent databases can co-exist on the same LIMS 2000 computer.

Within the database are logical files called data sets. Each data set defines how information is to be stored in the database and contains individual records of related information. A maximum of 99 data sets can be defined by the user, and up to 50 items, each of up to 1000 characters long, can be stored in an individual record in a dataset.

Some data sets will be common to all laboratories and are therefore included in the LIMS 2000 standard database definition, whilst others will be unique to a particular installation. For example, every sample that arrives in the laboratory is logged into the system, creating a record for that sample in the sample data set. The test data set contains the information on the tests the laboratory can perform; these are assigned to each sample during the log-in procedure. Each sample/test pair creates an analysis record in the analysis data set. When an analysis is completed, the results are entered into the results data set. In addition to these data sets, there are a number of others that store ancillary information such as users, submitters, charge numbers, and service groups.

Configuration of the database is flexible and data sets can be defined to meet the needs of the users. This is achieved by the system manager using a configuration program: all of the LIMS 2000 programs use the information specified in this module. Each manager is led through a series of easy implementation steps by the applications utility that essentially customizes the system for each individual laboratory.

Modifications, however, are controlled by a security system that prevents unauthorized changes, as all additions to, and modifications of the system are logged. At any time, any data set can be made subject to an audit trail. Thus, LIMS 2000 helps a laboratory comply with regulatory agency guidelines by maintaining a discrete record of database changes.

18.4 The LIMS 2000 Software Modules

The basic system, designed for use by a large number of laboratories, is described in the next three sections and gives the reader an overview of how the system works. This is followed by a presentation of some of the additional programs that are available.

18.4.1 Sample Management

This module is used for entering sample information into the sample data set. When a sample is entered into the LIMS, it is assigned a unique number, either manually or by the system, by which it is possible to follow its progress through the system. Additional information such as the source, the submitter and the test(s) assigned to it are also entered at this stage. Each sample-test combination is termed an analysis and a record for it is created in the analysis data set.

A series of pending analyses can be assembled into a worklist. This is achieved by using templates, that have been defined by the operator and stored in the worklist template data set, and a list of pending analyses. The templates, and thus worklists, can include duplicate samples (DUP), spiked samples (SPK), and standards (STD). Once a worklist is created, it can be printed out and given to an operator or chemist who will then analyse samples according to this worklist.

Using worklists in this way enables a manager to ensure that the correct analytical procedures are followed. It is also possible to trace the analytical results back to the raw data by reference to the worklists used.

18.4.2 Results Entry

Results can be entered manually in two ways. The first is on a sample by sample basis where the appropriate result is entered via a screen for each sample. The second is to use a worklist, where results for more than one sample can be entered on a single screen. This is used where a laboratory assays multiple samples for a test.

Results entry by worklist is accomplished via a spreadsheet format with freedom to move from cell to cell. The main feature of this approach is that results can be calculated from other results, eg. the concentrations of an analyte in samples can be calculated from standards run in the same worklist. If the relationship among items on the worklist changes, results can be re-calculated easily and inserted into the spreadsheet.

```
                         LIMS/2000 DATABASE
            ==========================================
                                      12/10/84 11:05    PAGE    1
          STATUS SUMMARY FOR SAMPLE NUMBER : 84000001

                          SAMPLE                        ANALYSIS
SAMPLE                    STATUS        ANALYSIS        STATUS

84000001                  RE-WORK       FL100           COMPLETE
                                        GC100           LOGGED-IN

          STATUS SUMMARY FOR SAMPLE NUMBER : 84000002

                          SAMPLE                        ANALYSIS
SAMPLE                    STATUS        ANALYSIS        STATUS

84000002                  IN-PROCESS    FL100           COMPLETE
                                        GC100           LOGGED-IN

          STATUS SUMMARY FOR SAMPLE NUMBER : 84000003

                          SAMPLE                        ANALYSIS
SAMPLE                    STATUS        ANALYSIS        STATUS

84000003                  IN-PROCESS    FL100           COMPLETE
                                        GC100           LOGGED-IN

          STATUS SUMMARY FOR SAMPLE NUMBER : 84000004

                          SAMPLE                        ANALYSIS
SAMPLE                    STATUS        ANALYSIS        STATUS

84000004                  IN-PROCESS    FL100           COMPLETE
                                        GC100           LOGGED-IN

          STATUS SUMMARY FOR SAMPLE NUMBER : 84000005

                          SAMPLE                        ANALYSIS
SAMPLE                    STATUS        ANALYSIS        STATUS

84000005                  IN-PROCESS    FL100           COMPLETE
                                        GC100           LOGGED-IN
```

Figure 18.2 Sample status report

18.4.3 Reports Management

Once results have been entered into the LIMS 2000, reports can be obtained. Reports management is used for short, standard searches of the database or a summary of the status of the samples; an example of the latter is shown in Figure 18.2.

User report management allows customized searches based on criteria defined by the operator. Status management permits the user to manually change the status of any sample or analysis to one not automatically assigned by the system.

18.3.4 Graphics and Statistics (GS 2000)

This is an application program that provides mathematical analysis of chromatographic data acquired using the CLAS system (see Section 18.5.1). GS 2000 consists of five modules. Four of these allow chromatograms to be normalized, added, subtracted, de-convolved, differentiated or smoothed. The fifth module, graph generation and statistics, allows the user to input sets of x,y data. Once the data have been input, GS 2000 creates a graph of the data and lets the user fit various curves to it. An example of the output is shown in Figure 18.3.

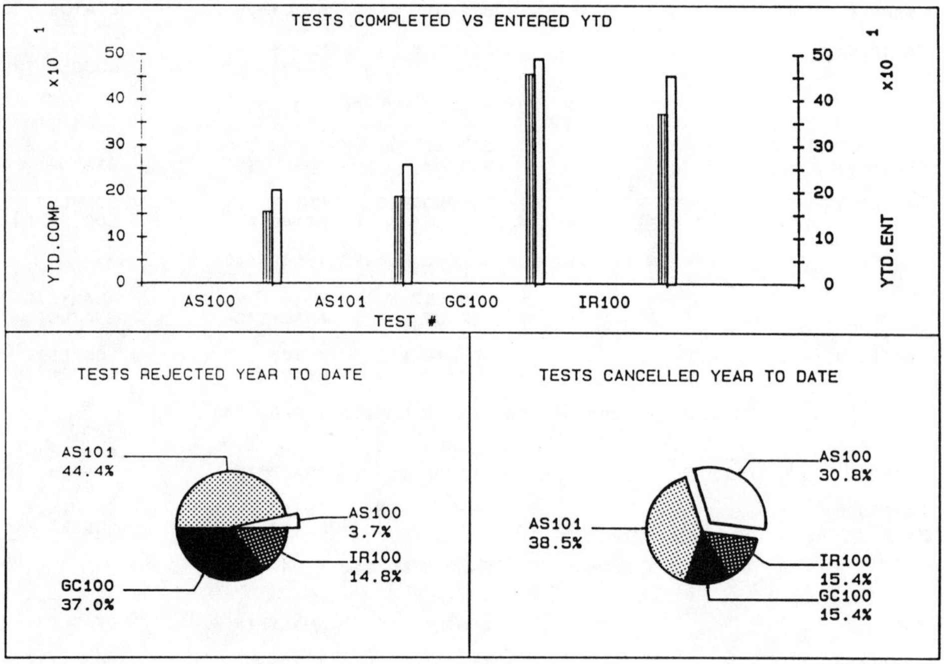

Figure 18.3 A variety of charts generated by GS 2000 from the same data

To make the graphics output device-independent, the module has been separated into two parts: core graphics (native calls) and device drivers. Subroutines are used to plot graphs, bar charts, or pie charts. Therefore, the operator need not be concerned with where the graphics will eventually be shown or printed out.

18.4.5 Laboratory Quality Control Applications

Today's analyst must not only collect data from samples, but must also collect information indicative of the quality of the data collection process. A possible method of dealing with this increase in data includes the use of an electronic spreadsheet for performing QC calculations. A program, Export, can convert a data file derived from the LIMS 2000 database into one that can be transferred into either Lotus or dBase. This is not an optimal solution, however, because the spreadsheet does not reside within the LIMS computer, This shortcoming was resolved with an enhancement of the results entry module.

The problem of performing inter-record QC calculations has been solved by incorporating a spreadsheet calculation capability within LIMS 2000 via results entry. The system takes into account the fact that most QC laboratories perform analyses on groups of, rather than on single, samples.

An important advantage of the QC enhancement is that it imposes a structured work-flow onto the activities of the laboratory and incorporates all the relevant data into the database. Furthermore, all of it can be retrieved, plotted, listed, or processed in any manner needed. For instance, the trend of any QC parameter can be displayed using a Shewhart chart (see Figure 18.4).

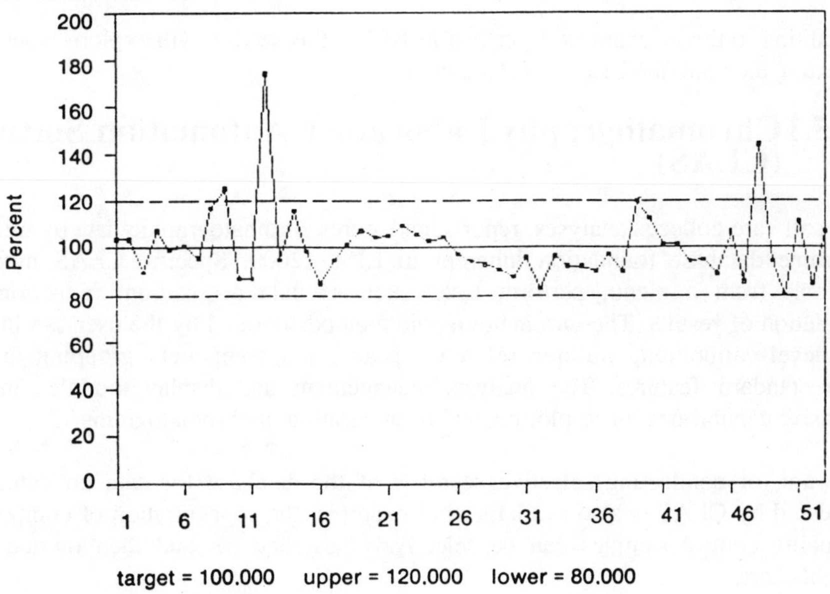

Figure 18.4 A Shewhart chart

18.4.6 Product Quality Control (QCLIMS)

To aid the laboratory manager in monitoring product quality, Perkin-Elmer has designed QCLIMS, a program that emphasizes the use of worksheets as an application on top of LIMS 2000. The worksheets are designed to assist in the process of obtaining and comparing results against the product specification. When a sample is entered into the system, it is assigned to the appropriate predefined product specification against which results for individual tests are checked. The worksheet produced by QCLIMS can be generated by sample or by test; that is, it can list either all the tests for one sample, or all samples to be analyzed by one test. Each worksheet lists the test, sample, and batch numbers, along with an indication from the product specification of the way in which any results will be checked.

As results are noted on the worksheet the analyst can manually indicate compliance or noncompliance with product quality, or let the software do it. The worksheets are designed to match the results entry screens, this reduces transcription errors and helps make results entry as error free as possible. Automatic specification checking will then highlight on the results entry screen any results that do not comply with the specification. Once the analysis is complete, comments can be stored in the database along with the sample data.

18.5 Data Acquisition

In addition to the mechanism described in 10.3.3, this section will explore other ways of getting data into the LIMS 2000 database.

18.5.1 Chromatography Laboratory Automation Software (CLAS)

This software collects, analyses, reports, and stores chromatographic data by utilizing the same database foundation inherent in LIMS 2000. Specific CLAS modules, available from a menu, perform tasks such as data acquisition, reduction and calculation of results. The chromatographic methods defined by the user can include multilevel calibration, multiple reference peaks, and component grouping, among other standard features. The analysis management and display modules include extensive capabilities for re-plotting and re-integration of chromatograms.

Software for conducting selective searches of the database for data collected and processed by CLAS is also available. For example, the concentration of components in quality control samples can be selectively searched for and then plotted on a control chart.

CLAS can be combined with LIMS 2000 to form an integrated Chromatography Automation Laboratory Information Management System. With the integrated system the results obtained from CLAS can be transmitted automatically into LIMS 2000 and can be correlated with the results from other testing techniques.

START THE BASELINE TEST: (THIS TEST WILL TAKE 2 MIN.)

MIN = 122875; MAX = 124483; AVG =123663
STANDARD DEVIATION OF BASELINE IS: 423.174

***** BASELINE ACCEPTED *****

C F RUN ON NITRATES: CHANNEL ONE
****

CUP#	T SAMPLE #	PEAK HT	CONC MG/L	CHARC	LIMITS
1	X PRIMER	551882	0.000		
2	X STD 1	117238	0.000	PEAK	
3	X STD 2	232931	0.000	PEAK	
4	X STD 3	344454	0.000	PEAK	
5	X STD 4	448700	0.000	PEAK	
6	X STD 5	557399	0.000	PEAK	
7	W WASH	6294	-0.018	FLAT	
8	S 8400000100	59147	0.175	PEAK	-
9	S 8400000200	117776	0.385	PEAK	0
10	S 8400000300	176180	0.602	PEAK	0
11	S 8400000400	225573	0.782	PEAK	0
12	S 8400000500	287816	1.009	PEAK	0
13	S 8400000600	336459	1.187	PEAK	0
14	S 8400000700	392977	1.393	PEAK	0
15	S 8400000800	458447	1.632	PEAK	0
16	S 8400000900	502131	1.791	PEAK	0
17	S 8400001000	561823	2.009	PEAK	+
18	W WASH	5257	-0.022	FLAT	
19	S 8400001100	57923	0.170	EDGE	-
20	I NBS244	455862	1.622	PEAK	
21	W WASH	5917	-0.019	FLAT	

REAL TIME PLOT

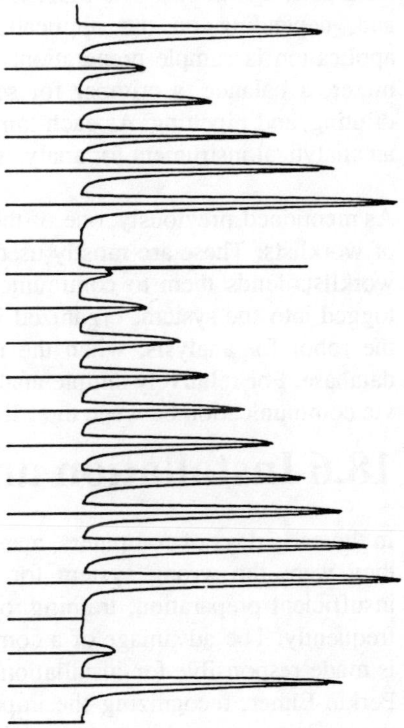

Figure 18.5 Real-time report from CFDA with analogue signal and the calculated results

18.5.2 Continuous Flow Data Acquisition (CFDA)

Continuous flow analysers are widely used in many laboratories, here computers have proved useful in analysing the data produced and in monitoring instrument performance. The speed and power of LIMS 2000 has been used in the design of the CFDA module.

This program can collect and process data from up to 64 CFA instruments simultaneously at a maximum rate of 20 samples an hour per channel. Real-time reports can be produced, as shown in Figure 18.5, if the output device is a Perkin-Elmer GP-100 plotter. Here, the analogue signal is plotted alongside a tabular report containing the sample identity and the calculated concentration. Post-run calculations on the sample concentrations in the peak file are also available.

18.5.3 MasterLab Robotics

One new route to increased productivity in the laboratory is robotics which can decrease turnaround time and relieve the staff of time consuming, boring, or hazardous operations. The MasterLab robot is designed to interface with LIMS 2000 and, depending on the application, with a variety of instruments. A common application is sample preparation. Instruments for this application include a vortex mixer, a balance, a crimper for sealing sample vials, and syringes for dispensing, diluting, and pipetting. As each sample is prepared, the robot can move it directly into an analytical instrument for analysis.

As mentioned previously, one of the methods of managing work is by the generation of worklists. These are mostly used by laboratory personnel, however, the nature of worklists lends them to communication with a robotics system. Thus, samples are logged into the system, organized into logical groups and the instructions passed to the robot for analysis; when the results are available they are passed back to the database. For relatively simple analyses the work of the laboratory can be automated via communication between the LIMS and a robot.

18.6 Installation and Training

In the early days of computers, many systems were purchased but never used: perhaps they were the wrong system for the company or the problem might have been insufficient preparation, training, or support. This still happens today although less frequently. The advantage of a completely integrated system is that one organization is made responsible for installation and can be called upon for training and support. Perkin-Elmer, recognizing the importance of this, has designed a series of training courses that normally are held either in-house or on the customer's premises. New training courses are added to the curriculum as new applications are added to LIMS 2000.

To aid the personnel involved in installation of the system a thorough implementation handbook has been developed. This contains worksheets to be filled in as tasks are completed.

Described below is an installation which shows the benefits of careful planning and implementation. In 1982, a large industrial laboratory reviewed its management system; although the system was considered workable, pressure for increased production at competitive cost caused serious strains. Specialists were spending too much time administering jobs. The first step was to set up a four-member automation task force; two members were managers and two were bench chemists. One of the latter, who had some computer experience, became the full-time team director, the other members worked part-time on the project. The laboratory's management believes that this was an important step.

The task force made a combined feasibility and return-on-investment study of a LIMS. To keep the margin of error in the company's favour, they deliberately underestimated the capacity of automation; despite this the study promised a dramatic return on investment.

Because of highly specialized requirements, one option that was considered was for the company to write its own software. The task force's studies showed, however, that the cost would be 1.3 - 1.5 million dollars and would require several man years. This effort would have drained off badly needed professional skills from the laboratory's daily operation. After many months of evaluation, the task force purchased LIMS 2000. Installation, involving a series of carefully staged steps, began in July 1984 and was completed at the end of 1985. The slow pace was deliberate; the task force felt it would give the staff time to assimilate the new methodologies.

The package met their specifications, but because of the company's special needs, it required customization (1,2). Even though the prior management system was maintained during and even somewhat beyond the installation, the total process was completed two months ahead of the lab's self-imposed schedule.

18.7 References

1. Bassett, D.O., 'Automation and Data Management at Calgon', *American Laboratory*, 17, 91-93 (1985)

2. Bassett, D.O., 'A Laboratory Automation and Data Management System', *American Laboratory*, 18, 88-90 (1986)

Chapter 19

LABSAM

S.B.Crisp and J.Mollner

19.1 Design Goals

Laboratory System for Automation and Management (Labsam) had as its design goal the provision of a highly flexible software system readily adaptable to the needs of the majority of laboratories, without user programming. This was idealistic as it was recognized that no system could meet all the needs of every laboratory. Thus, the design had to provide the convenient means to allow users to attach additional programs where the standard software was insufficient.

It was envisaged that Labsam development would be on-going, with new features added over its lifetime, so great importance was attached to providing consistent interfaces to all of the users' equipment and programs from revision to revision, allowing all users to take advantage of new enhancements.

Since a LIMS must accept data from all types of instruments, both analogue and digital, Labsam had to acquire data from a wide range of hardware interfaces, signal types and digital data, the latter including Laboratory Automation System (LAS).

Often taken for granted, but crucial in the design phase is database performance and flexibility and also disk space efficiency. Fast response times are a key element in the acceptance of a LIMS; flexibility is necessary to cope with the addition of new items to an established database without the need for structural changes; and storage efficiency, measured in terms of disk space, is a major cost consideration when purchasing a LIMS.

Finally, the system and the data had to be secured against misuse and illegal access and for compliance to GLP.

19.2 Hardware Required

The requirement for the computer was a multi-tasking CPU capable of supporting up to 40 simultaneous users, and at the same time interfacing to many peripheral devices, some demanding priority interrupt handling for data capture. As laboratories vary widely in terms of numbers of staff, instruments, workload and the amount of data generated, a solution which could easily be expanded or changed was also needed.

The final requirement was a system giving a good price/performance ratio now, with the guarantee of continually striving for improvements in the future.
The HP 1000 family met these requirements, ranging from the smallest A-400 CPU with 0.5 - 4 Mb memory and six I/0 parts, to the largest A-900 with up to 24 Mb of memory and 16 I/0 ports. Furthermore, a complete system, optimized for performance, can be configured from HP's comprehensive range of disk drives, terminals, printers, plotters and other peripherals.

The operating system, RTE-A, is uniform across the range, giving total portability of data and programs and a fully structured product. RTE-A is real-time and multi-tasking and provides for the multi-user environment with additional data security to that provided by Labsam.

19.3 Description of Software

The Hewlett-Packard LIMS software is provided in five distinct packages:

1. *Labsam* is the basic package which provides the standard functions of laboratory data management including the internal routines for writing to, and reading from, the data base.
2. *Labquest* is a data retrieval package with searching, sorting, listing, and statistics capabilities.
3. *Labtools* provides standard read and write routines to the database and file structure of Labsam. These may be included in user written programs which write to, or read from, the database or Labsam files.
4. *Labtrail* is an integral part of Labsam which provides full audit trail capability of all sample data within the database. Whenever a result is altered, Labtrail records the old result or results, who made the change, when and why. Although an integral part of Labsam, this function may be enabled/disabled by HP, allowing customers the choice of whether or not they want an audit trail.
5. *Contributed library* is a growing collection of user written programs which provide added functionality to Labsam. The contributed library software is a formal way of sharing programs between users. The library is updated approximately every six months and is distributed free of charge.

19.4 Database Structure

The following discussion covers the most important elements of the LABSAM database and its structure.

19.4.1 Defining Data Sets

To provide a flexible data structure, Labsam uses a definition file to describe all sample data entered; this file will vary according to the needs and type of the laboratory. Every data set defined has two keys or titles, a main key of sixteen characters and an alternate key of eight characters. A data set definition is referred to as an IDC (identification code) and comprises two parts: the whole description and the two keys, the latter are referred to as the Main and Alternate IDCs.

IDC Definition

Main IDC	Alternate IDC	Record No. of IDC file

Reference table for IDCs

Main IDC	Alternate IDC	Descriptive Data

IDC File

Figure 19.1 Structure of an IDC file and reference table

Seven different types of IDC are available for differing data handling needs:

1. *Single IDC* defined for a single test and its result.
2. Joint IDC defined to calculate a final value from the results of several single IDCs.
3. *Master Data IDC* defined to hold descriptive data for a sample, ie. sample type, customer, sampled by whom.
4. *LAS IDC* defined to schedule a LAS (Laboratory Automation System) GC or LC run for a sample.
5. *Text IDC* defined to allow entry of a comment during result entry or validation.
6. *Group IDC* allows the grouping of several single, joint or group IDCs into a single menu (this used only in a test request).
7. *Class IDC* defined to schedule a variable group of tests,eg. heavy metals.

Table 19.1

A typical set of IDC descriptive data for a test result

HEADER	
01 Main IDC	: pH
02 Alternate IDC	: pH
03 Description	: pH value
04 Data type	: S (type of IDC)
05 Status	: 1 (active/inactive/incomplete)
FORMAT CONTROL	
10 Format (final result	: F21
11 Units (final result)	: PH units
12 Input mask (final result)	: nn.n
13 Help number	:9004
14 Format (prelim. value)	: F21
15 Units (prelim. value)	:
16 Input mask (prelim.)	: nn.n
LOGICAL CHECKS	
20 Sample type	:
21 Sample preparation	: Liquids
22 Analytical procedure	:
23 Normal limits	: 2.5 - 9.0
24 Detection limits	: 2 - 13
25 Alarm limits	: ALARM, PH STYPE
26 Plausibility number	: 0
27 Code file name	: (allows short form entry - file decodes)
OPERATIONAL DATA	
30 Invert IDC	: No
32 Multiple preliminary values	: Yes
33 Delete preliminary values	: No
34 Validation level	: 2
35 Algorithm	: (Allows further calculation)
36 Laboratory	: Water laboratory
37 Instrument	: PH1
38 Sequence	:
39 Work station	:
40 Account code	:
41 Time due	:
42 Evaluation routine	:
45 Start address in codefile	: 1
46 Delta-Limit, Time:	
47 Hold Prevalue?	: No
ORGANIZATIONAL DATA	
60 Comment	:
61 Days to retain	: 30 (Time the result stays in data-base until deletion)
62 Predecessor IDC	
63 Date created	: 850408
64 Created by	: System manager
65 Date last modified	: 870305
66 Last modified by	: System manager

Item 25 shows different alarm limits stored in the alarm file, @ALARM, triggered by different sample types stored in a file @STYPE for the TEST PH

This combination of seven types of IDC is fundamental to the high degree of flexibility within Labsam. The advantage of this approach is that no definitions are duplicated in the database and that it is very easy to define new data or edit existing entries without reconstructing the database in any way.

19.4.2 Indirect Definition of Data

In many cases, single values cannot be defined for a sample; items such as limits may vary for different sample types. For instance, the limits for the pH value, sugar and alcohol content of wine will vary according to its colour and cultivar. Labsam gives the capability of defining multiple values for limits and most other parameters in a separate file with appropriate selection criteria.

To make an entry into the pH IDC for variable normal limits as shown in Figure 19.2, the limit file must be referred to in the IDC definition (Fig. 19.1) at line 23 - normal limits.

```
23 Normal limits: limit, pH, colour, cultivar
```

Indirect definition is possible for most entries in an IDC.

IDC	Colour	Cultivar	Normal Limits
pH	Red	Pinotage	2.0 – 6.0
pH	White	Muscat	3.0 – 7.0
pH	Red	Pinot	1.8 – 5.0
Sugar	Red	Pinotage	2.0 – 15.0
Sugar	White	Muscat	20.0 – 100.0
Sugar	Red	Pinot	1.5 – 12.0

Figure 19.2 A Labsam indirection file of test limits

19.4.3 Sample Organization Requirements

It is usual practice in a laboratory to identify samples by a unique number and/or lettering system, in Labsam this is called the External Sample Key System. To cover all the different numbering possibilities there is a user defined structure of up to sixteen characters. The key elements of this may include internally generated sequential numbers, external strings, date, laboratory code, check sum, and fixed strings. Many laboratories wish to operate on a batch, sub-sample system; here, the external sample key may be configured as a hierarchical two-level system where the first key level represents a batch, and the second level the sub-sample within the batch.

On log-in of a batch of samples, data common to the batch may be entered for the batch key and data unique to each sub-sample may be entered for the second level key.

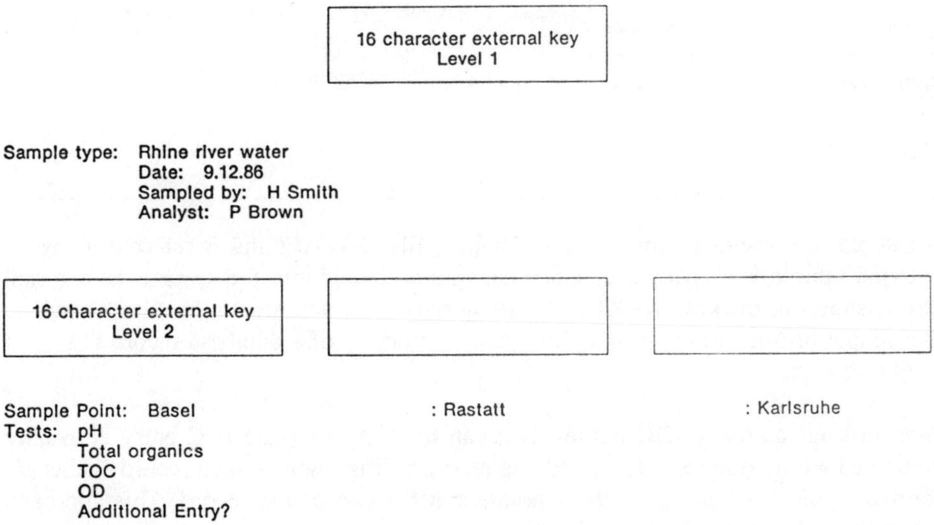

Figure 19.3 A two-Level external sample key

19.4.4 Storing Sample Data in the Database

The database files are handled by a common program which performs every transaction. This allows user written programs to access the same routines as the Labsam core programs, thus making program development straightforward whilst ensuring database integrity. The relationships of the database are shown in Figure 19.4.

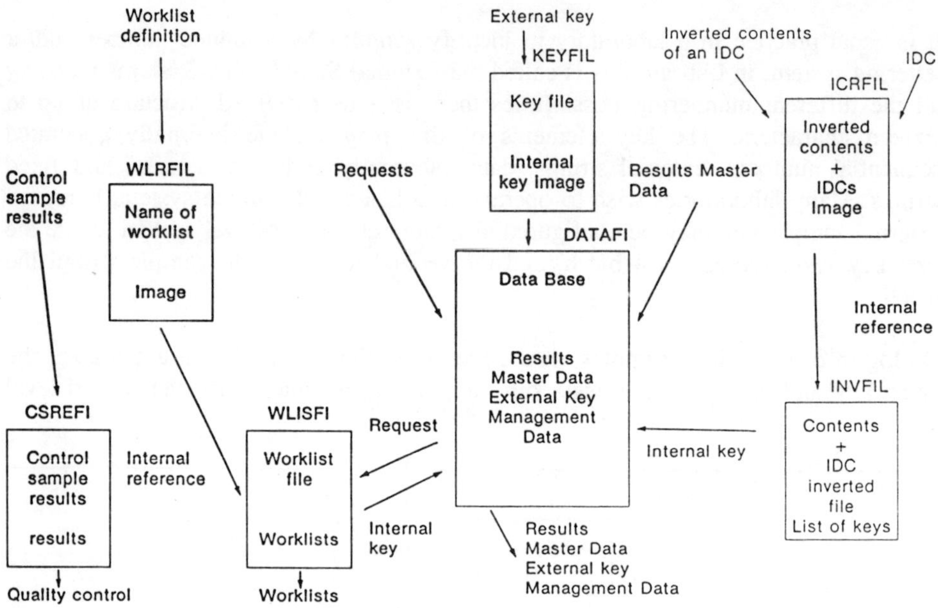

Figure 19.4 Labsam database structure

A sample in Labsam is stored in the database file DATAFI and is referred to by its external sample key. Whenever a new sample is logged into the system the external key is stored in the key file KEYFIL along with the reference to the data file where the sample information is stored. The general structure of a database record is shown in Figure 19.5.

A record has as many IDC entries as it can hold. A complete IDC entry is usually contained within one record. Should one record be insufficient, then record chains are formed in the database with chain pointers at the end of the record. This approach meets the disk space efficiency objective.

Finally, in a two-level (batch and sub-sample) database, record relationships are as shown in Figure 19.6.

19.4.5 Inverted File Structure

As outlined in the previous section, Labsam writes all data for one sample sequentially to a record. Individual records for one sample are not sequential but chained; this enables fast access to the information as searches for a sample first access the key file for the sample record number.

Fixed part	Fixed IDCs	IDCs for requests, results and Master data	Seperator	Chain Pointer

The fixed part of a sample record is held in the first record of the chain and has the following entries:

Internal key for sample; counts over the whole life of the system
Last request date for sample
Last result date for sample
Record number of first level (if sample is part of a batch)
The setting of bits 0-3 of the last 8 bits indicate:
00000001 out of alarm limits
00000010 out of detection limits
00000100 out of normal limits
00001000 audit trail exists

Bits 4-7 are reserved for future use (binary 0).

A set of internal data (fixed IDCs) is written automatically after sample Log-in to allow sample tracking. The fixed IDCs include:

External Sample Key 1
External Sample Key 2
Log-in date
Closing date (blank at Log-in)
Status

Any result or sample information (Master data) stored together with the reference to the description in the IDC definition file - Fig. 1, and a set of control data for the result as is shown in the following IDC structure.

Length of entry in words
Flag to indicate audit trail entries
'0' NO AUDIT TRAIL FOR THIS ENTRY
'1' ACTUAL ENTRY, BUT AUDIT TRAIL EXISTS
'2' AUDIT TRAIL ENTRY, NOT FOR ACTUAL LISTS
Type of IDC
IDC record number
Prevalue number
Measurement number
Record number of IDC inverted file or content inverted file
Result date (packed)
Result time (packed, 2 seconds resolution)
User number (result)

-------------------------------end of master data entry-------------------

- Request date
- Request time (2 seconds resolution)
- Validation dates - 5 validation levels
- User number (request)
- User number (validation), 5 validation levels
- Indicates out of limit
- Request status
- Out of limit flag

----------end of request entry----------

- Master data or result with variable length

----------end of result entry----------

- Separator shows length of audit trail
- User number for audit trail entry
- Date for audit trail entry
- Time for audit trail entry
- Comment for audit trail; variable length (max. 80 characters)

----------end of audit trail entry----------

Figure 19.5 General structure of a database record.

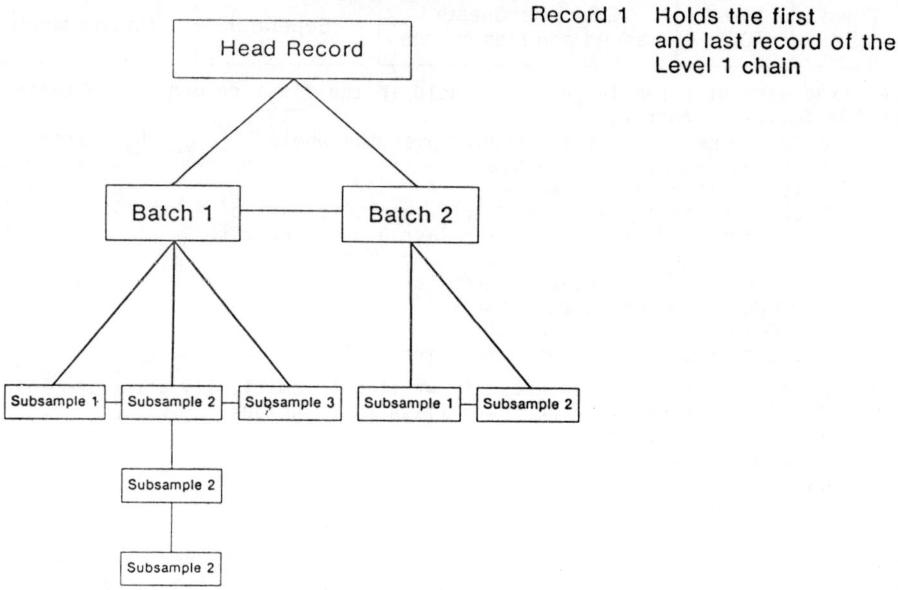

Figure 19.6 Record chains in a two-level external sample key database

For access by data rather than by sample a similar approach is taken. A second file, ICRFIL, similar in structure and function to the key file, contains the IDC record number and the first and last record number of a chain in the inverted file INVFIL (Fig. 19.4). This contains, for each IDC, a list of record numbers representing samples in the database (DATAFI), see Figure 19.7.

ICRFIL	IDC RECORD	VALUE	FIRST RECORD IN INVFIL	LAST RECORD IN INVFIL		
INVFIL	IDC RECORD NO.	VALUE	DATAFI RECORD NO.	DATAFI RECORD NO.	DATAFI RECORD NO.	CHAIN POINTER

Figure 19.7 Inverted file structure

By inverting an IDC, it becomes possible to locate specific record numbers in the database pertaining to specific IDCs, ie. to look up all pH values, the file handler program reads DATAFI locations from INVFIL and returns a list of all sample keys with pH values. Inverting an IDC results in fast access times to data in the database. In cases where you wish to have fast access not only by IDC, but also by Sample Master data (sample information) it is possible to invert the master data IDC contents. This result is then held in the ICRFIL - value - (Figure 19.7). If fast access to all samples of Rhine Water is required the master IDC identifying the river name would be inverted.

19.5 Auxiliary File System

Outside the sample database and file structure, an auxiliary file system (AFS) exists as a common utility for handling definition files, entering, modifying and listing of definitions and for providing definitions, to other Labsam programs. This allows you to handle all definitions in a common manner, providing wide flexibility and ensuring uniformity and integrity of the system. Functions such as worklist definition, sample log-in definition, coded entry, master menus, test menus, indirection definitions, and text definitions, are all handled by the AFS. Users have the possibility to define private files for access by their own or Labsam programs.

19.6 Functions of Labsam

An overview of the system is presented in Figure 19.8; each section is discussed separately.

19.6.1 Monitor

The systems manager establishes Monitor for individual users by defining which programs, modules and menus they can use. This program administers user passwords, provides security, access to I/0 devices, department security, and defines the sample validation authority levels.

Monitor also defines the layout of the home screen and the module names used for each individual user. For some programs, a control file is assigned in Monitor which defines the program functions, ie. the sample log-in procedure may differ according to the type of samples being received.

19.6.2 Log-in Test Entry

Log-in is the way in which samples are entered into Labsam by providing the appropriate user-defined screens, the information input is then passed to the database. Users may predefine sets of master data prompts to be used for all sample types and use the data entered through this routine to call other menus to capture further data specific to individual samples. This entire array of master data may then be used to trigger selection of predefined test menus, with the possibility to edit the final menu and to assign priorities.

Once samples are logged in, label printing, if desired, takes place. Users may freely select the format (text or bar code), content and number of labels. If required, a worksheet, giving all tests for a sample may be printed at log-in. To aid the speed of log-in, a 'ditto' function copies a set of master data from one sample to the next; also, fixed defaults may be defined for use during log-in.

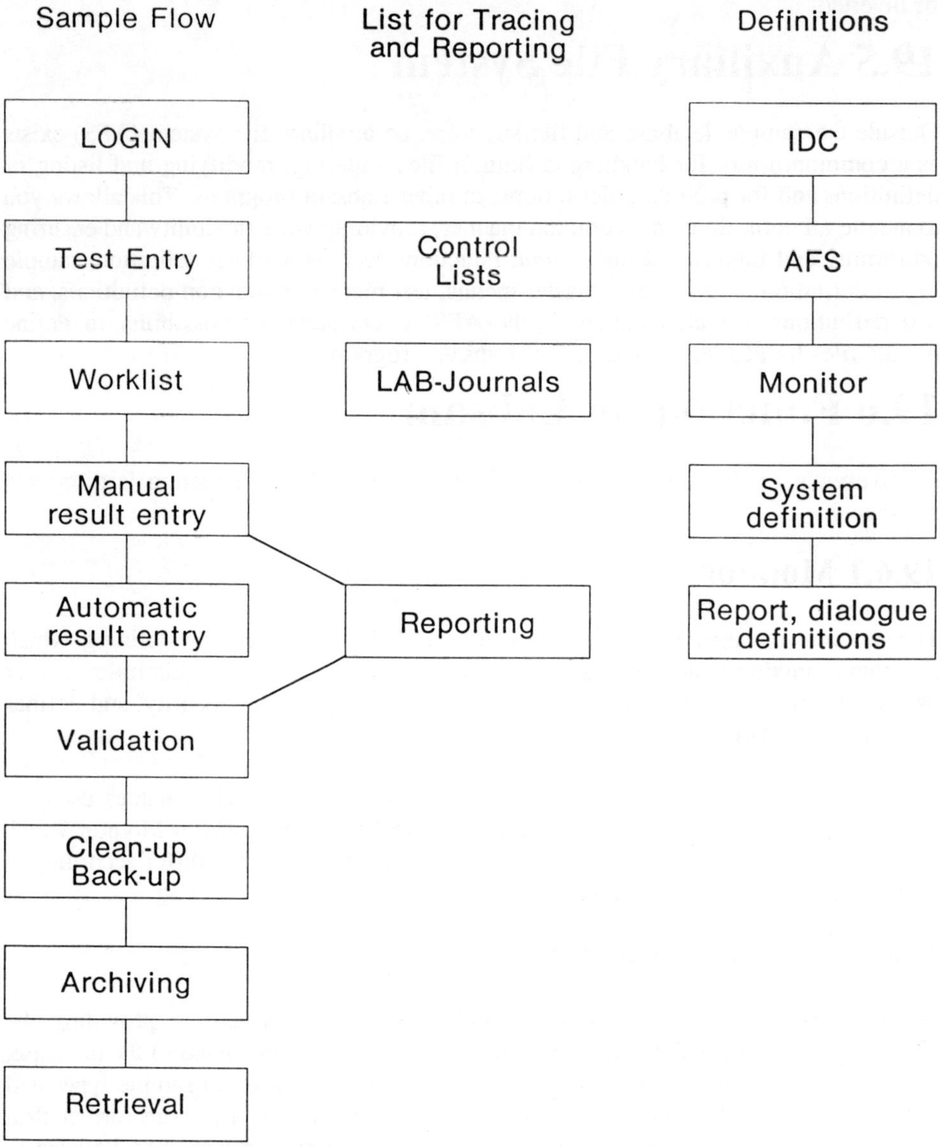

Figure 19.8 Labsam system overview

User written evaluation routines may be defined and triggered automatically from log-in. These can be used to check data and key integrity and to add further data in the background. Users may wish to use these routines to print specific labels or worksheets or to log-in, programatically, batches of tests received regularly. These evaluation routines are triggered by the input of a master data item.

19.6.3 Worklists

A worklist shows all samples waiting for a test or set of tests; information pertaining to each sample may be printed; the amount is defined by the user. Selection of what samples appear on a worklist may depend upon the test requests and sample master data. Samples are sorted according to priority, the date of test request, and the sample information. Worklists also have the ability to handle control samples without the need for manual entry. Control sample frequency and position are user predefined and will appear in the correct position in the analysis sequence.

Worklists have several different applications internally within the system. The same worklist format may be used to capture results either manually when analysts perform tests in the order of the worklist, or automatically if the analysis is carried out on an instrument interfaced to Labsam; ie. the same worklist will appear on a terminal with columns for results entry, or in the case of on-line instruments, the results received will be written to the database in the same order as they appear on the worklist.

In conjunction with LAS, the worklist is converted to an analysis sequence which can be used directly by the LAS computer. When the assay is complete, the sequence, with the sample keys and the results, is transmitted back to the Labsam database.

19.6.4 Result Entry

Results can be entered into the system either manually or automatically, each is be considered separately.

Within Labsam there are four possibilities for manual result entry, namely:

1. *Worklist oriented,* Section 19.6.3, where results are requested in the order of the selected worklist (all samples for one test).
2. *Worksheet oriented,* where results are requested in the order of the worksheet (all test results for one sample).
3. *Dialogue oriented,* where results and/or sample master data are requested in a user predefined order.
4. *Sample oriented,* where analysts enter sample key, test name and result.

To assist analysts during result entry, Labsam provides immediate warnings if a result is outside alarm or detection limits (these alarm messages are also written to a file and may be printed on a user specified printer).

Instruments connected to the system may automatically transmit results directly without operator intervention. Where instruments are unable to handle the Sample Key, all analyses must be worklist oriented, ie. Labsam controls the order of analysis and receives results through a digital interface in the same order as the worklist, or with a reference to a position in the worklist.

For instruments able to handle a sample key, two methods of analysis order and result capture may be used. First, Labsam may transmit sample key numbers across the interface and the instrument sends back this together with the result. Second, instruments that are unable to receive the sample key must have this typed into the instrument by an analyst. The instrument can then transmit the sample key plus result to Labsam. Obviously, these automatic dialogues differ from instrument to instrument and users must write interpreters to handle the different data strings and formats. For both automatic and manual entry, results may often be entered without final calculation or combination with other results. Labsam will then allow the inclusion of user specified algorithms for every test so that the final results may be mathematical derivatives and/or combinations of different tests. The algorithms may be chained so as to combine two or more calculated results to a final result or results. An accessible file contains calculation constants which may, for instance, be derived from daily instrument calibrations.

User written evaluation routines, triggered by result entry, allow for highly complex calculations, or may evaluate a result and derive a further result which may be numeric or a character string, eg. good, bad, overlimit, underlimit, released, and not released.

19.6.5 Validation

When tests are initially defined on the system, a required validation level for the result is set; the more complex or critical tests will usually require a higher level for final validation. The users of the system who have to validate data are assigned a validation capability of one to five. Labsam offers both automatic and/or manual validation.

If automatic validation is selected, Labsam checks the validation level of the person who enters the result. If this level equals or exceeds that required by the test, the result is flagged finally validated. If not, the result is flagged as validated to the user's level. Automatic validation may be set up to recognize out-of-limit results which will not be validated and will remain at the status 'result exists - not validated'.

In the case where all results from a sample have reached final validation level, the sample itself will be finally validated automatically. Again, with all results entered but with an out-of-limit value or any result having insufficient user capability, the sample will reach the status 'Results exist - not finally validated' and the result and sample must be manually validated by a person of sufficient authority.

Two modes of manual validation are provided, either validation for single results or for the sample as a whole. The program shows the sample number, status of the results and sample, the actual result and any out-of-limit messages. The validation program can provide users with a list of all samples of similar status, eg. all samples with results complete or all samples validated to level X.

Some functions, not directly concerned with validating samples are, for convenience, available to the user who is validating results. Should a sample or result require additional tests or retesting, manual validation allows the user to request an analysis and may prompt the user for additional master data related to the sample. At this point, new labels and a new worksheet may be printed if requested.

The ability to add, modify or delete a result from validation is available. Users must have a validation capability higher than the level to which a result is already validated before any change can be effected. Changes to the results from validation will, if implemented, trigger the audit system which will then require a reason for the change.

Upon completing a sample, a user written program may be called (depending upon the sample status) which further processes the sample data. The action of these programs depends upon user requirements but often the need arises to pass sample key information to other programs. This might, for example, be lists of samples finally validated and subsequently transmitted to another computer. Another function available upon leaving the sample, is the ability to append a comment automatically. This comment, now a part of the finally validated sample data, may be searched for by Labquest or used as a Master Data trigger for other functions and programs such as Automatic Reporting.

19.6.6 Reporting

Report presentation and content is individual and the system has an extremely versatile report formatting capability. This allows definition of multiple report formats with automatic format selection based on sample data.

Reports may be printed in background automatically, triggered by either the sample status reaching final validation or by any sample master data item, usually appended to the sample during validation.

Up to 200 different report formats may be predefined by the user and the report format he wishes to use may be selected from the monitor menu. The samples for inclusion in the report must be extracted from the database. To achieve this search, the user may input up to three sample master data items including a wild-card, eg. Report all samples that are:

```
:Sample type = Rhine Water
:Date        = 09/12/86
:Analyst     = P Brown
```

A report, using the specified format of Rhine water samples received on the 9th December 1986, and analyzed by P Brown will be printed. Wild-card entry -/12/86 would report all samples for December 1986. Report formats may include predefined text elements, sample master data items and their descriptions, formatting instructions and standard header information.

19.6.7 System Maintenance

Labsam includes a maintenance program (CLEANUP) which takes care of the worklist file, deleting those with no open requests, and repacking the worklist file. Information on the utilization of files and the remaining disk space is printed on termination of the program CLEANUP. System back-up provides users with facilities to copy the database plus files to a second disk or more usually to magnetic tape. A list of the users and the programs accessed from the monitor, together with the date and time of use, is kept in a file, DAYFIL, from which it is possible to review use of the system.

19.6.8 Archiving and Retrieval of Data

Archiving of samples is done on a user defined time window and is to tape or disk. As each medium is headed with a directory, it is possible to use the same media for several archiving runs.

Only completed samples can be archived, and where the same entries remain in the database they are flagged as archived and the data locked to ensure the integrity of the archive. Sub-samples are always archived with their batch; a batch is never deleted from the database as long as a sub-sample for the batch exists. Deleting of data from the database is an optional function of *ARCHIVING*. Minimum time in the database for each test is predefined. On running *DELETE*, a sample is deleted when every test has fulfilled the minimum time requirement.

Every archived file is identified in the directory by the date of the archiving. In retrieval, the main reference is to the archive date. Selective retrieval from an archived file is performed in a similar manner to a database search, by entering single or multiple selection criteria.

The selected data is written to either the actual 'live' Labsam database, or to a second database for retrieved data. The advantage of using a second database is the ability to purge the whole file. When retrieving to the live database, only logical deletion is possible, which takes more time. Data in the second database may be searched by Labquest and data from several different archive files may be combined into one database. Retrieved data has the identical structure to 'live' data in the database except for a flag which prevents modification of the data.

19.6.9 Overview Lists

To allow users, supervisors and managers to monitor the performance of their laboratories, Labsam provides several modules for producing lists of workload and work status, results, and for tracking samples through the laboratory.

1. *Login journal* Shows all samples logged in during a selected time window. The information includes the sample identification key (sample number), four items of sample master data plus the overdue and closing dates for the sample.
2. *Test request per sample and Samples per test* requests This shows all open tests for a given sample or batch and a list containing all samples for a given test where no results exist.
3 *Master data, Result and Dialogue oriented result* lists and Laboratory journal These lists show sample master data and results, either all master data and results or a predefined set of the same.
4. *Status list* Two lists may be produced to give sample status, ie. logged in, request exists, result exists, result complete, result validated to level X, finally validated, reported. Either all samples at a given status within a time window or, for one sample, the status of all requested tests plus date, time and user's names, for each status, may be produced.
5. *Remaining and Overdue list* These lists show either all samples with open tests or overdue tests. The sample number is accompanied by the relevant list of open tests. Overdue may be defined for the whole sample or on an individual test basis.
6. *Other lists* These are available to assist in the basic configuration of Labsam. It is beyond the scope of this chapter to provide such detail, suffice to say that it is possible to organize anything user definable into a consolidated listing.

19.6.10 Labquest

Labquest is a data retrieval package specific to the Labsam database; all information can be searched either singly or in combination within one statement. The combination may be 'and', 'or', 'exclusive or', wild-card, greater than, less than, equal to, not equal to, or, without, and can be used singly or in combination to qualify data. Sets of commands used frequently or optimized for economical searching may be stored in command files. Simple English abbreviations are used in the search definition, eg:

```
SE, DB, FIL1, pH.GT.5,1/9/87-2/9/87
```

is the Labquest abbreviation of:

Search for all pH values between 1/9/87 and 2/9/87 with value greater than 5,0.

The results of the search of the database DB should be placed in a disk file FIL1.

The system will optimize search efficiency where all commands are in the same statement and these may be chained together for added efficiency. Several search outputs may be merged into one file for further use. Also, the output may be sorted, according to user selected criteria and listed to a terminal, printer or a disk file which may then be transported externally.

Statistical analysis may be performed on the results of a search for inclusion in the Labquest report. Mean, SD, RSD, number of entries and minimum/maximum are available, whilst for more complex calculations, a statistics package may reside in the system. Access to and from this statistics package is via a disk file, and graphical display of results is handled in the same manner.

19.6.11 Quality Control

As a means of measuring instrument performance, quality control samples are frequently included in analysis sequences. Labsam will automatically include these samples in worklist sequences based on a user defined schedule.

```
Central Labs                                                     Enviv. GP.
10/08/86                                                              11:17
                                CONTROL STATISTICS

Time window:   10/07/86 - 10/08/86

Main IDC         : pHTEST        Control Sample Type:        P
Alternate IDC    : TEST1         Batch Number       :     1112
Instrument       : INS1          Control Sample     : pH24

Normal Value     : 2.3           Acceptable Dev.[%] :    20.20
Std. Deviation   : .2            Variation Coeff.[%]:     8.70

Date   Value  Dev.   Name   -3s    -2s    -1s    Norm  +1s    +2s    +3s  St
                            1.70                 2.30                2.90

10/07  2.4     .1    Herb    :                  :   *              :  9
10/07  2.3     .0    Herb    :                  : *                :  9
10/07  2.3    0.0    Herb    :                  *                  :  9
10/07  2.2    -.1    Herb    :                * :                  :  9
10/07  2.1    -.2    Herb    :              *   :                  :  s
10/07  2.0    -.3    Herb    :           *      :                  :  s
10/08  2.1    -.2    Tom     :              *   :                  :  s
10/08  2.2    -.1    Tom     :                * :                  :  9
10/08  2.3    0.0    Tom     :                  *                  :  9
10/08  2.4     .1    Tom     :                  :   *              :  9
10/08  2.5     .2    Tom     :                  :      *           :  s
10/08  2.6     .3    Tom     :                  :         *        :  s
10/08  2.4     .1    Ed      :                  :   *              :  9
10/08  2.4     .1    Ed      :                  : *                :  9
10/08  2.3     .0    Ed      :                  :*                 :  9
10/08  2.3     .1    Ed      :                  :   *              :  9

Number of Samples : 16    Standard Deviation :    .2
Average Value     : 2.3   Variation Coeff.[%]: 6.77
```

Figure 19.9 Control sample statistics

To aid the user in tracking instrument performance and to ensure quality results, a program in Labsam provides a semi-graphic statistic (Figure 19.9) for every instrument running control samples. Statistical values for sequential analysis are calculated so that limits may be set. The mean value is set along with allowable limits; trends (user definable) can be recognized and results of the control samples outside the limits will be flagged.

The 'Name' column displays the name of the user responsible for each control sample measurement; this column may be suppressed if required. Each quality control list applies to a specific test and a specific control sample. Should a change of batch number occur within the selected time window, a new quality control list will be generated automatically. This ensures that each control sample will have an independent set of statistics.

19.7 Implementation

As has been shown, Labsam is extremely adaptable to the needs of an individual laboratory. The system may be configured by a user without the need for programming, but many hooks are available (user written evaluation program calls) for more specialized functions.

A successful implementation of the system starts with a careful survey of all laboratory functions to be included in the system. An 'Application Fit Analysis' may be carried out to ensure that Labsam can be configured to meet most of the laboratory's needs without programming. Once a good fit is ascertained, and the system purchased, an action plan should be defined. The need for user written programs will emerge during this analysis.

This plan should include: sample flows and sample identification needs, list and report formats, instrumentation interfacing requirements, sample volumes and numbers of tests, priorities for data and data searches, a layout of laboratory personnel, their responsibilities, capabilities, functions, and the required security. In addition, detail for every test (including the sample master data items) that must be captured by the system is required. Tests must be fully characterized for limits, units, special calculations, names and formats. Existing laboratory dialogues including abbreviations for tests, and sample master data must be obtained. Required report formats, including special dialogues, type and content of information, must be defined along with detail of which reports are needed where.

Once all the data are obtained, the system's physical layout must be defined and the user may begin an orderly, planned configuration of LABSAM. Users may begin to write programs or to specify program performance and to draw up program reference specifications.

19.8 Once Labsam is Running

Assuming that the implementation was a carefully planned and executed operation, there remains only the need to keep Labsam matched to the changing requirements of the laboratory. Changes in the system are possible during system operation and in general it is not necessary to shut the system down to perform additions it.

New test procedures, different sample types, new users, or changes in users' capability are a few examples of the many parameters which can be added or modified by those authorized, without interrupting a running system, and no reconfiguration of the database is required.

19.9 Labtools

It only remains to discuss the possibility for users to write their own programs to add specific functions to their system. There is a set of subroutines: Labtools, which provides the read/write mechanism for both the database and certain files. Labtools may be used, in conjunction with other software packages, eg. with a communication package for the transfer of data to and from other computers. Data may be transferred from Labsam and sample data may be received from the external computer and written to the Labsam database.

The user library is a collection of programs written using Labtools. The use of these procedures ensures that all user written programs remain compatible with all new releases and updates of Labsam.

19.10 Summary

This concludes the chapter on Labsam as it stands today. It is not an exhaustive list of all the features, it is intended rather to give the reader a general overview of the system and a feel for the degree of flexibility inherent in it.

Chapter 20

The VG Manager System

Bob Hillhouse and Michael Head

20.1 Concepts

The VG Manager LIMS is designed to take advantage of the benefits offered by the VAX computer architecture, manufactured by DEC. Some key features of the system are the modular and flexible applications software, distributed data acquisition and fast data access coupled with good data security. This is augmented by the use of an industry standard operating system and database.

VG Manager consists of five major software modules which can be operated as either stand alone systems or as a fully integrated LIMS: Sample Manager, Data Manager, Instrument Server, Multichrom and Bridge Communications.

Layered software can be installed and still allow the user all the usual system utilities that are available. The importance of this is that other vendors' software products can be used on the same computer without modification. Whilst the LIMS is operational, other users can concurrently use any other software application that runs under the installed operating system.

An advantage of the VMS operating system is that Sample Manager can run on the complete range of VAX computers which may be connected together using DECnet. PCs can be added to the network to provide work stations for individual users by using third party terminal emulation packages or via DECnet-DOS which allows an integrated approach to the connection of PCs to the DEC environment.

20.2 System Architecture

The overall system structure is shown in Figure 20.1

20.2.1 Chromatographic Data Acquisition

The present chromatographic data acquisition system, Multichrom, runs on a MicroVAX. It has been redesigned from an earlier version (Minichrom) that ran on a PDP 11 computer. Data are acquired by a Remote Interface (RIF) which contains the A/D circuitry which gives a one micro volt resolution in a one volt signal and is controlled by a PDP 11 processor; one RIF can capture the data from eight channels at 50Hz or 60Hz simultaneously. For the MicroVax to handle these data, a DEC KXT11 board also with a PDP 11 processor, is used to communicate with the RIF to accept and store data directly onto a computer disk.

A 16 channel chromatography system comprises two RIFs and one KXT11 board to support the interfaces. A maximum of 64 channels of data can be acquired by one MicroVAX with four KXT11 boards installed. The software is menu driven and is shown in Figure 20.2.

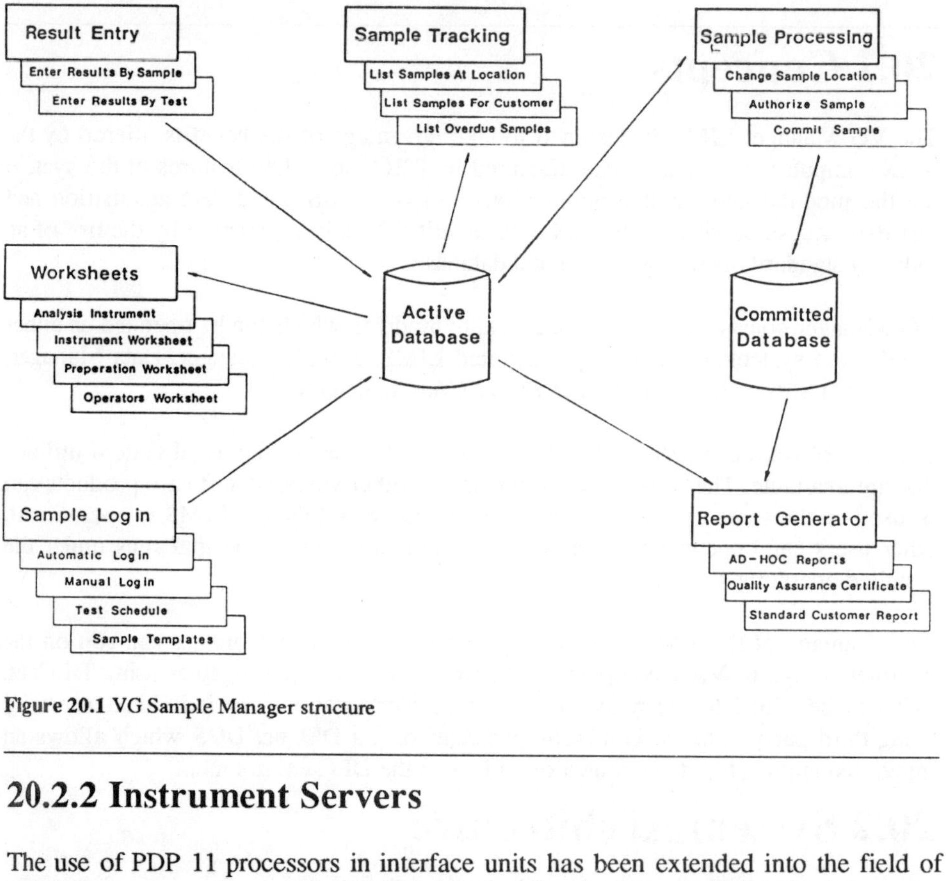

Figure 20.1 VG Sample Manager structure

20.2.2 Instrument Servers

The use of PDP 11 processors in interface units has been extended into the field of serial data capture. As used in a VG Instrument Server (Figure 20.3), data from up to eight serial input devices can be collected and transferred to the LIMS. Information in the form of a string or a file is acquired and sent to the VG Data Manager package residing on the MicroVAX. This package relates the sample ID to the result and stores the result in the relevant database.

20.3 An Overview of Sample Manager

The various software packages and hardware interface units, together with Sample Manager, form the basis of VG LIMS. Sample Manager has been designed to provide a sample management solution for a wide range of applications in production, process and research laboratories. Consequently, it is supplied as both a turnkey and a general package which may be customized using a number of application utilities.

Figure 20.2 Multichrom display with menu

Sample Manager has the ability to log samples into a database, assign tests, enter results and produce reports. The logging of samples into the database can be simplified by using a sample template, where these templates contain predefined information relating to the individual sample itself, or the sample type.

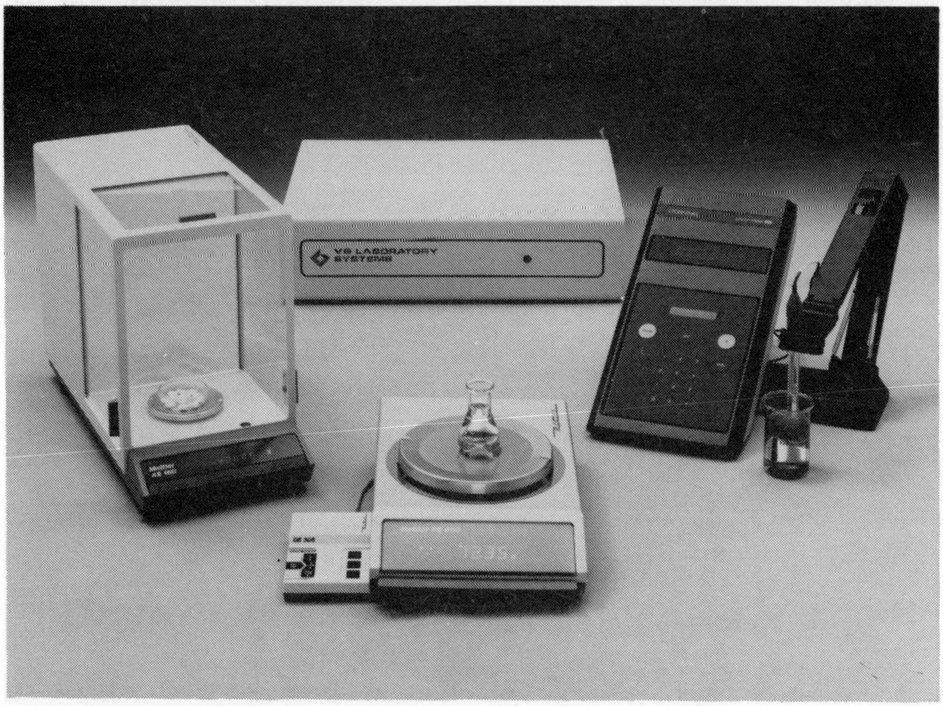

Figure 20.3 VG Instrument Server

Information can be pre-defined including the time when results will be expected for the sample, specified in days and hours; the type of report format, if a special report is required; a hazard, initial preparation and special instructions or comments; when the sample is logged in a project and job identification can be defined, together with sample name. To further increase the speed of entry a sample name can be pre-defined with an incremental value, hence subsequent sampling from a valve could have sample names Valveabc001, Valveabc002, Valveabc003; as this field is searchable a report for these samples can be obtained.

Bar code labels, which significantly reduce identification and transcription errors, can be used for single or job entry of samples, or for system commands. A new feature of Sample Manager is the use of Psion Organiser portable terminals, which provide a low cost solution for data display in the laboratory. A Psion and bar code reader (Figure 20.4) may be attached to the Instrument Server for each analytical instrument. Additionally, there is remote data entry into a Psion, a hand-held and portable device, which allows information to be gathered off-line and then transferred to the VAX when convenient.

Figure 20.4 A Psion hand held terminal with a bar code reader

Organizations can be grouped into departments, laboratories and process areas, for example, and database information can be seen through 'windows' provided by these groups. A user in one laboratory will only see his methods when browsing with his identifier, but another may see all methods within the organization. Sample Manager can be supplied with a facility to stop different groups accessing different parts of the database. This feature is important where more than one department is using the system, or in applications where it is not desirable that analysts can access different test information on the same sample.

20.4 The Database and Software Structure

An efficient relational database cannot be purchased: it requires careful design. None of the current relational database products fully conform to the relational model. Indeed, the relational model can be implemented in a totally satisfactory manner using DEC RMS indexed files, and the real issues are related to creating a relational model for mapping onto a suitable file structure. In this section, a description is given of the way in which VG has solved this problem for the laboratory information environment.

Sample Manager has been designed using the standard DEC RMS indexed file structure, a major advantage of which is the easy integration of other software packages into the LIMS, eg. RS/E and All-in-One. Two report writers are provided with a standard system: a tabular report generator for fast efficient data retrieval and a 4GL SQL-like report language. Alternatively, if the purchasing company's computer department has a knowledge of Datatrieve, then this can be used simultaneously whilst laboratory users are operating the other VG report facilities.

A laboratory performs work on two types of samples: active and completed. Active samples are those undergoing data entry and testing, and completed samples are those for which all testing is completed, approved and reported. To provide the best possible response for the user, the database has been split into two parts: active and committed, see Figure 20.5. The organization of the database allows searching of each separately or both together. Daily routine operations such as sample log-in, test result entry, validation and approval can be performed quickly on the active database.

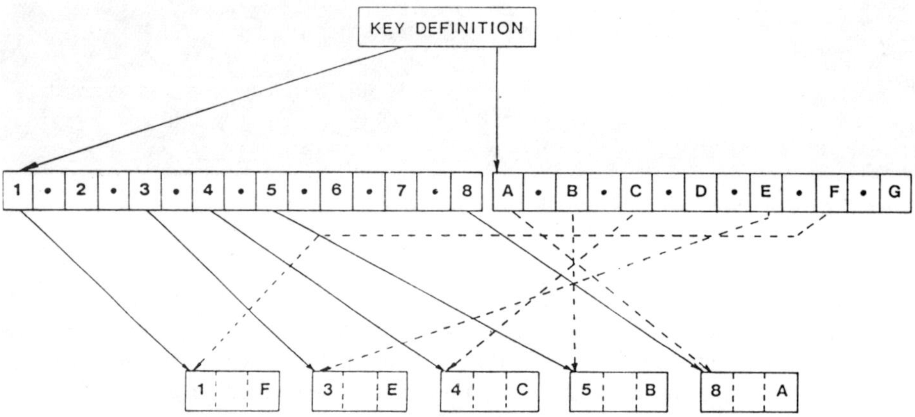

Figure 20.5 Organization of the System Manager database

An audit trail option is provided that records all modifications made to the database, this facility requires additional software overheads which usually require additional CPU capability and fast disk access. Again, extensive use is made of standard VMS features to provide both audit trail and journaling facilities within Sample Manager.

In applications where database integrity is of the utmost importance the database disk may be shadowed by a second disk by taking advantage of the VMS shadowing software.

By accessing RMS files directly, Sample Manager avoids the overheads of using additional software. Information is retrieved from indexed RMS fields using sequential and keyed access modes with variable length records. The file structure and access methods have been designed for the optimum storing, retrieving and manipulation of information in the laboratory environment.

Browse provides instant access to database entries, which eliminates the need to refer to hardcopy printouts of libraries. Browse has been implemented throughout the LIMS and is activated by placing the cursor on any system parameter and pressing the 'find' key. If this is pressed once, the information is browsed at the bottom of the screen by continually pressing the spacebar. Alternatively, pages of information may be obtained by pressing the 'find' key a second time and using the 'next' and 'previous' page keys. An example of the Browse command is given in Figure 20.6. Search criteria may be used as the beginning of a Browse search routine.

VG Sample Manager **Version 2.**

Option > MIL

Instrument Definition Menu

HPLC01	High performance liquid chromatography
HPLC02	isocratic hplc system
HPLC1	isocratic hplc system
HPLC2	gradient hplc
HPLCPHARMA	gradient hplc
HPLC_PLAS	High performance liquid chromatography
ICCH26	Dionex Ion Chromatograph
INFRARED	V.G. Infrared Spectrophotometer
IONCHROM	Dionex Ion Chromatograph
IR_PLASTIC	V.G. Infrared Spectrophotometer
JFTOT	Jet Fuel Thermal Oxidation Tester
KARLFISHER	Metrohm 547 Karl Fisher Titrator
KEROLAMP	Kerosene Burning Lamp
KEVEX0700	Kevex 0700 Energy Dispersive X-Ray
KF	Fisher ASTM Colorimeter
KT7BAL	Mettler KT-7 Balance
LECOS	LECO Sulfur Analyzer
MANUAL	Manual Test Method

Figure 20.6 Example of a browse on an instrument database

In this way, the valid entries for the fields in a record can be displayed easily and transcription errors can be avoided. Browse can be used to select and group together similar samples requiring processing and reporting and, if required, it can act as an on-line help facility by displaying the options available.

Additional sample information may be added to the database structure using simple on-screen editors. In many cases, the organization will change its requirements from one year to the next, or indeed, between the date that a system is ordered and the date the system is installed. To simplify the addition of new sample information various set-up templates are available as shown in Table 20.1. Alternatively, any further information can be added to the system, either numerical, date or text. The provision of the sample set-up templates allows the new data to be integrated into the system with the use of the Browse facility. During log-in the operator is asked for the normal entries and then the additional information from the template.

Of special interest is the field type IDENTITY which allows the creation of known values for a particular class of sample information. For example, a new field called COLOUR of type IDENTITY is created and defined in a sub-template. When a value is requested the operator can enter a value (red, green, blue, etc.). If this does not currently exist as a value for this field, it can be added to the allowed list. When subsequent samples are logged in the operator can use Browse and the correct list of allowed values is displayed. Each piece of information is stored against the sample ID, with keys allowing the following searches:

all samples with colour = red
all samples with colour = any (for samples which colour was requested).

Table 20.1

Sample information templates

Analysis	Sample
Boolean	Sampling point
Customer	Syntax
Date/Time	Templates
Group link	Test schedule
Hazard	Text file string (4 chr)
Identity	Text file string (8 chr)
Instrument	Text file string (20 chr)
Location	Text file string (30 chr)
Numeric (integer)	Text file string (40 chr)
Numeric (real)	Text file string (44 chr)
Personnel	Text file string (50 chr)
Preparation	Text file string (60 chr)
Product	Text file string (234 chr)
Project	Time interval
Report format	Time offset

20.5 System Features

Sample Manager provides a flexible working environment for the analytical chemist in production, process and research environments. System commands can either be entered using computer mnemonics or using a menu structure. Each system manager or user, if allowed, can optionally customize a menu structure using quick on-screen editors. The descriptive text, computer mnemonic, position of each function on the screen and the position of a function on a particular menu page, can all be selected in the customization process. If the complete menu structure requires changing then the text message file may be edited. The VG data acquisition and information systems are presently supported in English, German and French.

When a new user is allowed access to the system, the system manager uses the same menu procedure to select a matrix of allowable operations and groups, before issuing the new user with a User Name and Password.

20.5.1 Sample Parameters

The organization of the database allows samples to be grouped together in various ways. Projects (or studies) allow the operator to set up and associate sample analysis with a specific study. For example, a customer may be involved in the development of a new product where samples are periodically submitted for analysis; the operator can ensure that all information relating to this product is stored in the database under a specific project name, see Figure 20.7.

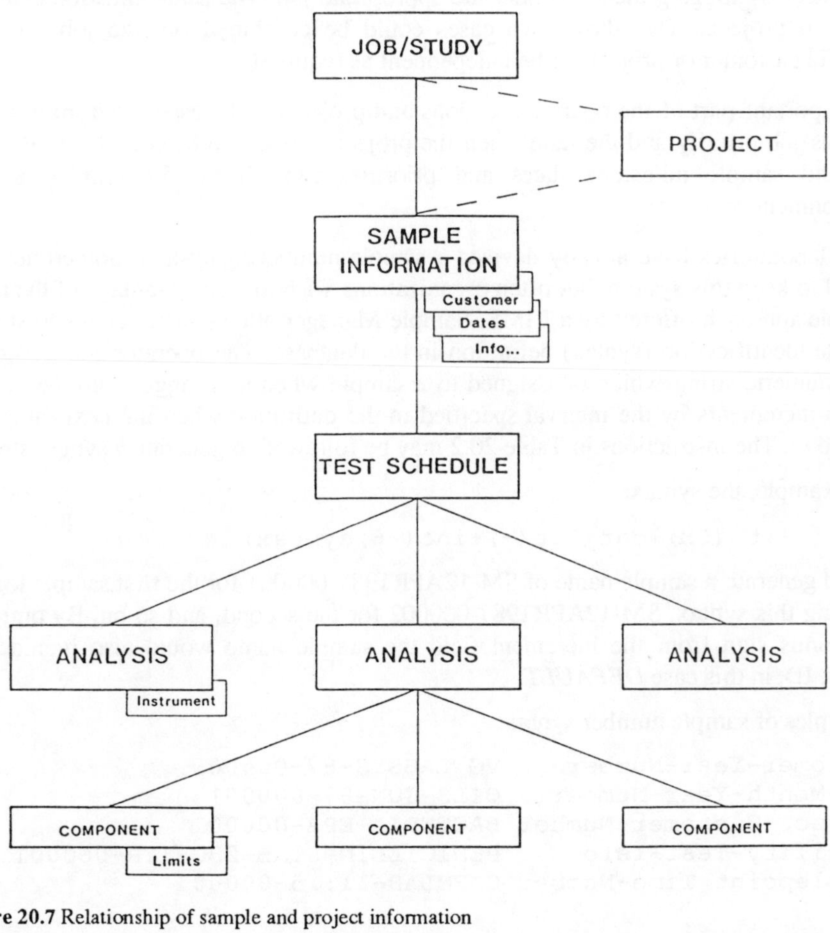

Figure 20.7 Relationship of sample and project information

In many situations, a sample contains a number of different aliquots which all require log-in, result entry, validation and reporting in a simple and efficient manner. The system allows samples to be treated as sub-sample sets of information. For instance, a laboratory may be analyzing batches of oranges from different sources. An orange is selected from each batch, separated into the various constituents and tests are carried out. One such test may be HPLC, where the column elutent corresponding to a peak may be further analyzed using techniques such as infra-red or mass spectrometry. Other examples are the kerosene constituents of a crude oil which only appear after the oil has been processed in the refinery, and products that are produced from an initial raw material in a process plant. Different sample templates can be set up which allow quick and easy log-in and entry of data for this type of situation.

Samples can be organized into jobs. This can be a series of related samples arriving at the laboratory together and logged into the Sample Manager as one job. Alternatively, the operator could create a job that is dedicated to a particular customer, so that each time samples arrive from the customer, the operator can segregate them from all of the others by logging them in under the appropriate job. The same situation can also apply to projects. The above two cases could be combined (ie. the job serves a specific customer or project) or be independent as required.

An important part of the operation of jobs and projects is the associated information such as job priority and the date when the project tests are to be completed. Project, job and sample milestone dates and priorities can all be different in a real environment.

Most laboratories have already devised a sample numbering system. Sometimes it is useful to keep this system, but often organizations wish to take advantage of the more flexible approach offered by a LIMS. Sample Manager allows the operator to store a sample identification (syntax) definition in the database. The operator can design an alphanumeric string which is assigned to a sample when it is logged into the system which increments by the interval specified in the definition when the next sample is logged in. The instructions in Table 20.2 may be followed to generate a syntax string.

For example, the syntax:

```
lit (SM)+dat+lit(-)+inc(-6,syntax)
```

would generate a sample name of SM-12APR1987-000001 for the first sample logged in using this syntax, SM-12APR1987-000002 for the second, and so on. By omitting the minus sign from the increment field the sample name would also include the syntax ID, in this case *DEFAULT*.

Examples of sample number syntax:

```
Customer-Year-Number       VG LABSYS-87-000001
Job-Month-Year-Number      OILS-JUN-87-000001
Project-Customer-Number    BACTERIA-EPA-000001
Stability-Test-Info        PENICILLIN-STAB-23C-1YR-000001
Samplepoint-Time-Number    CHEMLAB-11:15-000001
```

Table 20.2

Numbering system syntax

Full Name	Abbreviation	Arguments	Comments
Literal	LIT	(text)	text inserted in name
Full date	DAT		DDMMMYYYY insert
Year	Y	(2)	YY inserted into name
		or (4)	YYYY inserted
		or (1)	Y
Month	M	(3)	MMM inserted (text)
		or (2)	MM inserted (numeric)
Day	D		DD inserted
Hours	HRS		hh inserted into name
Minutes	MIN		mm inserted
Day of week	DOW		eg., Wednesday
Samplepoint	SPT		From sample record
JobID	JOB		From sample record
Customer ID	CID		From sample record
Project	PRJ		From sample record
Product	PRD		From sample record
Incremented number	INC	(n,director)	numerical field of
		or(n-,director)	of width n

Increment Directors

Customer	-	increments for new samples against each customer
Prodspec	-	increments for new samples against each product
Project	-	increments for new samples against each project
Samplepoint	-	increments for new samples against sampling point
Syntax	-	increments each time this syntax is used

20.5.2 Sample Information

Like a child learning, a computer will only become knowledgeable when it has been taught about a given laboratory environment. Library utilities are provided that allow this environment to be defined during the initial stage of an installation. These same utilities are available on a day to day basis as the system begins its regular routine use.

Laboratory table information describes the contents and work of the laboratory. This information is, in effect, a reference library for the laboratory. Twelve database tables are used to store laboratory information, additional tables may be using the system utilities. The names of the standard tables are shown in Table 20.3.

Table 20.3

Names of the standard laboratory tables

Table		Description
Customer	-	customers or sample sources
Location	-	locations around the laboratory
Preparation	-	sample preparation methods
Sample point	-	origin of sample
Hazard	-	hazards or special handling methods
Format	-	reports
Personnel	-	laboratory or managerial staff
Instrument	-	laboratory instruments used for analysis
Syntax	-	sample identity definition
Projects	-	on-going studies
Analysis	-	Analytical operations or methods
Prod spec	-	product specifications

20.5.3 Test Schedules

A test schedule is a list of analyses which may be carried out on a sample, together with the instruments required; these are used to speed up the assignment of analyses to samples. A schedule consists of a number of pre-defined entries each representing an analytical method. During sample log-in, if a test schedule is specified, the operator may assign the required tests to the sample from those listed, or remove entries that have been assigned as standard tests, if inappropriate. Selection of the various tests is provided by cursor control within a simple user friendly interface (Figure 20.8).

```
VG Sample Manager                                   Version 2.0
              Edit screen for tests assigned to sample

Sample name. . . . . :  SM-22MAY1987-000018
Schedule ID. . . . . :  BACTERIA
Schedule Description :  Coliform and Nitrate Tests on Potable Water

Analysis ID  Instrmnt ID  Status       Analysis ID  Instrmnt ID  Status
COLIFORMMF   INCUBATE35   V            NO2_NO3      DIONEX_IC    V
NO_NO2_AIR   PE597        V            PAH          VGTRIO2      V
PAH_GCMS     VG7070       V

Analysis ID  Instrmnt ID  Std. test    Analysis ID  Instrmnt ID  Std. test
COLIFORMMF   INCUBATE35   Yes          NO2_NO3      DIONEX_IC    Yes
NO_NO2_AIR   PE597        No           ODOR         NONE         Yes
PAH          VGTRIO2      No           PAH_GCMS     VG7070       No
PCDD_DF_HR   VG7070       No

Press DO to save edits on this page, EXIT to abandon page.
```

Figure 20.8 A Test Schedule

An interesting aspect of non-standard tests is the ability of the system to schedule tests which have yet to be assigned. In many situations, the analyst knows that a number of tests can only be specified after the results of previous tests. The use of unassigned tests allows flexibility within the system and permits laboratory workloads to be realistically evaluated.

Sample Manager can also be driven from the result entry of the laboratory testing procedure rather than the sample log-in. For example, a sample can be measured at a pH station and other test and sample information can be entered at a later date.

Both test limits and product or legal specifications can be entered and used within the system. Product specifications can be assigned to groups so that in the case of the oil industry, for example, different grades of kerosene may be grouped under a particular specification, whilst in the pharmaceutical industry, different specifications may be applied to a given drug preparation by different countries.

20.5.4 Sample Registration

Samples can be registered on the system in a number of different ways. Single, batch and multiple samples can all be registered in a simple manner with the minimum of keyboard entry. Templates can be set up which hold common data and test schedules which contain both the standard and non-standard tests that are to be assigned to each sample type. In cases where each sample requires a different data entry, the information can be selected by using the cursor during a sample template modification procedure, before each one is logged into the system.

```
VG Sample Manager                                    Version 2.0

                      Sample request login screen
Request id . . . . :  BACTERIA
Description  . . . :  Basic Job Header For Bacteria

Sample No.              Sample Name               Test Schedule  Description
       603  _AITKIN_FE-87-000001                  BACTERIA       12 University Avenue
       604  _ALEXA_DIES-87-000001                 BACTERIA       67 Snelling Road MN
       605  _ALFA_LAVAL-87-000001                 BACTERIA       10 Pascal Road MN 10
       606  _BADGER_CON-87-000001                 BACTERIA       43 Keys Road MN12738
       607  _BCED-87-000001                       BACTERIA       100 Albert Road MN 1
       608  _BFI-87-000001                        BACTERIA       561 Sutherland Avenu
       609  _CALEDONIA-87-000001                  BACTERIA       102 Southhampton Way
       610  _DAHLASC-87-000001                    BACTERIA       Wellington House Bro
       615  _ERA-BOHL-87-000021                   BACTERIA       12 Washington Drive
       620  _BADGER_CON-87-000002                 BACTERIA       43 Keys Road MN12738
       626  _ERA-BOHL-87-000023                   BACTERIA       42 Clarinda Road MN1

Template ID  Repeat count                   More
__________
```

Figure 20.9 Sample registration

Automatic registration of samples can be initiated using the VG Watch-dog Timer module which can be programmed to routinely register samples, for example, at various sampling points on a process plant or oil refinery.

20.5.5 Validation

Flags are used as a means of rationalizing the current status of a sample, result and test. These flags are in the form of a single character which represents a particular status or condition and are stored and updated in the appropriate record. Sample Manager validation procedures have been designed to cope with the stringent pharmaceutical environment as well as the general laboratory.

In pharmaceutical laboratories, the validation procedures can be implemented at various stages of the process, whereas, in the general testing laboratory, the system can be configured so that only the final report needs to be authorized. Authorization can occur in a number of steps with one user, say a supervisor, who is able to authorize all the results and another user, say a laboratory manager, authorizing the sample itself. This procedure may be too restrictive for some applications; the system can also be set up so that a sample is authorized when there are outstanding tests, which for some reason will not be done. In this case an authorization text string will be requested. Once authorized, the sample can then be committed, ie. transferred from the active to the completed database.

```
VG Sample Manager                                        Version 2.0

Option > $TAB
                        Tabular Report Generation
-----------------------Sample record selection fields------------------------
   Columns selected :          1  Total column width :          32

   Job Name
   Project Id
   Product Name                         Use cursor keys to move to item
   Sampling Point                          INSERT to add item to table
   Preparation Id                          REMOVE to delete item from table
   Location                                SELECT on upper column to change
   Batch Name                                     sample selection criteria
   Login Date
   Date Results Reqd

   Numeric Id              Recieved Date              Hazard Id
( 1)Text id                Date to Start Tests        Priority
   Status                  Date Results Availabl      Container
   Suspended               Login by                   Customer
   Cancelled               Sample Type                Report Id
   Sampled Date            Sample Name                Test Schedule
```

Figure 20.10 Tabular report writer - command page

To produce reports from Sample Manager, there are two reporting mechanisms. The first is the Tabular Report Writer that enables users to make selections from a number of criteria such as: job, project, log-in dates, costomers, etc. Using the cursor along with the powerful Browse feature allows the possible choices to be displayed interactively as the search criteria is developed as shown in Figures 20.10 and 20.11.

```
VG Sample Manager                                    Version 2.0

Option > $TAB
                         Tabular Report Generation
 ----------------------Sample record selection fields--------------------------
      Columns selected :            2  Total column width :          30

      Job Name                : BACTERIA
      Project Id
      Product Name
      Sampling Point
      Preparation Id
      Location
      Batch Name
      Login Date              :  1-JAN-1987 00:00 - 21-JUL-1987 00:00
      Date Results Reqd

 ( 1)Numeric Id             ( 2)Recieved Date             Hazard Id
      Text id                   Date to Start Tests       Priority
      Status                    Date Results Availabl     Container
      Suspended                 Login by                  Customer
      Cancelled                 Sample Type               Report Id
      Sampled Date              Sample Name               Test Schedule
```

Figure 20.11 Tabular report writer - set-up

The database is then searched and all the tests available for selected samples are displayed on the screen, see Figure 20.12. The user can select which of these tests will have results reported, as well as the order in which they appear. Finally, the report is formatted as presented in Figure 20.13 and a choice of output location is given, either printer, display, test file or VAX mail box. Many users take advantage of the VAX mail box and security system, and allow customers to dial the computer via a modem to collect their analytical reports.

```
VG Sample Manager                                   Version 2.0

Option > $TAB
                         Tabular Report Generation
  Test          Components                  Total column width :     127
COLIFORMMF      ( 3)Coliform Bacteria
NO2_NO3         ( 4)Nitrate + Nitrite (a
NOX             ( 5)Oxides of nitrogen
NO_NO2_AIR      ( 6)Nitrate & Nitrite     ( 7)comments
                ( 8)Within FDA limits?       colour of vapour
ODOR            ( 9)Threshold Odor
```

Figure 20.12 Tabular report writer - result display

VG Sample Manager Version 2.0

Sample information selected by:-
Job Name is BACTERIA
Login Date from 1-JAN-1987 00:00 to 21-JUL-1987 00:00

20-JUL-1987 10:55

Numeric Id	Recieved Date	COLIFORMMF	NO2_NO3	NOX	NO_NO2_AIR			ODOR
		Coliform B	Nitrate +	Oxides of	Nitrate &	comments	Within FDA limits?	Threshold
168	23-MAY-1987 13:45	100	10	20	35	Solution was cloudy	yes	5
481	1-JUN-1987 10:05	100	100	23	27		yes	6
584	5-JUN-1987 09:54	100	254	24	29		yes	6.5
585	5-JUN-1987 10:00	2300	124	27	31		yes	6
586	5-JUN-1987 10:05	10000	20	29	30		yes	5.5
587	5-JUN-1987 10:10	0100	100	35	32		yes	6.1
588	5-JUN-1987 10:48	100.00	100	37	23	Low nitrate level	yes	6.3
589	5-JUN-1987 11:35	100	100	40	24		yes	7.0
590	5-JUN-1987 11:36	1000	100	39	30		yes	7.2

Figure 20.13 Tabular report writer - output

The second report writer provides a 4GL which allows a report to be formatted to an exact specification, as shown in Figure 20.14. Extensive facilities provide for the following features: sideways reports, the inclusion of ASCII text files, and graphical representation of the results. The header and footer of the report can be easily defined and there are word processing features available. Interactive Browse is also a feature of this report writer.

Organic Analysis Data Sheet

Sample Number

Laboratory Name : Pacific Analysis Inc.	**Case No** : AD-76
Lab Sample ID No : 2730	QC **Report No** : 1
Sample Matrix : Soil	**Contract No** : 99-99-99 S1
Date Release Authorized by : ____________	**Date Sample Recieved** : 3-OCT-1986

VOLATILE COMPOUNDS

Concentration : Low _x_	Medium ___
Date Extracted :	3-OCT-1986
Date Analyzed :	3-OCT-1986
Conc/Dil Factor : 1	pH :
Percent Moisture: (Not Decanted)	

C.A.S. Number		ug/kg	C.A.S. Number		ug/kg
74-87-3	Chloromethane	125.0	78-87-5	1,2-Dichloropropane	45.6
74-83-9	Bromomethane	223.0	10061-02-6	trans-1,3-Dichloropropene	30.1 u
75-01-4	Vinyl Chloride	60.1 u	79-01-6	Trichloroethene	30.1 u
75-00-3	Chloroethane	60.1 u	124-48-1	Dibromochloromethane	30.1 u
75-09-2	Methylene Chloride	154.2	79-00-5	1,1,2-Trichloroethane	30.1 u
67-64-1	Acetone	98.2	71-43-2	Benzene	45.6
75-15-0	Carbon Disulfide	60.1 u	10061-01-5	cis-1,3-Dichloropropene	30.1 u
75-35-4	1,1-Dichloroethene	30.1 u	110-75-8	2-Chloroethyl Vinyl Ether	60.1 u
75-34-3	1,1-Dichloroethane	265.3	75-25-2	Bromoform	30.1 u
156-60-5	trans-1,2-Dichloroethene	45.2	108-10-1	4-Methyl-2-pentanone	30.1 u
67-66-3	Chloroform	30.1 u	591-78-6	2-Hexanone	30.1 u
107-06-2	1,2-Dichloroethane	30.1 u	127-18-4	Tetrachloroethene	30.1 u
78-93-3	2-Butanone	30.1 u	79-34-5	1,1,2,2-Tetrachloroethane	30.1 u
71-55-6	1,1,1-Trichloroethane	30.1 u	108-88-3	Toluene	30.1 u
56-23-5	Carbon Tetrachloride	4.5 j	108-90-7	Chlorobenzene	30.1 u
108-05-4	Vinyl Acetate	60.1 u	100-41-4	Ethyl Benzene	30.1 u
75-27-4	Bromodichloromethane	30.1 u	100-42-5	Styrene	30.1 u
				Total Xylenes	30.1 u

Data Reporting Qualifiers

U - Compound was analyzed for but not detected. Reported value is the minimum detection limit for the sample.
J - Reported concentration is estimated based on a 1:1 response or is between 0 and the Detection Limit.
C - Pesticide identification confirmed by GC/MS.
B - Analyte was also detected in the blank so reported value may be due to contamination.

Form 1 - Page 1 / 3-OCT-1986

Figure 20.14 An example of output from the 4GL report writer

Sample labels can be customized using the 4GL report writer and bar code labels may be included on the label using simple commands.

20.6. Application Areas

20.6.1 A Physical Testing Laboratory

In physical testing, the user has the problem that all samples are unique and might range from the contents of a refrigerator coolant to the constituents of a plastic swimming pool. The incorporation of user-definable input screens, not only for result entry but also for sample log-in, enables the user to enter relevant information quickly. As the fields in the user definable log-in screens are searchable, information relating to earlier samples of the same type can be retrieved with the minimum of delay. There is no limit to the number of pages that can be defined for sample log-in.

20.6.2 Environmental Protection

An area of great interest to many people is monitoring the levels of pollutants, pesticides and herbicides in environmental samples. The Environmental Protection

Agency (EPA) is the government body that monitors pollution in the USA and laboratories either because they manufacture herbicides and pesticides or because they are involved in contract work for the EPA, need the ability to produce reports to the EPA standard. The format of the report is very precise and to obtain the maximum benefit of a LIMS package, reports must be written to this standard. To this end the Report Writer in Sample Manager is able to faithfully reproduce the EPA forms. In addition, when the sample is logged in the specific EPA form is identified in the template.

20.6.3 Oil Refineries

In the typical application of a tanker transporting crude oil to a refinery, the crude being cracked and the subsequent final products such as jet fuel or plastic polymer shipped. The complete process is continuous. Thus, the concept of a sample entering the process is misleading, instead the introduction of sampling points has evolved. These are at all the stages from the tanker, unloading crude at the jetty, through the various cracking processes to the final product.

To ensure that a production process is being maintained, whenever a result is entered manually the last five or so results can be displayed. This quick manual check ensures that if a process starts to go out of control it is noticed and acted upon quickly. The final product is then checked against a product specification, and if found to fail, can either be re-routed back into the refinery or used for another product. As refineries can process millions of pounds of crude oil daily, the essential parameters of speed and accuracy are paramount in ensuring efficient plant operation.

VG have systems of this type installed, and communication of analytical results to the process computer is also part of the installation. Again, the report writer specifies the format of the information being transferred.

20.6.4 Pharmaceutical Industry

Due to the pressure of the regulatory authorities, the pharmaceutical application is unique in that it requires a full audit trail. This audit trail monitors any changes to the information in the database by storing the original value, the new value, the date and time the change was made and the operator ID. As each instrument entry in the database has calibration and service information, a calibration standard can be automatically logged on every day, week, month or year, to be run as part of a test sequence. Also within the Instrument Entry there are as many pages as required for freeform text to define the Standard Operating Procedure (SOP) for that instrument. The same is true for all analyses. In this way, whenever a worksheet is produced, a full SOP can also be printed out to ensure that correct procedures are followed.

One important aspect of any pharmaceutical quality control workload is that of stability testing. The use of the background timer in Sample Manager allows the full five or seven years of a stability program to be defined with the various sampling dates and intervals. On a sampling date, Sample Manager will automatically log-in the sample, produce a worksheet, the number of bar code labels required and describe the physical location from where the sample should be taken.

20.6.5 General Production Installation

The background or Watch-dog Timer can be used for routine applications such as sample log-in, worksheet generation and automatic disk back-up even while data are being put to the disk. This ensures that many routine tasks occur in the background with no operator intervention. As the timer is activated by VMS there is no need to reset or restart the timer if ever the VAX is turned off.

As a production system is more likely to have automatic data collection, the display of the last entered results may not be convenient. The inclusion of automatic results into a Shewart Quality Control Chart ensures the system maintains its own checks on the quality of data.

20.7 Summary

VG Sample Manager and Multichrom are both stand alone packages and together with the full VG Manager LIMS, which includes all the modules described, are installed in a number of different application areas in Europe and North America. The products described are available exclusively for the VAX range of computers with the VAX/VMS operating system.

Chapter 21

LIMS on a PC

Harmon W.Brown and Gloria Lawler

21.1 Introduction

Typical laboratory information management systems run on a multi-user computer such as a VAX and cost upwards of $100,000. These are designed to keep track of data for tens of thousands of samples. There is also a need for a simpler system that might keep track of a few thousand samples, and cost an order of magnitude less. Nelson Analytical has developed such a product called RLAB that runs on the IBM PC class of computers. It is designed for the smaller laboratory, or a section of a larger department that wants to maintain its own database.

One of the advantages of using a PC is the wealth of software that is available. There are several excellent database packages, one such is R:base System V from MicroRim. Their product provides the programmer with a very powerful toolkit with which to develop an application. Nelson Analytical has taken R:base and created a comprehensive LIMS product that will allow the chemist to keep track of chromatographic and other kinds of data. It also has other features such as the ability to prepare worklists, backlog reports, and a variety of types of information.

21.2 What Is RLAB?

RLAB, the Nelson Analytical database product, is a powerful laboratory information system (LIMS) designed especially for the analytical chemistry environment. It enables the analyst to track the progress of laboratory testing from the time a sample is logged into the laboratory until its testing is complete. Although RLAB offers features that apply specifically to chromatography, it can be used in any analytical laboratory.

RLAB runs on MS-DOS-based personal computers (PCs) equipped with a hard disk and at least 640K bytes of memory. The system can be operated on a single PC, or several PCs can be linked in a network to create a multi-user system.

The database allows storage, retrieval, and reporting of information typically generated in analytical chemistry laboratories: background information about samples, information on the tests used to analyse them, and the test results themselves. In addition, RLAB allows laboratory managers to create job lists for analysts and to approve test results.

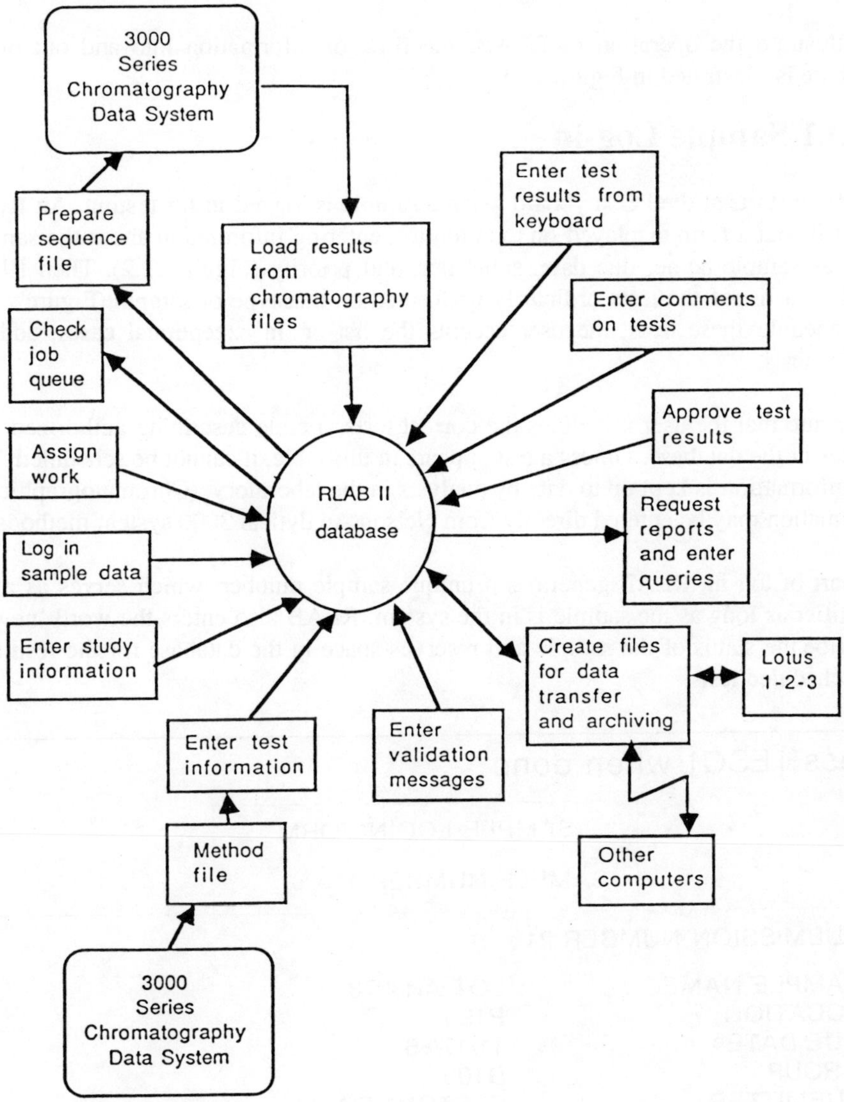

Figure 21.1 The flow of information in the RLAB LIMS

RLAB also communicates with other systems developed by Nelson Analytical and by other software manufacturers. For example, it writes sequence files that can be used by the Nelson 3000 Series Chromatography Data System. Conversely, it extracts information from files created by the 3000 system. RLAB also writes files that can be read into Lotus 1-2-3 for spreadsheet and graphic analysis.

21.3 How RLAB Works

To illustrate the operation of RLAB, the flow of information into and out of the database is illustrated in Figure 21.1.

21.3.1 Sample Log-in

The flow starts at the PC keyboard when a sample is logged in for testing. An RLAB user fills out a form displayed on the monitor, entering information about the sample, such as sample name, due date, submitter, and priority (Figure 21.2). Then RLAB displays a list of the tests ordinarily performed on that type of sample (Figure 21.3). To schedule these tests, the user accepts the list or, in exceptional cases, adds or deletes tests.

To ensure that the user schedules the correct tests, a table describing authorized tests resides in the database. Unless a test appears in this table, it cannot be scheduled. This test information is kept up to date by analysts in the laboratory. (Chromatographic test information may be entered directly from Nelson Analytical 3000 system methods.)

As part of log-in, RLAB generates a unique sample number, which serves as a key identifier as long as the sample is in the system. RLAB also enters the word 'new' to describe the status of the sample and reserves space in the database for the results of the scheduled tests.

Press [ESC] when done

SAMPLE LOGIN FORM

SAMPLE NUMBER 113

SUBMISSION NUMBER 21

SAMPLE NAME	LOT AR3973
LOCATION	P195
DUE DATE	11/17/86
GROUP	D100
SUBMITTER	R. STONNER
DEPARTMENT	QA
PROJECT	LX45
SOURCE OF SAMPLE	VAT 144
PRIORITY	3
COMMENTS	2 OF 50 CAPSULES BROKEN
SAMPLE TYPE	MULTI

Figure 21.2. Form used for logging samples

Delete	Add new	Previous	Next

TEST NAME	ANALYST
LC1	JUDY
CATION1	DOUG
PH	LINDA
LC2	JOHN

ANALYTE	UNITS
NA	PPM
K	PPM
CA	PPM
MG	PPM

Figure 21.3. Form used for scheduling tests

For convenience, a predefined group of samples that make up a 'study' can be logged into the system in a single step. Information about the testing is stored ahead of time in the database, see Table 21.1. Log-in is then virtually automatic. Entry of the study name and a few additional facts triggers RLAB to read the information and log the entire group of samples. Figure 21.4 shows the entry form for studies.

Table 21.1
Static tables in the RLAB database

Table Name	Description of Contents
TESTS	Stores information on the tests performed in the laboratory. Each row in this table contains a unique test name. Columns are set up to hold related information such as instrument name, settings, and the name of the analyst who normally performs the test. The table also contains columns to hold chromatography parameters such as injection volume and internal standard amount. These columns can be filled automatically with parameters transferred from a chromatography method.
TESTCOMP	Stores the names of the analytes determined by the tests listed in the TESTS table. Each row contains a unique test-and-analyte combination. Entries in a column named Samptype define which tests are used on different sample types. In addition to Testname, Analyte, and Samtype columns, the table includes result specifications for each analyte: high and low detection limits, physical test limits, and logical test limits. A Flag column contains integers that invoke a validation message whenever a result falls outside these limits. Validation messages are defined in the VALIDS table.
TESTSTUDY	Stores all the information necessary to log-in a study. A study consists of one or more samples, each subjected to the same or different tests at certain intervals. All the samples are logged and the tests scheduled in one step. The table columns hold the study name, sample names, sample types, test names, and replicate numbers to identify repeated tests on a given sample. An Interval column holds values that specify, for each test, the number of days between log-in and testing.
VALIDS	Holds validation messages associated with 'flags' on test results. When a result falls outside the limits set for a given analyte, RLAB flags it with a number specified in the TESTCOMP table. The purpose of the VALIDS table is to store definitions for each of these flag numbers. RLAB uses these definitions as message in reports containing the flagged result.

```
Press [ESC] when done

        STUDY LOGIN
        STUDY NUMBER 3
        STUDY NAME     water1

DUEDATE
SUBMITTER
SOURCE
PRIORITY
COMMENTS
```

Figure 21.4 Form used for logging studies

At the end of the log-in procedure, RLAB has the options to print log papers (Figure 21.5), bottle labels (Figure 21.6), and analysts' job lists (Figure 21.7).

```
      SAMPLE NUMBER:          123

SAMPLE NAME:  LOT AR397    PRIORITY:  3
CURRENT LOCATION: P195
LOG DATE: 10/27/86      LOG TIME 10:35          DUE DATE: 11/17/86

PROJECT: LX45     STUDY NUMBER:          SUBMISSION NUMBER:  4

SAMPLE TYPE: MULTI      GROUP:
SOURCE OF SAMPLE: VAT 144

COMMENTS:

                 TEST NAME            TEST DATE
                 ---------            ---------
                 FSHPLC               10/27/86
                 COLORIMETRIC         10/27/86
                 WSHPLC               10/27/86
```

Figure 21.5 Log paper for a sample

```
Sample Number:              123 Logdate: 10/27/86
Sample Name: LOT AR397
Submission Number:                4
Study Number:
Comments: NONE
Source: VAT 144
Submitter: R. STONNER
Project: LX45               DUEDATE: 11/17/86
Type: MULTI
```

Figure 21.6 Bottle label for a sample

JOB LIST FOR MOLLY TRAVIS

12/01/86

	SAMPLE NUMBER	TEST NAME	PRIORITY
TEST DATE: 11/18/86	109	WEIGHT	3
TEST DATE: 11/19/86	110	FSHPLC	2
TEST DATE: 11/18/86	22	FSHPLC	2
TEST DATE: 12/18/86	94	WSHPLC	2
TEST DATE: 11/17/86	6	FSHPLC	3
	8	FSHPLC	2
TEST DATE: 12/18/86	65	AAS	2
TEST DATE: 11/17/86	9	FSHPLC	2
	10	DISSOLUTION	2
	10	FSHPLC	2
	10	VISUAL	2

NUMBER OF SAMPLES: 9

Figure 21.7 Job list for an analyst

Analyst Job Assignment

Edit Reset Previous Next

Analyst JOB Assignment

Job	Sample	Test	Date	Analyst
45	FRAMEWORK4	WEIGHT	11/17/86	MOLLY TRAVIS
46	FRAMEWORK5	VOL.DISP.	11/17/86	MOLLY TRAVIS
47	FRAMEWORK6	DISSOLUTION	11/17/86	MOLLY TRAVIS
48	C-128	FSHPLC	11/18/86	MOLLY TRAVIS
50	C-130	FSHPLC	11/18/86	MOLLY TRAVIS
53	AA102	AAS	11/18/86	MOLLY TRAVIS
55	AA104	WEIGHT	11/18/86	MOLLY TRAVIS
56	AA105	VOL.DISP.	11/18/86	MOLLY TRAVIS
57	AA106	DISSOLUTION	11/18/86	MOLLY TRAVIS
58	AA456	AA-MG	11/18/86	TINA ROSS
59	AA457	AA-MN	11/18/86	TINA ROSS
60	AA458	AA-CU	11/18/86	TINA ROSS

Figure 21.8 Form used to assign work

21.3.2 Assigning Work

Once samples are logged in, the laboratory manager has the option to assign tests to analysts. This procedure is not mandatory, because the analyst who usually conducts a test can be named in the database and then is automatically assigned the job. Job lists based on these automatic assignments are available as soon as a sample is logged in. The manager's intervention is needed to redistribute the workload or to accommodate absences and departures from the routine. The manager can review the entire workload on-screen, and allocate work as needed. Figure 21.8 shows the form on which tests are assigned to analysts.

21.3.3 Checking the Job Queue

As the first step in their daily routine, analysts check the system's job queue to find out which tests are on their agenda. Each analyst can print out a copy of their job list (Figure 21.9).

TEST STATUS REPORT

10/27/86 08:08:04

Number	Sample Name	Study	Submis	Test	Analyst	Status	Testdate	Proj
2	LOT CB504		2	COLORIME	TINA ROSS	NEW	10/27/86	DF-23
4	LOT AR397		4	AAS	TINA ROSS	NEW	10/27/86	LX45
4	LOT AR397		4	COLORIME	TINA ROSS	NEW	10/27/86	LX45
5	LOT FW289		5	AAS	TINA ROSS	NEW	10/27/86	TF783
5	LOT FW289		5	COLORIME	TINA ROSS	NEW	10/27/86	TF783
6	LOT NS347		6	AAS	TINA ROSS	NEW	10/27/86	TS67
7	LOT MS189		7	AAS	TINA ROSS	NEW	10/27/86	CD347
9	LOT CB504		2	COLORIME	TINA ROSS	NEW	10/27/86	DF-23
9	LOT CB504		2	AAS	TINA ROSS	NEW	10/27/86	DF-23
123	LOT AR397		4	COLORIME	TINA ROSS	NEW	10/27/86	LX45
124	LOT AR397		4	COLORIME	TINA ROSS	NEW	10/27/86	LX45
124	LOT AR397		4	AA-P	TINA ROSS	NEW	10/27/86	LX45
125	LOT FW289		5	COLORIME	TINA ROSS	NEW	10/27/86	TF783
125	LOT FW289		5	AA-P	TINA ROSS	NEW	10/27/86	TF783
125	LOT FW289		5	AA-MG	TINA ROSS	NEW	10/27/86	TF783
125	LOT FW289		5	AA-FE-1	TINA ROSS	NEW	10/27/86	TF783
126	LOT NS347		6	AA-CA-2	TINA ROSS	NEW	10/27/86	TS67

Total samples in report: 17

Figure 21.9 Report from the laboratory job queue showing the tests scheduled for an analyst

21.3.4 Preparing Sequences

Before running automated analyses on a batch of chromatography samples, analysts can use RLAB to prepare a sequence file. The sequence, shown in Figure 21.10, is formatted for use by the Nelson Analytical 3000 system. As it uses the unique sample names and numbers entered at log-in, the sequence links chromatographic results with the correct sample records in the RLAB database.

SAMPLE NAME	METHOD NAME	DATA FILE	AMOUNT INJECTED	INT.STD. AMOUNT	DILUTION FACTOR	SAMPLE WEIGHT
1 AUTOCAL1RE	TEST	BIO	1.0000	1.0000	1.00	1.000
2 AUTOCAL2RE	TEST	BIO	1.0000	1.0000	1.00	1.000
3 SAMPLE1	TEST	BIO	1.0000	1.0000	1.00	1.000
4 SAMPLE2	TEST	BIO	1.0000	1.0000	1.00	1.000

Figure 21.10 Sequence used to guide a series of chromatographic analyses.

21.3.5 Entering Test Results

Keyboard entry of test results. The testing phase can produce results in many forms. All types of test results can be keyed into the database individually; Figure 21.11 shows the screen form that analysts fill out to enter test results manually.

Edit Previous Next Quit

RESULTS UPDATE FORM

SAMPLE NUMBER	TEST NAME	ANALYTE	NUMERICAL RESULT	TEXT RESULT	UNITS
1	PH	PH			PH UNITS
2	PH	PH			PH UNITS
7	PH	PH			PH UNITS
11	PH	PH			PH UNITS

	PHYSICAL LIMITS	DETECTION LIMITS	SPECIFICATION LIMITS
UPPER	14.	10.	8.5
LOWER	0.	4.	6.

Figure 21.11 Form used for entering test results.

Batch entry of test results is far faster, more reliable and convenient than manual entry when large numbers of samples are tested. RLAB accepts results in bulk from two sources and has potential links with virtually any laboratory instrument.

One source of bulk data is the Nelson Analytical 3000 system. This software runs on the same PC as RLAB. The two systems work as an integrated package, communicating through compatible files. Using sequences generated by RLAB to guide batch data acquisition, the 3000 system collects and analyses data from chromatographs and stores the results in files on the PC disk. Analysts then use the

Nelson Analytical PEAKSUMMARY software (which comes with RLAB) to transfer the values in these result files to the RLAB database.

If defined in the database, a result must fall within limits. If outside these limits the result is flagged, and an appropriate user-defined message appears on standard reports that include the result. Such messages might include the acceptable result range, for example, 'pH should be 6.0 to 8.5'.

In addition to numeric test results, RLAB accommodates narrative information about a test performed on a given sample. Analysts can express results in 'essay' form or enter comments about unusual conditions, as shown in Figure 21.12

Press [ESC] when done

TEST COMMENTS ENTRY

SAMPLE NUMBER	123
TEST NAME	LC1
COMMENTS	CHROMATOGRAM CONTAINED UNKNOWN PEAKS. CONTAMINATION SUSPECTED.

Figure 21.12 Form used for comment entry.

21.3.6 Test and Sample Status

While testing is in progress, RLAB automatically keeps track of the status of each test. When a sample is logged in, RLAB gives each test a 'new' status. When results for a test enter the database, its status changes to 'done'. If the laboratory manager rejects the results for a test, its status changes to 'error'.

Concurrently, RLAB tracks sample status. At log-in, samples are labelled 'new'. Their status changes to 'active' as soon as the results of the first test enter the database. When all test results are in, the sample status becomes 'complete'. If a result is rejected by the laboratory manager, RLAB gives the sample an 'error' status. RLAB assigns the final status, 'approved', when the laboratory manager gives approval for all the results.

21.3.7 Approval

After all tests on a sample are complete, it becomes eligible for approval. The laboratory manager examines the test results and approves or rejects them individually. If a test result for a sample is rejected, RLAB puts the test back into the laboratory job queue so it can be repeated. When all test results are approved, the sample receives an 'approved' status.

21.3.8 Reporting

While a sample is in the laboratory, any RLAB user can check its status by requesting a standard report. The sample status report, shown in Figure 21.13, is one of seven that the system produces. Other standard reports provide information on the tests conducted in the laboratory, list tests scheduled for a group of samples, list test results, show the locations of samples, and list the analysts' workload. Output can be displayed, printed, or routed to a file.

SAMPLE STATUS REPORT

10/27/86 08:08:04

Number	Name	Study	Submis	Duedate	Status	Prior	Project	Location
123	LOT AR397		4	11/17/86	ACTIVE	3	LX45	P195
124	LOT AR397		4	10/27/86	ACTIVE	3	LX45	P195
125	LOT FW289		5	10/27/86	ACTIVE	3	TF783	F56
126	LOT NS347		6	10/27/86	NEW	4	TS67	NL34
127	LOT FW294		5	11/27/86	NEW	3	TF783	F56

Total samples in report: 5

Figure 21.13 Sample Status Report

Besides requesting standard reports, users can pose their own queries to retrieve information in report form. By specifying a sort order and selection criteria, the user can define the report content and the order in which the content is presented. For example, RLAB responds to queries like this one: Print a summary report sorted by sample name, where the test is pH and the result exceeds eight.

21.3.9 Information Transfer

RLAB enhances the value of test results stored in the database by allowing further analysis. The system automatically writes compatible files and offers menu access to Lotus 1-2-3. Users can apply all the computation and graphic capabilities of this package to the laboratory data (for example, see Figure 21.14).

Results in the database can also be sent to other computers. RLAB offers the option to load data selectively into files in a variety of standard formats. By using commercial terminal emulators, users can send the files to other PCs and to larger IBM, Hewlett-Packard, and DEC computers.

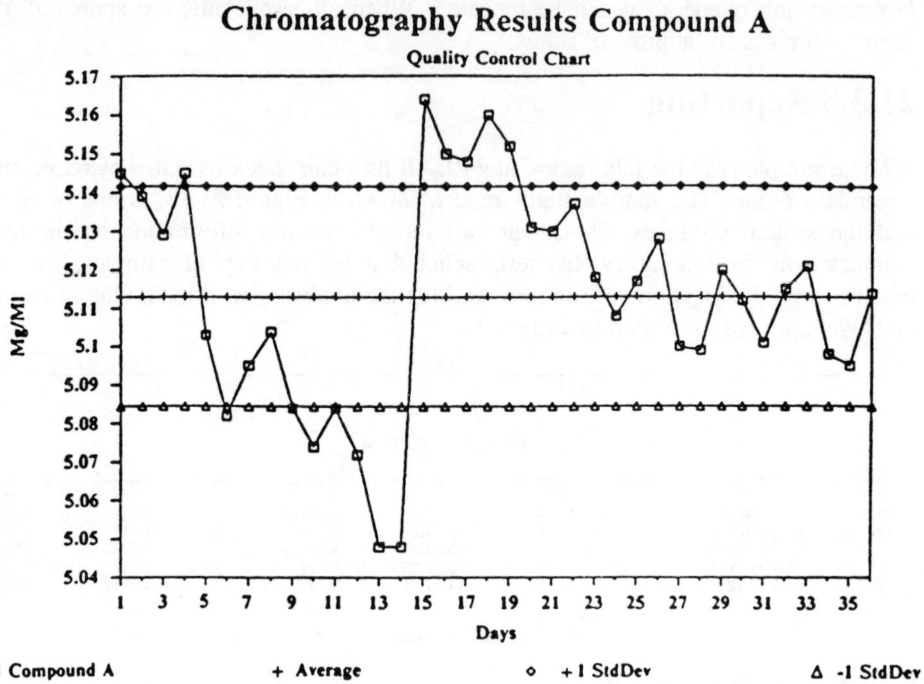

Figure 21.14 Example of laboratory data plotted by Lotus 1-2-3

21.3.10 Archiving and Back-ups

As samples are logged, tested, and approved, information accumulates in the database. These historical records may be useful for a while, but eventually they become obsolete and need to be deleted. RLAB allows the system manager to archive selected data by transferring it to a long-term storage device. RLAB also allows periodic database back-ups.

21.4 The RLAB Database

The nucleus of the RLAB LIMS is its database. It is designed to store the types of information typically encountered in an analytical chemistry laboratory, particularly one using chromatography. The RLAB is a relational database, containing tables to organize data: sample number, sample name, submitter, and submission date are just a few examples associated with each column is a format specification. All data entering a column must have the prescribed format: text, note, integer, real number, monetary value, date, or time. In addition, data must conform to certain range requirements:

- Text data are limited to a maximum number of characters.
- Note data, also consisting of text characters, have no specific length limits.
- Integers can range between plus and minus 999,999,999.
- Real numbers can range between plus and minus 9×10^{37}.
- Monetary values can range between plus and minus $99,999,999,999,999.99.

In addition to the basic format range requirements, laboratory managers can define acceptable limits for result data and 'rules' to control data entry. At delivery, the tables in the RLAB database contain a few rows of practice data. These data are provided so that the users can try out the system before adding rows of real laboratory data.

21.4.1 The RLAB Database Tables

In using the database to store and retrieve data, the user encounters two types of table: static and dynamic. Static tables hold information that remains the same over a relatively long period; it is modified only when laboratory procedures change. For example, tests the laboratory performs are stored in a static table, see Table 21.1. In contrast, dynamic tables change constantly. Typically, data are added daily and archived periodically. For example, test results are stored in a dynamic table, a list of these dynamic tables is shown in Table 21.2.

The system manager, responsible for database maintenance, will encounter a third set of tables that define the system's standard reports, data-entry forms, and data-entry rules. These tables (as well as the static and dynamic tables) can be modified to accommodate the specialized needs of the laboratory.

21.4.2 Relations Among Tables

The database tables are related in two ways:

1. Key columns are duplicated in two or more tables and contain data in common (for example, see Figure 21.15).

2. Entries in certain key columns of dynamic tables cause the system to read and extract information associated with duplicate columns in static tables (for example, see Figure 21.16).

Table 21.2
Dynamic tables in the RLAB database

Table Name	Description of Contents
SAMPLES	Stores background information on each sample. Each row contains a unique, system-generated sample number and a unique, user-supplied sample name. Among the other types of data are due date, priority, sample type, and submitter. Most data in this table are entered at log-in; however, the log time and date come from the computer, and the sample status is assigned automatically by RLAB.
JOBQUEUE	Stores information about the tests to be conducted on a particular sample and the names of the analysts assigned to conduct the tests. Most of the information enters the table when the sample is logged in. Each row in the table contains a unique sample-and-test combination, which is assigned a system-generated job number. These job numbers are used in the COMMENTS table to identify narrative associated with a sample-and-test pair. Besides job numbers, RLAB enters test status (new, done, or error), test time (the time a result enters the database), and test date (the date a result enters the database).
RESULTS	Stores all test results. Each row accommodates results in both numeric and text form for one analyte determined in one sample. The sample is identified by a sample number, which must also be found in the SAMPLES table. Whenever a result enters the RESULTS table, RLAB records the system date in the COMPLETE column. When a result enters the RESULTS table for the first time, RLAB places the value 0 in the AUDIT column. If the result is changed for any reason, the AUDIT value becomes 1. Rows in which an audit value of 1 appears are not reported in the standard Sample Report nor in the Summary Report. When a result is outside the specifications stored in the TESTCOMP table, RLAB enters the appropriate flag number in the FLAG column. This number invokes validation messages from the VALIDS table.
COMMENTS	Stores comments on sample-and-test pairs. Each pair is represented by a job number in this table. Comments can be any length needed. They can report anomalies encountered in testing or describe test results in narrative form.

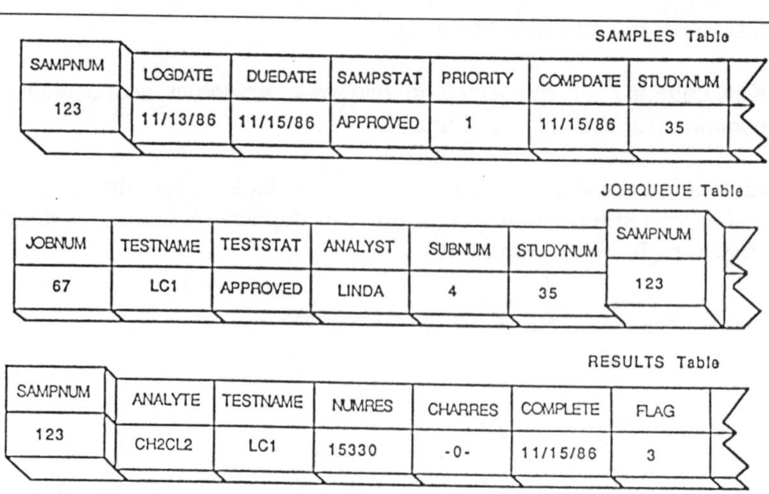

Figure 21.15 The sample number is a link between database tables

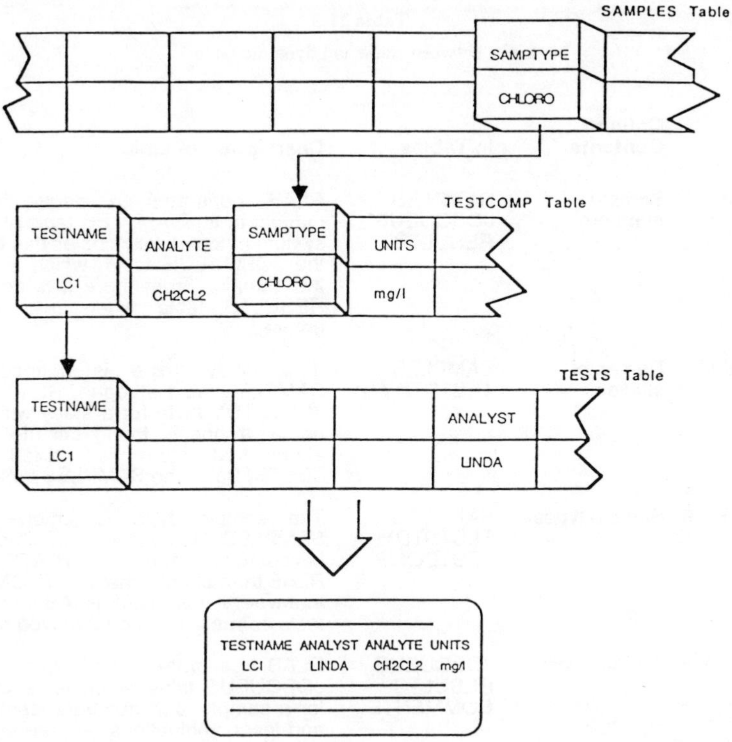

Figure 21.16 RLAB retrieves test data associated with a sample type code in the SAMPTYPE column

Table 21.3 lists some of the key columns in the static and dynamic tables and shows how they link the tables.

21.4.3 How RLAB Functions are Organized

The jobs possible with RLAB are distributed among four levels of users. Passwords for each level allow access to separate menus offering different sets of commands. These levels are hierarchic: users above the first level have access to all commands on lower levels.

User access levels

Level 1 - Users The most limited level of access is the 'user' level. Users in this category can do the following jobs:

- Log samples into the system.
- Log studies into the system.
- Request standard reports showing sample status, test status, and test results.
- Query the database to retrieve selected data. Create files containing result data for analysis by Lotus 1-2-3.

Table 21.3

Links between static and dynamic tables

Key Column Name	Column Contents	In Tables	Description of Link
SAMPNUM	Sample numbers	SAMPLES JOBQUEUE RESULTS	RLAB generates a sequential sample number in the SAMPLES table at log-in. The system automatically copies this number into the JOBQUEUE table, which lists tests for the sample. From there it is copied to the RESULTS table when test results are entered.
STUDYNAME	Names of studies	SAMPLES TESTSTDY	If a study name is entered into the SAMPLES table at log-in, RLAB checks the TESTSTDY table for a study with the same name. If one is found, all the information about that study is copied from the TESTSTDY to the SAMPLES table.
SAMPTYPE	Sample types	SAMPLES TESTSTDY TESTCOMP	The sample type is entered into the SAMPLES table at log-in, either from the keyboard or from the TESTSTDY table. RLAB then checks the TESTCOMP table for that type. If it is found, RLAB displays all the tests normally done on that type of sample.
JOBNUM	Job numbers	JOBQUEUE RESULTS COMMENTS	RLAB generates job numbers in the JOBQUEUE table when tests are selected for a sample. Job numbers identify sample-and-test combinations and are later used by the system to identify results and comments.
TESTNAME	Test names	TESTS TESTCOMP JOBQUEUE TESTCOMP RESULTS TESTCOMP	Test names are unique identifiers for tests performed in the laboratory. Both the TESTS and tables contain information describing the tests. Information from the table is displayed at log-in to allow selection of tests for samples. Once this selection is made, RLAB puts the test names and corresponding sample numbers into the JOBQUEUE table. Before studies are logged in, test names pre-selected for samples, reside in the table. At log-in, these test names also enter the table. Finally, the test names, associated analytes, and units are copied to the RESULTS table when results are entered.
FLAG	Number	TESTCOMP RESULTS	Whenever a result enters the indicating table, RLAB checks RESULTS out-of-range, VALIDS the TESTCOMP table to see results whether limits are specified. If the result is out of limits, the flag number associated with the analyte is copied to the RESULTS table to flag the out-of-range result. Each flag number is related in the VALIDS table to a warning message that RLAB prints in standard reports containing flagged results.

Level 2 -Analysts The next access level in the hierarchy is the 'analyst' level. Users in this category can do jobs typically performed by laboratory analysts:

- Check the laboratory's job queue for work assignments.
- Build chromatography sequences for use with the Nelson Analytical 3000 Series Chromatography Data System.
- Query the system for sample locations.
- Enter test results and test-related comments.
- Enter and edit test descriptions.

In addition, analysts can perform all the 'user' functions.

Level 3 - Laboratory Managers In the third level of the hierarchy are 'laboratory managers'. Their functions are as follows:

- Assign jobs.
- Approve test results.
- Browse test comments.
- Enter specifications for a study.
- Enter messages that appear when results are outside specifications.

In addition, laboratory managers have access to all functions on the two lower levels.

Level 4 - System Maintenance The broadest access is assigned to level 4 users. Tasks at this level focus on database maintenance rather than laboratory operations. Users at the system level can:

- Customize RLAB by modifying data-entry rules, data-entry forms, standard reports, database tables, and passwords.
- Back-up and restore the database.
- Pack the database to improve utilization of disk space.
- Archive obsolete records.
- Unload selected data to a file for transfer to another system.
- Input previously unloaded data.

In addition, system users have access to all functions at lower levels.

This task-oriented organization scheme integrates the operation of RLAB with the work in the laboratory. It also provides for the security and integrity of the database.

21.4.4 The RLAB Main Menu

The Main Menu of RLAB reflects the system's organization (see Figure 23.17). It offers direct access to the user, analyst, and laboratory manager functions. Access to system functions is available indirectly through the laboratory manager's menu.

```
<<< Nelson Analytical RLAB II Sample Management System >>>
        (1)  Perform USER Functions
        (2)  Perform ANALYST Functions
        (3)  Perform MANAGER Functions
        (4)  EXIT RLAB II
```

Figure 21.17 RLAB main menu

21.5 Summary

RLAB is a powerful relational database product for the analytical laboratory that runs on the PC class of computers. It is designed to keep track of information on thousands of samples and will produce a variety of reports.

Chapter 22

How to Evaluate a LIMS

D.C.Mattes

22.1 Introduction

When a system is purchased, the importance of evaluation is amplified because there are not the options available to modify it which would be available for an in-house system. It is often difficult to recover from errors, and this should be a concern of the LIMS project team who are selecting the system.

Unlike many other types of computer systems and laboratory automation, the full benefits of a LIMS should not be expected until several months after the system has been introduced into the laboratory. This time lag means the impact of making a wrong decision is greater because time and effort have been wasted on the wrong system. To back out of this awkward situation is very costly, so the need to make the right decision the *first* time is very important. The success or failure of a LIMS to meet the needs of the laboratory will reflect on all those involved in the decision process.

An evaluation is very subjective, this chapter is based on the author's experience in dealing with the evaluation of LIMS as well as commercial computer solutions for a scientific environment. The discussion focuses on three aspects: the objectives and expected benefits of the system, an outline of topics to be investigated, and a plan for the execution of the actual system evaluation.

22.2 Objectives of the LIMS Evaluation

It is important to discuss and agree the objectives of a LIMS evaluation prior to it taking place. Moreover, these must be presented in a concise fashion to the users, to build a consensus regarding the importance of this process.

The overall objective of any evaluation is to ensure that the chosen system meets the requirements defined by the laboratory. So it is very important to define and document the needs and the expected benefits of the system. Without careful definition of the requirements, and the evaluation criteria of the LIMS with respect to those requirements, the acquisition of the LIMS may well prove to be an exercise in futility.

To specify the objectives in a clearly defined statement may be difficult, but if the expected benefits of the evaluation are thought of at the start of the process, this may clarify things. These are:

- To build confidence about the ability of the system to contribute to the laboratory productivity. This should reinforce the expectations or the benefits expected.

- Introduction of the system to a nucleus of the users to help the overall acceptance for the LIMS (see Chapter 6).

- Ensure that the requirements which have been defined for the system are valid. The introduction of a *real* system into a *real* environment for evaluation tests the quality of the requirements as well as the capability of a system to meet those requirements.

The key issues for the successful evaluation of any LIMS are first that the technical considerations of the LIMS and the procedures used for the implementation of the system must meet the requirements of the organization. This is the most obvious and sometimes only consideration in the evaluation process. Second, the LIMS must be subjectively acceptable and in compliance with the attitudes of the staff in the laboratory.

22.3 Remember the Requirements

In evaluating any LIMS applications the initial action on the part of the evaluator must to be to remember the specification documents which have been prepared. These are too often only considered in the light of developing a LIMS either in-house or through a software house. However, it is equally important to carefully document the system requirements when planning to acquire a LIMS from a vendor. The reason is that in order to evaluate or select a LIMS package it is necessary to fully understand the requirements of the laboratory and to have those who will be involved in the system all agree upon the system requirements. The preparation of specification documents will ensure that the requirements and functions of a LIMS have been defined.

It is important to understand from the outset that the products which are supplied by the various LIMS vendors are not the same. Each product has its roots in a unique history which has been responsible for the functions programmed into the system. The input into a system may be the result of training or work experience of the system designer, or the desired market for the product which focuses on a particular type of laboratory environment.

There are several issues which should be addressed in the requirements specifications which may not be discussed when considering custom development because of basic assumptions made or implied by the development team. These assumptions should not be taken for granted in the evaluation of a commercial LIMS package. Following personal experience, it has become increasingly clear that functions which may appear to be obvious or necessary for a LIMS to function effectively may not be available in some commercial systems. Therefore, it is important to obtain a clear definition of *all* issues important to the implementation of a LIMS. The evaluation process is intended to provide this information with a level of assurance to support the decision to implement a specific system.

If a requirements document is defined with the assumption that the LIMS will be developed under the control of the laboratory, it should be reviewed before evaluating vendor supplied systems. It is common to make assumptions when a system is developed 'in-house' which cannot be assumed of a vendor supplied LIMS. Therefore, issues such as the following, may need review and resolution before beginning the evaluation process.

- Who requires access to the information?
- Where are those that require access physically located?
- What type of computer systems are currently in place?
- Will the system integrate with the relevant software systems?

22.3.1 Current EDP Environment

A primary consideration will be directly related to the overall environment in which the LIMS will operate. If a LIMS is developed in-house, the issues of integration are normally not considered in the same sense because the development process will, by default, involve the generation of a system which is compatible with the current systems in the laboratory.

This issue can be reduced or removed from the evaluation process if the following situations are true for the environment for which the LIMS is planned: if the laboratory in question is an independent operation and the information which is generated within the laboratory does not require any interfacing with other areas of the organization, or if the LIMS is being installed in a 'virgin' EDP environment where future systems and products will be required to interface with the LIMS, but no current systems are in place.

A major benefit from the integration of a LIMS into a laboratory is the availability of information in an organized format. This benefit will *not* be realized if the LIMS does not integrate into the same environment as the people who require the information produced by the system. The areas of integration which should be considered are:

- The use of system hardware which is similar or compatible with current EDP systems. The support of the system and the availability of data are greatly enhanced if all systems in the organization can be linked to a common network.

The use of some common development tools. A major issue here will be the selection of a DBMS which is used to implement the LIMS. As many organizations mature they are more concerned about the general handling of information which is a resource of the company. LIMS data is a major element of the overall information of many organizations. For this reason it may be deemed necessary for the LIMS database to be compatible with the existing information management infrastructure.

22.4 What to Evaluate

Once the objectives and the structure of the process are clear, the LIMS project team should focus on two areas. First, the evaluation of available products to assess their ability to provide the critical LIMS functions in comparison to the documented requirements. Second the vendor should be evaluated in order to ensure that the commitment of the laboratory to the LIMS will be reflected in the ability of the vendor to support that commitment over a reasonable lifetime. These two sections of the evaluation may be considered in either order, although in reality, there may be three phases of evaluation where the initial phase is very cursory and requires only a minimal investment in time.

22.4.1 Evaluate the Functions of the LIMS

The functional evaluation can be further divided into first, does a product provide the functions required and second the subjective evaluation process, which focuses on the 'value' of the product.

The most critical aspect of the objective evaluation is to prepare an evaluation checklist which lists each function that has been identified in the user requirements. It may be of value to rate these requirements and score each product; this however, is often not a realistic approach to the problem. An alternate method is to review all requirements and rate them as either required, desired or optional.

22.4.2 Categories for Functional Evaluation

The following categories are provided as a guide for the development of an evaluation matrix or checklist. Within each category there is a description of the objective of the evaluation and several specific topics which are applicable to the category are listed as examples. The areas are summarized in Table 22.1.

1. Environment The focus here is to understand the intended environment for the LIMS. This is important because it is essential for integration and providing the functions required. Issues which are related to the scope and implementation of the LIMS should be evaluated under this category. There are three major divisions which should be addressed under this category in the evaluation.

Hardware This should evaluate the hardware required to use the system. Is the system compatible with current systems? Is the hardware from a 'standard' computer vendor? Are communication options available or required for data collection? As noted above, this is a critical area of the evaluation if the system must integrate with an existing computer environment in the organization, or if the system is to be the foundation of an emerging computer environment.

Software The need here is to evaluate the robustness of the system based on the tools used for implementation. For instance, does the system use a standard or proprietary operating system? Is the system based on a well supported DBMS? Does the software provide for data sharing?

User environment The general user interface needs to be evaluated. Does the system use forms and/or menus? What are the requirements for a 'friendly' user interface?

2. Sample management functions All LIMS provide a basic set of functions which fall into various general categories. The sample management programs update sample progress in the system; these must be consistent with the collection and introduction of samples into the laboratory to reflect the way of working. Related issues are: sample information which should review the type and extent of information which is required for the LIMS to function or can be optionally entered; label generation, including the types of labels and bar code support, if required.

3. Data collection A major concern in the implementation of any LIMS should be the data collection methods available. This is important because it is an area where productivity gains can be dramatic. Also, it is an area which must continually be expanded or developed as the functions and/or instrumentation used in the laboratory change. In general, this is one area where flexibility is a major concern because of the dynamics of instrumentation and testing procedures in the laboratory. Typical areas for consideration are: instrument support, manual and automatic data entry, data collection interface, and communication with intelligent instruments. Under this category it is important to consider issues related to data validation. How will the system respond to data out of range, or to typographical errors?

4. Results Management The way the system organizes, handles and reports data must be evaluated. Two concerns in this area are related to the system's ability to adapt to procedures performed in the laboratory, and the data review and approval processes which are required. Some systems allow the user to define the forms and related information which are to be collected for each laboratory procedure, others are relatively limited allowing only a few fields for test results. Many systems provide a specific data approval hierarchy which cannot be modified; if this does not match the laboratory environment, the system will not be acceptable to manage the data. In general, does the flow of data in the system and its availability match the working practices of the laboratory to allow the system to be integrated into the current laboratory environment?

Table 22.1

An Evaluation Checklist for a LIMS

1. Environment		
	1. Hardware	System compatible with current systems Hardware from a 'standard' computer vendor Communications options for data collection and to corporate computers
	2. Software	Standard operating system DBMS used Data sharing capabilities Software tools used
	3. User interface	Forms Menus Command language
2. Sample management functions		Log-in of sample details Allocation of priority Alteration of priority Scheduling of work Batch log-in of samples Label generation Bar code support Sample receipt forms
3. Data collection		Interfacing problems Which systems Flexibility to meet change in practices and instrumentation Manual data entry Automatic data entry Chromatographic data acquisition Interfacing with intelligent instrumentation
4. Results management		Ability to mimic laboratory procedures Storage and organization of data Ability for re-definition Flexible approval system Reporting capabilities Flexibility of report writer
5. General management functions		Sample status Workload reports Customer billing Backlog reports
6. System support functions		Data back-up/recovery/archiving Training and other support functions Documentation for users and system manager Validation procedures Warranty
7. Product history		Age and version of product Number of installed sites References

5. General management functions In order to aid the management of the laboratory, the LIMS must provide basic management reports. These are generally important to monitor productivity as well as provide status reports to laboratory staff, and include workload reports, time tracking, and backlog reports.

6. System support functions These functions are important to manage the computer system in the customer environment. These tools should allow the system manager to monitor and manage the computer resources used by the LIMS to ensure the effective use of these resources. The areas in this category should include: data back-up/recovery functions, data archiving, and documentation.

7. Product history No evaluation would be complete without investigating the history of the product. This will help provide a basis for evaluating the stability and robustness of the system. This should include a discussion with users of the system in other laboratories specifically asking questions about any areas of concern which have arisen in the evaluation.

This has provided a brief overview of the areas which should be covered during the LIMS evaluation. In Table 22.1 a sample evaluation checklist is included. *Caution*, remember that the value of the system evaluation is directly related to the care which is taken to reflect the system requirements in the evaluation process.

22.4.3 Categories for Vendor Evaluation

In addition to evaluating the product it is important to evaluate the LIMS vendor. Unfortunately, there are products which are introduced into the LIMS market which appear to be the answer to a wide range of problems, yet the product never meets the expectations of the vendor. The result is that the product is not supported and the vendor may even leave the LIMS market. It is important to realize that the use of a LIMS is a long term commitment and that some of the benefits are not apparent until the system has been in use for a long period. Therefore, in order for the LIMS to be successful, the system must be supported by the vendor for its anticipated life.

Table 22.2

Evaluation checklist for a vendor

History	Age of the company
Size	Stability
Profitability	Target market of the product
Resources	Staff
	Expertise
	Background
	Facilities
References	
Plans for future products and/or growth of existing package	

Table 22.2 provides an outline of a checklist for the evaluation of the vendor. These categories will not be explained in detail because the issues are more subjective and have a common focus, which is that if a LIMS is purchased the laboratory enters into a relationship with the vendor where the latter has a direct impact on the effectiveness of the laboratory.

22.5 Performing the Evaluation

Once the objectives and scope of the evaluation process are understood, the actual evaluation process can begin. Although there are various methods which can be used, one evaluation plan is outlined here.

22.5.1 Select the Evaluation Team

A critical element with respect to the evaluation process is the selection of the evaluation team which should be composed of representatives of the LIMS project team. This is important not only to ensure the capability of the team to address all issues with respect to the system requirements, but also to ensure a level of confidence within the organization that the selected system will be successful.

22.5.2 Identify Options

The first step of the evaluation process should be to identify the products to be considered in the evaluation. This may include a broad range of products which potentially meet the requirements for the LIMS. A list of vendors and/or products can be gleaned from product literature in sources such as *Analytical Chemistry.*

22.5.3 Develop the Evaluation Checklist

The evaluation checklist should be developed at the start of the process and used to identify products which appear to meet the minimum requirements for the LIMS. This checklist provides a rational means to compare systems and to ensure that critical functions are not overlooked.

22.5.4 Paper Evaluation

An initial round of the product evaluations should be intended to reduce the list of potential products to a manageable set which meets the basic requirements of the system. This task can be performed by evaluating the documented functions of the system against the evaluation checklist. This process can be expedited by sending a copy of the evaluation criteria to each vendor and asking him to respond with respect to his products. An alternate method is to obtain documentation, and review each product via an interview with a vendor representative and system documentation.

Following this stage of the evaluation, the evaluation team should discuss their findings and select the most viable solutions, and narrow the initial list of products to two or three systems.

22.5.5 On-line Evaluation

Each product which is selected as a result of the paper evaluation should be evaluated with 'live' data in the target environment. This part of the process requires planning to ensure that the entire system can be evaluated to some level of detail in a short period of time. One mechanism to perform this phase of the test is to use historical data from a specified period of time in the laboratory, and to process this data through the LIMS for the evaluation. The test should also include interfacing a variety of instruments to the system if on-line data collection is required.

Following the execution of this phase of the evaluation, the project team should meet and form a consensus on the most viable system to meet the specified need. At this point, the decision can be made with an assurance that the requirements for the LIMS can be satisfactorily met by the proposed system and that the implementation of the system can be successfully completed.

22.6 Summary

The objective of the evaluation is to ensure the selection of a LIMS that meets the specific requirements of the laboratory and its staff. This process, though extensive and time consuming, is necessary because of the large commitment which is required by the organization to successfully implement a LIMS and achieve the anticipated benefits.

Chapter 23

Validation of LIMS

Adrian Hampshire

23.1 Introduction

A Laboratory Information Management System (LIMS) may come from one of three sources:

1. Supplied off the shelf or suitably modified from a company specialising in such systems.
2. A custom system is written by a software house and matched to the hardware.
3. The required elements of hardware may be purchased and the necessary software developed within the organization.

Whichever is chosen, it is essential that the final system meets the users' requirements fully, correctly and reliably. In order to achieve this confidence, all of the parties involved in the project have to ensure that the quality of development is maintained and controlled throughout the lifetime of the project.

Computer hardware and software systems are everywhere, however some still continue to fail, with damaging consequences. The techniques of quality control have been widely applied to the hardware for sufficient time to result in a substantial improvement in reliability and correctness. The same techniques have not been so extensively used in the area of software despite the fact that software represents the major cost component in the development of most computer systems.

The attention to quality in the area of computer hardware can be seen in the numerous agreed standards. Some standards do exist in the software world but they are not as effective as those for hardware. This reflects the comparative difficulty of defining standards for software versus those for hardware.

It can no longer be accepted that programs may, on occasion, fail in an irritating way and that this failure must be viewed as the price to be paid for advanced technology. Computer systems are all pervasive and are responsible for many critical activities, eg. air traffic control and banking systems. Although not life threatening, the example of a major banking network that failed on 29th February 1984, because it had not been programmed to take account of leap years, proved very expensive, not least in bad public relations.

23.2 The Need for Validation

A major reason for the apparent slowness in addressing problems of quality control in software is that, whilst hardware problems frequently occur as a catastrophic failure of the system, software errors are often obscure and are not immediately revealed when they occur. Many techniques have been developed which attempt to minimize the occurrence of errors, these include some of the newer languages, structured programming and proof of correctness. All have made contributions to improved systems quality (1).

The rationale for ensuring systems quality is clear:

1. *Legal liability* Most suppliers disclaim legal liability concerning the use of their systems. Case law that tests such disclaimers is vague and there is little guidance as to the validity of such claims. However, if developed in-house then it would seem obvious that legal liability relating to the use of the system cannot be disclaimed. The FDA considers the end-user to be primarily responsible for hardware and software validation. A simple certificate of validation by the vendor is not considered adequate to discharge the users' responsibilities for system validation. The FDA states that the vendor may, however, be held liable to some degree for drug product errors if the causative factors involve computer system defects. This statement has yet to be tested adequately.

2. *Cost effectiveness* Software errors or 'bugs' cost money to find and fix. The cost is both to the system developer and to the user. Any suitable strategy to reduce these costs should be readily accepted.

3. *Customer requirements* Many customers can exert sufficient leverage by the fact of their size and purchasing power to ensure that particular standards of quality are adhered to. Many industries are controlled by authorities which regulate their activities or products. Such regulatory bodies as the FDA (USA) or the DHSS (UK) enforce their particular quality standards.

The policy of the FDA in the area of computer handling of drug test data has taken some years to mature and remains, in many areas, still to be adequately formulated (2-5). However, the general principle applies that raw data should be maintained in a form that is accurate, reliable and safe from deliberate or inadvertent change. This is regardless of how the data are stored: whether manually or by computerized methods. These aspects are discussed in the next chapter.

However, all results must be derived consistently and correctly from the raw data before their acceptance and reporting. A level of confidence must be achieved to show that any derived results are a true representation of the raw data.

The step by which this level of confidence is achieved is called *quality assurance* and is aimed at providing evidence that a given process functions as it is intended to do.

Quality assurance is an integral step undertaken in the validation of a LIMS. Validation is primarily concerned with the provision of *documented* evidence that a system acts as it purports to do (6). Validation has three main components:

1. Documentation of what the system is supposed to do and any operational limits that apply.
2. Application of quality assurance tests to provide evidence that the system functions correctly.
3. Documentation of all evidence derived from the tests.

Validation must be applied at all levels of transformation of the data from the basic analytical signal to the final result. Although the analyst may be unable to investigate the generation of signals within a particular scientific instrument, as long as there is the requisite evidence that the supplier of the instrument has undertaken adequate validation of the various processes involved, the output of the device can be defined as raw data.

Likewise, any further manipulation carried out on the data received from this instrument to derive further results must also be validated.

Any LIMS is constructed of a hierarchy of processes. Validation of processes at each level up to the top of the hierarchy will result in a fully validated system. The overall aim is to demonstrate that the system behaves correctly and consistently. This should instil confidence in the system to both users and regulatory authorities. The means of achieving this confidence is through validation.

23.3 Quality Assurance in Systems Development

23.3.1 Designing for Quality

It is essential to consider quality assurance from the earliest stages of a LIMS development. By designing and developing systems with quality in mind the likelihood of encountering problems at a later stage is much reduced. Arguably, the most important step in the development process is the definition of requirements. Designing and developing systems that fail to meet the user's requirements is an obvious and unnecessary waste of resources that is still widely practised. As much fault lies with the user, in failing to define requirements clearly, as lies with the developer in failing to pursue and eradicate any ambiguities in the requirements specification.

For a system of any complexity it will be necessary to subdivide the problem into manageable parts. The requirements of each part should be closely specified along with full consideration of the interfaces with other parts of the system. This design step should lead to a thorough understanding of the module so that the tests required to validate it can be designed at the same time.

The test plan should be stated in the form of a validation protocol defined as a prospective experimental plan that, when executed, will produce documentary evidence of the validity of the process (6). This may be termed *prospective validation* because the validation protocol is defined before system development starts.

23.3.2 Developing for Quality

An impeccably designed system with a thorough validation protocol can be ruined very easily by software development that abides by no strict standards. Standards and operating procedures should be applied both to systems development and modification, whether in hardware or software.

These standards and procedures should be agreed by all concerned with the development. They should be laid down in formal documents which will be available to all the development team. It will be a prime function of the management of the development team to ensure that the standards and procedures are followed.

These procedures should recognise that no system is static either during or after development, so that control of change in the system is of vital importance.

The validation protocol should have laid down a number of tests that can be applied during the development stage. These tests will be applied and their results fully documented. The application of validation during the development stage is sometimes called *concurrent validation.*

23.3.3 Testing for Quality

At the completion of development, further rigorous testing is required to prove the correctness of the system. This may be called *retrospective validation.* For systems developed in-house retrospective validation will comprise the collection of all test data generated throughout the development cycle into a comprehensive document that establishes the system validity.

For systems purchased from a third party supplier retrospective validation will form the major effort in establishing system correctness. The users should insist that the supplying company provides them with documentary evidence of satisfactory application of prospective and concurrent validation, and should then apply rigorous testing to confirm the reliability of the supplier's assurances.

23.4 The Methods of Validation

The methods of validation fall into three categories described below.

23.4.1 Critical Examination

Design inspections, code inspections and 'walkthroughs' are manual checks applied by the software quality assurance team to the system design and the product. Critical examination of design documents and source code is undertaken in order to reveal errors in both the design of the system and the way this has been translated into computer programming code.

Such techniques are the only techniques available in certain situations. For example, human review is the only method for ascertaining that a system will meet the user's requirements. Although certain automated systems can assist in the administrative task, checking the documentation of development and validation of a system is a human task.

Many aspects of conversion of software source code to a running system are handled by the computer. Compilers create intermediate code and undertake extensive checking of the syntax used, linkers bring together the compiled modules to form a running system and check that all modules required are present. However, neither compilers or linkers are able to verify the logic of the system.

Where system design is undertaken using a formal requirements language, automated tools are available to assist the correct translation from requirement to source code. However, such systems are, as yet, not widely available and rely too heavily on a comprehensive and correct description of the problem in the formal language used by the analyser. Achieving such a description may be fraught with almost as many problems as constructing the correct source code from a comprehensive, though less formal, requirements definition.

If available, automated analysis techniques should be employed. Where such facilities are not available, inspection techniques should be employed by the quality assurance team. 'Manual' inspection techniques in which the code is reviewed and 'executed' by the inspection team as if by a computer suffer from two main problems:

1. Psychological problems may arise in the development team from the critical, in-depth examination of its work by another team.
2. Manual inspection techniques require investment of large amounts of time.

23.4.2 Dynamic Testing

The methods of critical examination described above are 'static' in that they are applied to the system in a non-operational state and do not address examination of the running system. The method of dynamic testing examines a system under operational conditions. It is undertaken by applying selected input data and then examining the resultant output to assess the reliability.

This type of testing does not prove a system to be correct because of the necessity of selecting the input data to be used. It can only reveal the existence or absence of errors in a particular test circumstance. It follows that the effectiveness of testing will only be as good as the quality of the design of the test cases. To achieve adequate system testing the following elements are required.

1. *A detailed test plan* This should define the complete range of tests that are to be applied to the system. The plan will be successful if treated as a hierarchy of tests to reflect the hierarchy of software modules in the system. Thus the function called *RECORD MAINTENANCE* might be subdivided into the elements *ADD RECORD, DELETE RECORD, MODIFY RECORD, DISPLAY RECORD*. The test plan should parallel this hierarchy and should describe how the levels of testing combine to establish a suitable level of confidence in the whole system.

2. *Test definitions* At each level in the hierarchy suitable tests should be designed. The purpose of these tests should be laid out in a formal definition of the aims and goals of the particular test. The operational limitations and any special test conditions must be defined and documented.

3. *Test data design* For each test to be undertaken suitable test data should be constructed and written instructions given for carrying out the test. It is not possible to design tests that will exercise and verify every possible route through the logic of even a small system. A simple 'IF condition a THEN DO action a ELSE DO action b' construct has two routes through its logic. If the actions themselves lead to further levels of choice and action pairs, which themselves lead to further choices and actions, then it will be readily appreciated that the number of possible routes through the logic very quickly expands to a level where testing of every route is impossible. In this situation no test plan can be expected to cover all logical routes through the system. However, many logical routes can be excluded on the basis of practical realities. For example, the logical route through a system that would lead one to test that the site of injection of a test substance was properly recorded in any one of four possible legs is a logical pathway that would not be tested if the subject species was a chicken. The design of test data should therefore be to test all reasonable routes through the system. The data sets should be constructed to cover both the range of expected values and also the handling of out-of-range data. If available, software tools can be used which record the logic paths through the system which have been exercised by a particular test.

4. *Documentation of execution and analysis of tests* All the tests identified by the test plan must be executed using all the test data sets. The execution of each test will be fully documented and the results analysed to reveal evidence of correct completion or errors.

The main disadvantage of dynamic testing is in its failure to test all possible logic routes through a system. Testing can obviously be applied during code development as well as to the final system, however some code development must take place before any testing can begin. Techniques that attempt to verify the system specification have the potential to reveal problems before investment in coding is undertaken.

It must be remembered that dynamic testing reveals only the absence or existence of errors in a given test situation; it does not prove correctness.

23.4.3 Proof of Correctness

This method of verification involves the formal proof of correctness of the program by proving the consistency between implementation and specification. As with static inspection methods, it can only prove that the system functions correctly as specified, not that the system design meets the user's requirements.

Instead of designing test data sets and applying these to the running system, execution involves the symbolic definition of inputs to the program which, when applied to the designed logic step-by-step, result in transformation to symbolic outputs which can be evaluated to relate input to output (7).

There is obviously potential for using a computer to assist in the construction and evaluation of program proofs which is otherwise tedious and error prone. Additionally, some effort has to be applied to the development of languages which are more amenable to automatic generation of proofs of correctness. Whilst ordinary testing can cover only a finite number of test situations, symbolic execution tests an infinite number of specific inputs and thus provides a proof of correctness for the test case under consideration rather than providing just output values for each test case considered.

As with dynamic testing the complexity of the logic in systems of the size of a LIMS may make methods such as symbolic execution impractical to apply. This technique may therefore be restricted to use only in the critical areas of the system where its disadvantages of complexity can be justified against its major advantage of a high degree of reliability.

23.4.4 Summary

No single technique of quality assurance covers all the requirements for validation of a system either for theoretical or practical reasons. It is therefore essential that a combination of all the techniques is used to inspire the highest level of confidence in the correctness of the system.

23.5 Documentation

All efforts in quality assurance are in vain if the activities planned and executed are not documented fully. Validation is the production of documented evidence to show that the quality assurance tests stated have been carried out and that the results obtained are the same as those expected. Without the full documentary evidence any inspecting authority will have no reason to accept that the LIMS operates correctly and reliably.

23.6 Validation in Practice

The user will need to undertake certain actions before and after delivery.

23.6.1 Before Delivery

The actions to be undertaken before delivery are:

1. *Establish the validation requirements* The user will need to discuss with the supplier what quality assurance techniques will be applied during the design, development and testing of the LIMS and how the documentary evidence is to be assembled to form the validation package suitable for inspection by the user and any regulatory authority. If an off-the-shelf solution is purchased then the validation package provided by the supplier will be historical documents over which the user has no influence. These must be assessed to see if they can adequately form part of a validation package. If not, more extensive acceptance testing may be considered than might otherwise have been necessary to achieve confidence in the system.

2. *Acceptance testing* As a customer, rather than a supplier, the quality assurance techniques of critical examination and proof of correctness will probably not be available. Therefore the user will need to concentrate on dynamic testing. Independently of the supplier, the user should draw up a detailed test plan, define the aims and goals of individual tests, document test execution, and analyse results as discussed above. Wherever possible the results of the testing should be compared with testing undertaken by the supplier, and any discrepancies should be explored. As well as providing essential documentary evidence for the validation package, the acceptance tests will, when satisfactorily completed, probably represent the contractual completion of the supplier's involvement in providing the system. Thereafter the suppliers role becomes one of support. It is therefore essential that the user is prepared to devote substantial time and manpower to the testing of the system.

3. *Establish the initial configuration* After delivery one of the most important areas of validation is the response to changes in the system, ie. configuration management or change control. In order to monitor any changes in the configuration, the exact initial configuration must be established. This exercise covers both hardware and software and will require collection of full documentation of model numbers, part numbers, serial numbers and version numbers for all elements of hardware and software.

23.6.2 Validation of Computer Hardware

Computer hardware, including the terminals and other peripheral devices, should be considered analogous to an analytical instrument. Preventative maintenance contracts should ensure optimum performance of the system. Analogue to digital (A to D) converters can be checked using peak simulator modules to monitor their output, although recently one manufacturer has introduced automatic validation of these units. Effective transmisson of the data from an analytical instrument should also be tested by sending the same results on several occasions. Further details of hardware validation are discussed by Motise (3).

23.6.3 Validation of Computer Software

The operating system software is such that its validation is too complex for the user to contemplate. Instead, the LIMS software should be tested thoroughly, as it uses the operating system, and these tests should indirectly validate part of the operating system. The validation protocol, embodying a dynamic test plan should be executed.

Tasks that form the LIMS fall into three areas, each of which has its own particular testing requirements.

1. Input related tasks must establish that only the data required for processing or output are accepted and that any other data are rejected. Only data within defined limits are accepted and all unacceptable data are rejected. Examples of this type of task are the entry of the 30th February as a date, or 59,6 rather than 59.6 as data. Only authorised individuals should have access to the system to input and modify data.

2. Output related tasks are to show that only the specified output in format, content and layout is produced and no extraneous data is observed, ie. information can be archived and retrieved efficiently and reliably or results are reported with no corruption.

3. Process related tasks are the most difficult to validate as they occur within the computer and require that the input and output tasks are validated. This aspect is concerned with calculations and internal data manipulation. The accuracy and reliability of any calculations used must be checked and formulae must be documented outside the LIMS. An assessment of data corruption must be made when passing files to and from other computers, whether on site or remote,as well as between analytical instruments and the central processor.

4. The relationship between the data sets must be validated to ensure that they have the correct logical relationship to each other.

23.6.4 After Delivery

The two activities related to system validation that must be considered are change control/configuration management and revalidation. Both activities are irrevocably connected.

Change control/configuration management Having established before delivery the initial configuration of the LIMS, any changes from this configuration must be monitored and documented. Thus, the complete history of all hardware and software elements can be established. Often there may be administrative procedures to be undertaken before a system enhancement can be ordered, it is a simple matter to alter the procedure to ensure that the enhancement is properly documented.

Revalidation Few systems are likely to remain in their initial configuration for long. They will be modified to take account of developing technologies, changing work practices or new requirements. Whilst the system is in the initial configuration then the original validation remains relevant to that system. If a change is made to the configuration then the original validation is no longer applicable. In this circumstance the new configuration should undergo revalidation to re-establish the required level of confidence that the new system again functions as it is meant to do.

The definition of a revalidation protocol according to Chapman (6) is repetition of the validation process or a specific portion of it, once the system is in a state of control. Thus, enhancements and new revisions to any part of the software will require revalidation. Revalidation tests are undertaken to ensure that if the LIMS is still functioning correctly and to document the fact.

In essence, revalidation will probably involve repetition of the same tests that were applied during the acceptance test phase to establish that the results achieved are the same. New tests may also be introduced to test the functioning of any new facilities introduced in the change in configuration. Revalidation is a matter of judgement: should a complete revalidation protocol be run to test a simple error correction or just the affected module? Even if there has been no change in the system it is strongly suggested that periodic revalidation of the software should take place (4). Furthermore, it would be preferable to partition the test data from the live data to ensure integrity of the latter.

There will be a substantial expenditure of time and effort in revalidating frequently changing systems. There is a need for such revalidation tests to be applied automatically by the system itself, and the output stored and analysed against results achieved from the same tests carried out during previous revalidations. Any discrepancies may then be automatically highlighted to draw attention to potential problems with the new configuration.

If the system is able to execute the revalidation automatically, then there is good reason to schedule revalidation tests to be run automatically at regular intervals, rather than only when change occurs. This will greatly assist in achieving the aims of validation which are to establish and maintain an acceptable degree of confidence that the system functions correctly and reliably.

23.6.5 Validation of Existing Systems

While most of this chapter has discussed validation of a new LIMS there are many existing systems that may not have formally undergone a validation procedure. The users or computer staff must then carry out a retrospective validation to establish if the system does what it purports to do. Documentation must be assembled that defines the system, together with test results, specifications, maintenance logs and failure reports (8). A test protocol is then established and executed to provide further evidence of the system's adequacy. Once validation is established, the normal change-control and revalidation procedures are used to ensure the system remains validated.

23.7 Automated Software Validation Techniques

Computer validation requires much time and effort to engender a high level of confidence in the system. It is ironic that, from the users point of view, this process can often involve the manual techniques that the computer was designed to eliminate. Software tools are becoming available that will help to automate this process (9). Examples of this are test coverage analysis to verify that every path of the program has been tried at least once. This does not test the logical paths through the program, but at least every program line has been executed. This is accomplished by a test coverage analyser (eg. Performance and Coverage Analyser, DEC) which inserts test points in each path of the program and maintains a record of what paths have not been executed. At the end of testing, the PCA program can produce charts indicating which paths in the program are not yet tested. Once the lines have been identified the programmer can decide what action to take.

Of more relevance to the user are tools such as DTM (DEC Test Management), which contains the descriptions of the tests to be carried out together with files of expected results. When a test has been executed a file of the results obtained can be compared with those expected, this is done automaticallyby the program, and a list of discrepancies produced at the end of the test. Thus DTM acts as a tool for creating and executing tests and verifying the results it still does not check that the logic of the operation is correct.

23.8 Summary

Documentation of validation procedures is all important. Say what you are going to do. Document what you said you would do. Do it. Document the fact you did it, record the results, and in the case of revalidation compare with the previous results.

23.9 References

1. Buckley, F.J. and Poston, R., 'Software Quality Assurance', *IEEE Transaction on Software Engineering*, SE-10 (1984) 36-41

2. Taylor, D.W., 'Inspection of Computer-Supported Toxicological Data Submitted to the FDA', *Drug Information Journal*, 18 (1984) 189-194

3. Motise, P.J., 'Validation of Computerized Systems in the Control of Drug Processes: an FDA Perspective', *Pharmaceutical Technology*, 8 (1984) 40-45

4. Motise, P.J., 'What to Expect when FDA Audits Computer-Controlled Processes', *Pharmaceutical Manufacturing*,1(1) (1984) 33-35

5. Figarole, P.L., 'FDA's Views and Experience: Inspecting Computer Supported Studies', *Drug Information Journal*, 17 (1983) 7-9

6. Chapman, K.G., 'A Suggested Validation Lexicon', *Pharmaceutical Technology*, 51-55, August 1983

7. Hantler, S.L. and King, J.C., 'An Introduction to Proving the Correctness of Programs', *Computing Surveys*, 8 (1976) 331-353

8. Agalloco, J., 'Validation of Existing Computer Systems', *Pharmaceutical Technology*, January (1987) 38-42

9. Anderson, W., 'Software Validation Techniques', *Drug Information Journal*,in press (1987)

Additional Reading

Hausen, H.L. and Mullerburg, M., 'An Introduction to quality assurance and control of software', in *Software Validation*, Ed.Hausen, H.L., Elsevier, Holland, 1984, pp 3-9

Chapter 24

LIMS and Good Laboratory Practice

D. C. Mattes

24.1 Introduction

During the acquisition or development of a LIMS it is extremely important to remember the constraints which are placed on the laboratory not only by its staff but also by outside organizations or agencies which are, in a sense, customers of the laboratory. One example of these constraints is found in laboratories in the pharmaceutical industry which must comply with FDA regulations and are subject to either GLP (Good Laboratory Practice) or GMP (Good Manufacturing Practice) regulations. Other laboratories can fall under similar guidelines imposed by other agencies such as the Environmental Protection Agency (EPA).

The requirements of all of these regulatory agencies are similar to those imposed by the FDA under the GLP regulations (1). Therefore, the impact of GLP upon a LIMS will be reviewed. Although GLP is discussed throughout this chapter, it is regarded as equivalent to Good Manufacturing Practice (GMP) or the equivalent local regulatory agency guidelines.

Taylor (2) has stated that: ....the biggest impact of GLP in the coming decade will be its influence on laboratory management's decisions on how far to carry computerization.

The FDA had recognised the growing trend towards computerized data acquisition systems and provided the following guidance:

1. Provision must be made so that only authorized individuals can make data entries.
2. Data entries must not be deleted and any changes must be made in the form of amendments.
3. The database must be as tamperproof as possible.
4. The SOPs must describe procedures for ensuring the validity of data.

The implementation of a LIMS in a regulated environment has shown the accuracy of Taylor's prediction. In general it is important to realise that the use of a LIMS does not overcome but exacerbates the problems of control which are required in the management of laboratory data.

In this chapter, the issues which Taylor identified will be investigated as they apply to the integration of a LIMS into a laboratory environment. This will begin with a review of some basic GLP requirements, followed by the issues of data integrity and data handling, security, audit trails, and system operations.

24.2 GLP Requirements

The GLP requirements (1) are specifically focused on the non-clinical laboratory and the studies which are undertaken during the development of a drug product. Although these regulations define the general handling of data within the laboratory environment, there are several specific references to computerized systems:

> In computer driven data collection systems, the individual responsible for direct data input shall be identified at the time of data input. Any changes in computer entries shall be made so as not to obscure the original entry, shall indicate the reason for change, and shall be dated and the responsible person identified.
>
> All raw data, documentation, protocols, specimens, and final reports generated as a result of non-clinical laboratory study shall be retained. There shall be archives for orderly storage and expedient retrieval of all raw data, documentation, protocols, specimens, and interim and final reports.

These two areas, in addition to the basic data integrity issues which are required in a laboratory, are the foundation for a series of points which need to be considered with respect to a LIMS.

24.2.1 LIMS Issues

One of the most basic issues in the regulations is the focus on procedures to ensure data integrity. For this reason there are strict guidelines with respect to the handling and accountability of data, specifically 'raw data'. Therefore, it is a primary concern to ensure the integrity of data collected, maintained, manipulated, and reported. The integrity of information which is maintained within a LIMS is a result of three major functions working together:

1. The basic data integrity features of a system.
2. The security aspects of the LIMS.
3. Audit trails which allow the system to track changes to information held in the database.

These three features allow a system to maintain data in a manner which is consistent with responsible laboratory data management and the requirements of regulatory agencies. Each of these features is examined in greater detail.

24.3 Data Integrity

Data integrity is the process used to ensure that the data which are collected or entered into the LIMS are stored, used, and reported accurately without corruption. Related to this are the data validation techniques which are employed within the LIMS. These techniques are generally more clearly defined than for manual procedures to prevent errors in data recording.

When looking at computerization it is important to consider the current manual procedures to ensure that automation does not destroy checks which have been carefully embedded into a laboratory's SOPs. Data integrity and security are no exception and yet these issues are often ignored in the development or planning of a LIMS. A short review of manual procedures related to the recording, security, and integrity of data will serve as a springboard into the consideration of computerized data storage methods.

24.3.1 Data Integrity and Manual Laboratory Procedures

The most common method for recording data is the laboratory notebook and a pen. This mode of data recording, which may be termed a 'non-volatile memory and storage system' by computer professionals, should be the model for recording data in a LIMS. The notebook has several basic properties which, in combination with standard practices, make it a model for the storage of information in the laboratory (3) and should be mimicked in a LIMS.

1. Notebooks are bound volumes with numbered pages which ensure that pages are not deleted or data added out of sequence. Data are then recorded with historical consistency and so that the development of ideas from experimental results can be easily reconstructed.
2. All entries in a notebook must be signed and dated. This ensures that the responsibility for the information is defined and that the responsible person is accountable for the recorded information. Also, the information must be dated relative to an absolute point in time.
3. All entries must be made in ink and no erasures may be made. The entries in a notebook are meant to be permanent, and any obliteration of data is usually against company policy as well as in disregard of regulatory requirements.
4. Corrections to data must be signed, dated and entered in such a way as to allow the original data to be readable.

These rules, which are common for notebooks used within laboratories independent of regulatory requirements, impose certain basic integrity features on the data:

1. The data are permanent.
2. The ownership and responsibility for the data is clearly identified.
3. The history of the data is clear from the initial recording of the information through any changes or calculations.
4. The notebook can be copies and stored as appropriate to provide for the orderly storage and retention of data as required.

The manual system, therefore, has a well-defined set of features which are considered standard for laboratory procedures. The penalty of the manual systems is the inability to search and use existing laboratory data in a convenient fashion. The use of a LIMS brings with it a wide range of benefits which have been reviewed in other chapters of this book. However, the benefits of a computerized system also bring problems associated with ensuring that the data are recorded and maintained in a fashion equivalent to manual systems. For instance it is difficult to 'sign off' on data displayed on a terminal.

24.4 Data Handling by a LIMS

The central issue is how data should be handled and maintained by a LIMS. This must be viewed in a broad context, because various types of data may be maintained in the database, including raw data and derived data including results. The requirements for each type are different and a LIMS may need to treat all forms the same as raw data, as it is the most restrictive case, or limit the types of data maintained within the system.

24.4.1 Raw Data

One of the major issues involved in the computerization of a laboratory environment is the definition of what constitutes raw data. GLP regulations define raw data as:

> Any laboratory worksheets, records, memoranda, notes, or exact copies thereof that are the result of original observations and activities of a non-clinical laboratory study and are necessary for the reconstruction and evaluation of the report of that study. Raw data may include photographs, microfilm or microfiche copies, computer printouts, magnetic media, including dictated observations and recorded data from automated instruments (1).

Although this may provide a formal definition of raw data, the actual determination of raw data is much more complex and often the automation of a laboratory raises many questions about this issue. Moreover, raw data must be defined by the laboratory or company management and documented in an SOP. For now, assume that raw data are the initial recordings or an exact copy of them and are not the result of any manipulations to create a new reading. This means that raw (primary) data which is collected by a LIMS through automated data acquisition or direct keyboard entry must be treated as raw data (4). That is to say that the authorship or ownership of the data must be identified and that the data may not be obliterated, or changed in any way when corrections are made which would render the *original* observations unavailable for inspection.

Taylor (4) has attempted to define raw data for computer-supported toxicology studies; the same definitions are also applicable to analytical data held or generated by computer. These definitions were presented for discussion and interpretation but nonetheless provide useful guidelines.

1. Where observations are converted to a written record, the written record becomes raw data.
2. Where observations are converted to written records which are photocopied, microfilmed or converted to magnetic media by input from a terminal, the raw data can be defined as the original material, microfilm or photocopied records. The information recorded within the computer is not raw data.
3. When a written record of observations is transcribed by terminal input into a computer and stored in a database. Only the written record is raw data.
4. When observations are input directly (either on-line or as batch records) to a computer database, the database becomes raw data.
5. Where observations are input directly to a database and reports are produced, the computer database can be raw data. Also, the resultant printouts must contain the collected raw data. However, if the printouts contain any derived data, they are not considered raw data (derived means summarised, calculated or manipulated). A key question is how can the laboratory demonstrate that the printouts faithfully contain the raw data?
6. In a situation where observations are recorded by a chart recorder and entries are also made by the operator in a manual log, only the chart recording is considered raw data.
7. Where the observations are recorded by a chart recorder and integrated or calculated values are produced on a printout, either the chart or the printout can be raw data, but not both. The decision as to which becomes raw data is left to the laboratory management and should be documented before any inspection.
8. Where observations are accomplished by equipment that produces printouts with integrated or calculated values, then the printout is considered raw data.
9. Where observations are entered directly as input to a computer, and the computer returns the input as incomplete or erroneous and the operator corrects or deletes data, and finally the observation is entered to the database. The database is considered raw data. Note that any changes to the data after

they have been entered into the database must be documented. If the data are further manipulated, reduced or plotted they are not considered raw data.

These definitions of raw data are useful for laboratories to use as guidelines or starting points for individual situations.

24.4.2 Processed Data

Processed or derived results are data which are the result of calculations, analysis, or some function which is performed using raw data as the input. Consequently, they are not subject to the same constraints which are normally associated with raw data because any system should be able to reconstruct results from raw data. Therefore, ability to reconstruct the history or background for any reported information is the key requirement in the handling of processed data.

A LIMS may handle these data in a variety of ways. Some processed data will be treated in a manner similar to raw data because of their importance, or the method by which they are derived. Some processed data may not even be stored in a LIMS based on the requirements of a particular laboratory. The following gives an example of each case.

A test requires that three individual measurements of a particular parameter are to be performed and the average of the three values reported. It is this average value which is tied to the product specification. In this case the three measured values are raw data and they are stored in a LIMS. When the data are reported the LIMS has a pre-defined procedure which averages the three values and automatically reports the mean. This mean is not stored in the LIMS but is reported.

A second case, which is very common, is the processing of chromatographic data by a LIMS. It is accepted in many laboratories that the retention of the processed analysis data such as peak heights, peak areas, and retention times, is critical, and that this is the actual 'raw' data. However, without related measurements from the standard samples it is meaningless. Therefore, once the standards are analysed and the sample peak areas have been converted into component concentrations they are stored in the LIMS as raw data even though they were derived from original observations. This is because the resulting concentrations cannot be regenerated easily.

24.4.3 Reported Data

Reported data is information which is consolidated into a convenient format for reporting to the submitter of the samples. This report may be organized in order to draw conclusions from the work, or to prove conclusions based on the information. A report is based on the assembly of retrieved information that has been stored in the LIMS by a well defined procedure. A LIMS, although able to generate reports, is not usually used to maintain an archive of generated reports. However, report generation procedures can be run any number of times and the same report should be generated consistently unless an approved change in the underlying LIMS data has occurred.

24.4.4 Data Retention Procedures

These are critical requirements in the GLP environment. A basic requirement of GLP is the 'orderly storage' of all historical data. This requires more than standard back-up procedures which are designed to provide for system recovery in case of a failure. Systematic data retention requires orderly archiving and retrieval functions within the LIMS. These functions should allow the user to extract from the system a specified subset of data, in relation to a logical entity within the laboratory, such as samples or studies.

When data is archived and deleted from the database, this should be recorded by the system, preferably via an audit trail, which maintains a record of the data archived for reference. The archive function should extract the data into a specified format, and after a successful comparison test, delete the data from the on-line system for efficiency. The archived data can then be removed from the on-line data storage facilities and archived. The system must also provide facilities to restore archived data back into the database in order to provide further reports or responses to regulatory review.

If archived data is not deleted from the database but maintained on-line for reporting and retrieval, the system must ensure that the data cannot be changed, in order to ensure the integrity of the archive.

24.5 Security

Security is focused on meeting the objective of making a LIMS, and the data contained therein, as tamperproof as possible. Security in a LIMS involves many issues. Before these can be discussed the scope of the security must be examined. What is security? How does it apply in general to a LIMS?

For the purpose of this discussion, security can be defined as those features within a LIMS which are designed to restrict access to functions or data. This broad definition of security can be divided into two major categories.

1. The protection of a LIMS and the data from those who may have access to a shared system environment but who are not known by, and are therefore restricted from access to the LIMS. This type of security is important in a data processing environment where the LIMS software is installed and running in a timesharing computer environment. This is opposed to a LIMS

residing in a local computer environment, where system and LIMS access limitations are one and the same.

2. Restricting access to functions and data within the LIMS based on the identification of users and their position within the laboratory organization. This is essential so that information in the LIMS is handled with the same level of integrity as manual systems which would be used in the laboratory environment, in its absence.

24.5.1 User Identification

Any limitation of access to functions and/or data within the LIMS is directly related to the LIMS ability to suitably identify a particular user. The mechanisms which are available to identify any user are an important factor within the LIMS environment. The identification of users in the LIMS must ensure uniqueness of each individual and should not be easily mimicked by other users. Features such as passwords or badge readers can be effective for this purpose. However, procedures should be established outlining restrictions to ensure that system users do not abuse and thereby nullify the identification process, eg. the writing of passwords on reminder pads should be prohibited and the regular change of passwords should be required. In addition, although each user will generally have only one identification, this may need to support various dimensions such as:

1. *Personal user identification* Many functions within a laboratory and therefore the LIMS require that an individual has responsibility for data. This requires that a LIMS can identify individual users.
2. *User group identification* There are many functions within the laboratory where a group of people need to be considered as equal. In general, this allows the LIMS to function in a way which is similar to the actual structure and organization of the laboratory.
3. *User responsibility identification* The laboratory is often a complex matrix of interactions which are related to a person's position in the matrix relative to others in the organization.

If the non-automated laboratory is reviewed, a clearer perspective of the issues to be addressed in this area can be seen. Without a LIMS, there are well defined procedures and responsibilities which ensure that data are appropriately collected, reviewed, and approved. The addition of a LIMS is not intended to avoid these practices and procedures but to make them more effective. The requirements for the ownership and review of data are not modified in the eyes of the regulatory agencies by virtue of automation.

24.5.2 Functional Security

One of the major security functions required within the LIMS is to prevent people from carrying out tasks which they are unauthorised to perform. These functions currently exist within the manual laboratory because each individual does not have an equal level of responsibility or authority. The ability to centralize functions within the LIMS and to make them *easily* accessible should not equate to making them *equally* accessible.

One mechanism which limits access to LIMS functions is often referred to as menu-based security. Each user may have an individual set of menu options which only allows access to approved functions based on the user identification. Although this method provides a basis for the protection of information from unauthorised access it has a limitation in that it only prevents or limits access when using the LIMS software. If the user can gain access to information in the LIMS through tools which are generic to the system, and outside of the LIMS control, then this security may not be sufficient.

At present, there are two main types of security used for a LIMS: class and heirarchy. In the heirarchical approach, the users and the software modules are assigned a numerical value, a user can access any module with a value less than or equal to their own but none with a greater value. In contrast, the class approach allows users to access the specific modules approved by the system manager. In either case, modifications to the user's or the software modules security rating can be made on-line.

24.5.3 Data Security

The responsibility for data security rests with the laboratory management; however, this is often delegated to the users. The manual methods in the laboratory provide an inherent security mechanism because data is physically passed from the owner to a manager in written form for approval. This makes the approval of data by inappropriate personnel essentially impossible because they do not have access to the data record. However, in a LIMS the data is stored in the system, and review and approval of data is handled electronically. This requires a careful definition of a security mechanism to ensure the integrity of the data in the system.

Therefore, the security procedures concerned with data are important to ensure that data are only approved by appropriate (approved) managers. Also, it is essential that data are not modified by a user without appropriate authority, or released before all necessary approvals have been obtained.

In determining the data security requirements for a LIMS, it is important to clearly define the flow of data which is required within the organization. It would be naive to assume that a LIMS which is installed in one organization would be appropriate for another. In this area, even a system which is 'GLP Compliant' may not be suitable.

An initial task is to determine the levels of approval which are required. In many QC or production laboratories there are only two: the first is the analyst or technician responsible for collecting and reporting the information, and the second is review and approval by the laboratory supervisor. However, in many R&D organisations there is at least one additional level of review, ie. scientific review to determine the impact of the data before they are approved for reporting or release outside of the laboratory environment.

Because of the many interactions in the laboratory in relation to data management it is important for a LIMS to have a tightly integrated system which involves the functional security, data security (approval) and data audit trails which will be discussed in detail in later sections.

24.5.4 Integration of LIMS Within a Laboratory

In the overall analysis there is one critical factor which must be addressed in the development of security mechanisms of all types within the LIMS. This is that the manual procedures of the laboratory should be accurately mirrored in the implementation of security within the LIMS environment. It can be handled from two very different and equally viable positions.

The first position is to base the future on the past. It is assumed that the current procedures within the laboratory are in compliance with GLP requirements. Using this as a working basis, the security mechanisms, including availability of functions, data ownership, the review process, etc. can be designed into the LIMS. For this to be successful, the LIMS must either be implemented in a way which enforces a security mechanism which is consistent with the laboratory, or the LIMS must have configurable security features to allow the system to be modified so that they become consistent with current laboratory practice.

The second position to meet this goal is to make the future correspond with the principle implemented within the system. Although this may seem a dangerous thing to do, there are many LIMS applications which have been implemented to ensure that the security features which are reasonable, workable, and can be enforced within the LIMS, are available. A review of these may lead the scientist to believe that the LIMS security features are equal or superior to the procedures currently in practice. This, in turn, can lead to the development of new procedures in the laboratory which are, as a result, mirrored in the LIMS.

In reality, a LIMS implementation involves a merging of the two extremes presented above. The security functions which are included in the application are based on current practices but tempered by what is technically feasible to implement within a computerized system.

24.6 Data Validation

Data validation features within a LIMS provide a broad range of data integrity features which ensure that information stored in the LIMS is compatible with expected results and data storage restrictions. These validation checks can cover a broad range of techniques which also improve the reliability and usefulness of data within the system. The use of validation checks within the system, especially if adequately flexible, will provide for various types of validation checks and provide significant data integrity benefits. These have been discussed in Chapters 4 and 9 from the viewpoints of the overall benefits of LIMS and as features of databases respectively.

24.6.1 Data Type Checking

This can be used to determine that the information entered into the LIMS complies with a pre-defined format. With user-defined data type checking the format and allowable characters can be defined and checked on data input, thereby reducing the typographical errors which may occur.

24.6.2 Data Range Checking

This can be used to ensure that the information entered into a LIMS is not only in a correct data format, ie. numeric data, but it also allows a preliminary check for erroneous data, or data outside of specifications. For example, in the recording of a pH value, the system can be configured to accept only values in the range of 1.0 to 14.0. An extension to this type of data checking is to allow the system to maintain several ranges of data. The initial range may be values which are legitimate results; a second more restrictive range may be values which reflect the specification limits for a material and test combination. Values outside of the range may be entered but should be automatically flagged for the review process.

24.6.3 List Checking

This is a checking function which should be available and used for validating the entry of text results. It addresses the problem of various LIMS users defining a similar result using different terms. An example may be a colour test. In a manual system, the review or search of data involves human interaction, and abbreviations or terms which are synonymous can be easily detected. However, in searching an electronic database for text results, data entered in different formats with the same meaning is difficult to detect. By checking all entered results against a *validation list* a system can ensure that a limited pre-defined set of terms is used for a particular test. This simplifies the data review and subsequent data searching processes.

24.7 Audit Trails

24.7.1 Purpose

Two of the most critical aspects of any LIMS from a GLP perspective are 'accountability' and 'historical reconstruction'. Accountability is critical in tracking who is responsible for data and its approval for release 'the individual responsible for direct data input shall be identified at the time of data input'(1). This also holds true for information which is entered to modify the initial data entry. In addition it is important not to 'obscure the original entry'(1) and to indicate the reason for change. It is this requirement in all data handling which allows the history of results reported to be reviewed in detail, and therefore allows the reconstruction of historical events.

The purpose of an audit trail is to meet these requirements, allowing a LIMS to provide a coherent and complete history of all information which is maintained within the database. The implementation of an audit trail can be achieved from a variety of perspectives. However, not all methods allow the system to meet these basic requirements for valid audit trail functionality.

The key requirements of any audit trail may be summarised as follows:

1. Record the original results for all data which is modified.
2. Record the identification of the person changing the data as well as the time the new data was entered.
3. Require the user to enter the reason for the change to the data.

24.7.2 Types of Audit Trails

With the basic requirements identified, there are still many types of audit trail used. The variety of these is generally dependent on the tools available and/or used for the development of the LIMS.

In reviewing the capabilities of an audit trail it is important not to confuse database journaling or the maintenance of a transaction log with a LIMS audit trail. Journaling is a database function which allows the system to record all changes made to the database, including additions of new information. Generally, a database system will provide a function referred to as *after image journaling* which records the state of the database after a complete set of database transactions. This allows the database to be reconstructed, generally to any specified point in time, if the database becomes corrupted. However, it does not allow reasons for change to be associated with updates to the system; moreover, it may overwrite records, leaving only the latest change on file. In addition, the journal file may not be in a format which allows a report of changes or a report of the original data to be generated. Therefore, this should not be considered a viable alternative to a system audit trail.

Audit trail functions must be under the control of the LIMS application software. In a database environment, the best way to ensure data integrity would be to restrict modification or update of data to the application software. In this way, when data is selected for modification the system can ensure that the audit trail function is invoked. The audit trail function must perform the following operations:

1. Enforce any laboratory restrictions on the modification of data which has been approved or validated.
2. Store the original data in an 'audit trail' either physical or logical.
3. Require the user to enter a reason for the data modification. A list of valid reasons selected by a code or menu should be optionally available.
4. Record the new information in the data record, using the current user identification and date.
5. Mark the modified data as an 'audited' data record, noting the existence of an audit trail.

Audit trail functions may be implemented inside or outside of the database environment. There are benefits to both ways. An audit trail within the database provides better reporting functions. However, an external audit trail can be more easily protected, increasing the overall integrity of a LIMS. A major concern is the protection of data from changes which are not made using the LIMS software which has been developed to ensure appropriate audit trails for all changes. From this perspective, the integrity of the LIMS audit trail needs to be carefully validated to ensure the integrity of the overall system, in addition to meeting regulatory requirements.

Another function which may be used to augment a data audit trail, but not replace it may be referred to as a system *activity log*. This is a log which is used to trace all access to the system whether it is for the purpose of modification or the standard entry of information. A log of this sort is capable of providing a sequential record of all system activity and is helpful in reconstructing the exact order data was entered and manipulated within the LIMS.

An important consideration within the scope of system audit trails is the ability to recover from system failures. Instances such as disk crashes or media failure will occur and the procedures to recover from these failures will need to be well defined. An audit trail or related logging procedure may be invaluable in determining what data may require re-entry into the system based on the dates of system back-ups used for system restoration.

24.8 System Operations and Maintenance

The final issue is that of operating procedures for the laboratory. The regulatory requirements are concerned that the operating procedures for any laboratory are clearly documented and followed (5). A review of a facility will often focus on the procedures of the laboratory and their compliance with the GLP regulations. The integration of a LIMS within the laboratory environment requires close examination of existing policies and a re-definition of the working practices in the areas where the system is to become an integral part of standard practice. This area is discussed in more detail in the next chapter.

24.9 Summary

The introduction of a LIMS does not change the regulatory restrictions which a laboratory is required to follow. Although GLP and related documents do not specifically address computerization in detail, the requirements applied to manual data handling must be translated to parallel requirements in a LIMS. A reasonable test is to require a LIMS to mimic as closely as possible the procedures which are currently enforced under the manual procedures.

24.10 References

1. Department of Health, Education and Welfare. Food and Drug Administration, 'Non-Clinical Laboratory Studies, Good Laboratory Practice Regulations', *Federal Register*, 43 (1978) No. 247, pp. 60018-60019

2. *Toxicology Laboratory Design and Management for the 80's and Beyond*, Tegeris, A.S. (editor), Karger, Basel, (1984), pp 77-83

3. Mattes, D.C., and Clark, B.J., 'Development of Systematic Data Access and Integrity Features for Pharmaceutical Computer Applications', *Drug Information Journal*, in press 1987

4. Taylor, D.W., 'Inspection of Computer-Supported Toxicological Data Submitted to the FDA', *Drug Information Journal*, 18 (1984) 189-194

5. Figarole, P.L., 'FDA's Views and Experience: Inspecting Computer Supported Studies', *Drug Information Journal*, 17 (1983) 7-9

Chapter 25

Procedures and their Documentation for a LIMS in a Regulated Environment

Elizabeth H. Brown

25.1 Introduction

In regulated industries, documentation is the only proof available to an inspector that the job of system development and maintenance is being done in an organized and acceptable fashion according to established procedures. Documentation should be considered one of the most important types of record that can be maintained as part of a computerized system. Complete documentation can maximize the return on the cost of the system since missing or inadequate documentation can cause, not only financial losses and time delays, but also regulatory citations.

The following is a list of the types of documentation that are important to the introduction and use of computer systems in regulated industries:

- Procedures for systems implementation
- Hardware and software descriptions
- Operating instructions
- Results of tests
- Training records
- Back-up procedures
- Approvals for development, release, change of the system
- Security
- Ongoing evaluation and maintenance schedules

The task, then, is to design and put into practice procedures which will guarantee that the necessary documentation will be produced in a consistent and organized fashion. These procedures may be categorized as follows:

- Security Procedures
- Change Control Procedures
- Training Procedures
- Quality Assurance Procedures
- Standard Operating Procedures

A Standard Operating Procedure (SOP) not only describes the documentation but also defines the procedures by which it is produced. The description of the procedures and the subsequent documentation are equally important.

It is irrelevant where the LIMS is situated or who maintains it, procedures must be in place in order to maintain control of the system, its software and to satisfy a regulatory review. Each of the categories of procedures noted above will be considered in detail in the following sections.

25.2 Security Procedures

Although there are many definitions of security, the type of security violation that is most likely to occur and against which reasonable defence can be established is the accidental destruction of electronic data caused by power failures, computer crashes, operator error, physical failure, and authorized users.

Destruction of data by intentional means or by natural disaster is less likely to happen than the accidental breach of security by an unsuspecting user of the computer system. From a systems point of view the issue of security can be addressed in four different areas: physical security, system security, data integrity, and data archive.

25.2.1 Physical Security

Issues of physical security include the handling of all types of computer hardware. The following should be considered in developing a security procedure:

1. Restricted access to hardware: both physical and telecommunications access to a central or laboratory computer.
2. Maintenance of the correct environment for hardware.
3. Correct handling of storage media: eg. generating back-up copies of floppy and magnetic disks, exercising magnetic tapes to maintain viability.
4. Storage of sensitive data off-site, perhaps in several locations.
5. Development of disaster recovery plans to provide alternate computing sources if required.

Documentation is required to support these procedures. This includes logs of computer access, chart recorders to monitor environmental conditions, and maintenance logs of data back-ups and tape exercising.

25.2.2 System Security

The area of system security involves those functions of the computer that can be applied to impose security. For example:

1. Use of account, password, and protection coding so that only authorized persons have access to the source code and data.

2. The extension of password security by causing passwords to expire automatically after a given time period.
3. Use of programming facilities to ensure that the minimum amount of data are lost if the computer should crash, together with the availability of compatible alternate systems to capture data.
4. Implementation of programming techniques to prohibit data modification, (except as designed into the system) eg. binary (uneditable) files, encrypted databases, or database passwords, should be described.
5. Removal of old user identities and passwords after staff have left the laboratory.

Procedures such as these should be described in the SOPs that outline system development techniques and security.

25.2.3 Data Integrity

There are several data integrity issues that apply to the development or acquisition of secure computer systems including LIMS that should be described in the SOPs for a laboratory.

1. The definition of raw data must be stated (several possible definitions have been given by Taylor (2) and in Chapter 24). This will enable it to be identifiable and steps taken to ensure that it is unalterable. Error correction must be done so as not to obscure the original data and must include date and time, who made the change, and why the change was made.
2. Data integrity is improved by the implementation of direct data capture when possible: eg. electronic balances, bar code readers, analytical instrumentation.
3. Programs should take advantage of the fact that the computer can date-time stamp activities that affect or alter raw data or results.
4. When appropriate, instrument calibration should be forced to provide additional data validation.
5. Records must be maintained regarding computer down time and hardware problems; this information can provide clues to answer questions that arise after possible hardware causes are forgotten.

25.2.4 Data Archive

There are two issues involved in data archive: the data itself and the computer programs that operate on the data.

Back-up procedures must exist so that programs and data can be retrieved after they have been archived. Archive and retrieval of data or programs must be consistent, correct, and dependable. Consideration must be given to the length of time data must be retained. The chosen storage media must satisfy the retention period. Records must be kept with details of the tapes used and their contents, in the case of programs the version number should be given.

As computer systems undergo regular modification, either by a vendor or by in-house staff, two considerations must be kept in mind:

First, although the current programs operate on current data, consideration must be given to the operation of programs on older, perhaps archived, data. If the new programs will not handle the older data format, then programs which will handle old data must also be archived. Differences between the old and new versions of the programs must be documented so that differences in results can be explained.

Second, any program modifications must take into account data which has already been collected and processed using older versions of the software, but which is part of an ongoing study or process. Handling of such data may require conversion to the new data format and subsequent validation to ensure that the conversion was done correctly.

In other words, the interaction of computer programs and data, both of which may be changeable, must be considered in order to demonstrate the retrievability and validity of new as well as old data.

25.3 Change Control Procedures

Change control is a procedure which ensures that modifications to a computer system are performed in accordance with established standards and performance criteria. It aids in preserving the validated status of existing computer systems.

Historically, there have been two major definitions of change control:

1. The handling of a modification (either software or hardware) to a computer system, either by the vendor or by in-house staff.

2. The evaluation of the cumulative effects on the operation of a computer system due to changes in software or hardware external to the computer application; eg. changes to hardware components such as floating point processors or software such as compilers, or purchased software such as statistical programs.

All computer professionals know that the mechanisms used to change computer software do not inherently provide any measure of change control. Depending on the organization's tolerance for risk, different procedural mechanisms can be implemented to provide the desired level of control.

The FDA realises the ease with which software can be modified and has recommended that all versions of a program be maintained so that a complete history of the program can be reconstructed. This can be accomplished by maintaining copies of modified routines on tape or other permanent media or by means of code management software which is available from several vendors.

Another area of concern to the FDA is the control of software maintenance and the continuity required to ensure that maintenance can be continued even after the original programmer is no longer available. The documentation produced to support the maintenance effort is therefore an important part of the system documentation which must be maintained by the vendor or by in-house staff.

25.3.1 Authorization

There must be a well-defined mechanism for initiating a change or enhancement to a computer system. This should handle any change, whether it is a scheduled upgrade or a change necessitated by unexpected failure. Preferably the change request should be written and approved by the user department prior to submission to the programming group.

Review of the impact of the change should take place among the users and the system staff before the change is implemented.

25.3.2 Implementation

Software must be secure from all except those authorized to change it. Most computer systems provide security schemes that can be used to protect the programs and source code. Special care must be taken that the source code is not readable by those not unauthorized to change it, restricting the reading of source code prevents copying and subsequent unauthorized modification.

Software undergoing change should be identified as to the current revision level. This is necessary so that an inspector can verify that the software being used is the latest version. There are various mechanisms available for determining the current revision dependent on the hardware and software being used. Some software allows an edit number or other designator to be attached to the source file and corresponding executable module. In other cases, dates of modification can be used to determine a relationship between the source code and production module. For interpretive languages there is no executable or production form for a module; in this case other means must be used to ensure that the module being used is the same as the module that was tested and validated.

The maintainer of software should make changes to a copy of the software, keeping a copy of the current version intact. As changes are made, tests that were applied to the current version of the software should be used to test the new version. Records of the tests performed should be maintained; if some tests are not done, the rationale should be explained as part of the testing documentation.

An implementation of new software can be aided by providing the user with both a current version of a program and the revised version. The user can thus help to test the program, and a more orderly implementation of the new program can be accomplished. Special care must be taken in such an instance, however, so that data or reports generated using unreleased programs are handled correctly and that the test documentation clearly describes the procedure followed in performing the tests.

As part of a change control procedure, mechanisms for identifying program changes should be handled in the same way as data correction: who, when, how, and why should be noted as part of the maintenance documentation. This is necessary so that SOPs or other references to the software reflect the software actually used to support a study.

Review of the changes should take place before the changed program is released to the users. The review process should be formal and should include any necessary approvals and sign-offs perhaps from management, users, and the Quality Assurance staff.

25.3.3 Impact

Changes to a program can have an impact on software in other areas, as often modules are developed and used across many different computer systems, as part of a good programming practice procedure. An error or a change in a module can affect the operation of another, perhaps unrelated, program. Therefore, some mechanism must be maintained that allows the computer staff to ascertain the effects of changing a module on other programs. Additionally, historical data must be considered so that software changes do not invalidate existing or archived data files.

Modifications can also have an impact on the users; part of the final review process must include an evaluation for additional training in order to implement the change correctly and efficiently.

25.3.4 Revalidation Triggers

Changes in equipment, system software, Standard Operating Procedures, and application software must be documented and evaluated in terms of the effect they may have on the functioning of the LIMS.

The FDA expects that once software is released, it is subject to periodic review. Such periodic evaluation, during which significant changes in the computer system are recognized and addressed, may trigger a phase of revalidation of the system as a whole. Within the pharmaceutical industry it is unclear whether such a periodic review should be based on time or modifications, but one or the other should be chosen, documented and followed.

25.4 Training Procedures

The GLP regulations state that:

...each individual engaged in the conduct of or responsible for the supervision of a nonclinical laboratory study shall have the education, training, and experience to enable that individual to perform the assigned functions (1).

The success of any computer system is affected significantly by the training of personnel.

25.4.1 The Developer

The developer or evaluator of a computer system must be trained and experienced in the job of computer system design and implementation. Furthermore, in order to evaluate or design a computer system for a regulated environment, the developer must also be familiar with the process which the computer system will replace.

The programmer or project leader should also have broad knowledge of the established programming practices within the company or programming group. He should understand the business that the company does. Also he must be aware of the requirements necessitated by adherence to GLP.

Unless the system developer is adequately trained, the software will not accommodate the functions and activities specified in the study protocol, nor will it satisfy the requirements of the programming staff and management.

25.4.2 The User

As with manual systems, the user of a computer system must be trained and supervized. Experience has shown that training in the use of a computer system should probably be repeated periodically in order to maintain the level of training desirable to perform the task.

Records of all training activities - including procedure reviews, technical training, and interaction with the computer staff - should be maintained.

25.4.3 The Auditor

The Quality Assurance auditor must have sufficient training in the use of the LIMS and its tools to perform complete audits. It is possible to give the auditor access to the system so that he may operate it as part of an inspection. The role of the auditor is discussed in more detail in section 25.5 of this chapter.

25.4.4 The Curriculum Vitae

The GLP regulations state that:

...each testing facility shall maintain a current summary of training and experience and job description for each individual engaged in or supervising the conduct of a nonclinical laboratory study (1).

Considering the involvement of computer professionals in the handling of laboratory studies on the computer, this regulation should be extended to include the training of the programming or other support staff. One way to accomplish this task is to maintain a set of Curricula Vitae (CVs) to detail the educational and employment experience that is relevant to an individual's current responsibilities. Records for programmers at a vendor site may be more difficult to obtain, but assurance that adequate records are kept by the vendor may be obtained in writing as a substitute.

CVs should be reviewed periodically and updated whenever necessary, including a review when an employee leaves the company. The CVs should be accessible to an FDA inspector if required.

Information on a CV should include the employee's education, employment experience including positions at the current company as well as other jobs, professional affiliations, qualifications and societies related to the current work, and continuing education such as technical meetings and special courses.

25.5 Quality Assurance Procedures

The Quality Assurance Unit (QAU) should maintain its traditional role of auditing and critically evaluating systems and operations. This should include the development, operation and maintenance of computer systems. The QAU should not, however, attempt to become expert in the field of computer systems. Its role is to inspect and audit the procedures in place to ensure the quality of those systems. It must understand the process and the role of the software so that it can assess the system specifications, conformance to the specifications and adherence to the regulations.

25.5.1 Audits and Inspections

The regulatory mandate for the QAU is:

to assure management that facilities, equipment, personnel, methods, practices, records, and controls are in conformance with the regulations (1).

To that end the QAU should inspect and audit not only the use of the computer system, but also the computer operations and the system development or evaluation stages.

The expertise of a QAU in the evaluation of the seven areas below (Gittleson, personal communication) should provide them with a good background for auditing and inspecting computer systems.

1. *Facilities* Environmental conditions and physical security.
2. *Equipment* Functional and technical specifications for hardware, maintenance procedures, power considerations and log books.
3. *Personnel* Training, qualifications, levels of access for various personnel within the systems staff.
4. *Methods* System development life cycle, historical retention of data and software, documentation and SOPs.
5. *Practices* Inspections of software development and use with actions and findings documented.
6. *Records* Identification of raw data, archives, audit trails, storage of magnetic media in conformance with rules for storage of paper.
7. *Controls* Access to source code for the purposes of modification, data modification, handling of archived data, back-up procedures, revalidation schedules, authorization of program changes.

The QAU should develop an audit checklist listing the items to be inspected during an audit and to facilitate the retention of audit records. Above all the QAU must assure itself that the development and maintenance process for the computer system is in control.

25.5.2 Retrospective Validation

For computer systems that have been in place for some time, it is often impractical to perform a prospective validation because much evidence for the quality of the system already exists. If the QAU has been performing audits of the use of such computer systems, the information in its records can be used as a starting point to generate a retrospective validation of the system.

A retrospective validation is the gathering of documented evidence to show that an existing system does what it purports to do. More information regarding retrospective validation can be found in the article by Ettorre (3). Such information may include the following:

1. The determination of whether development or acquisition of the existing programs have followed current standards and practices.
2. Gathering of information to support the fact that validation did occur although documentation was not kept, including monthly reports, personal notes, and memos.
3. Inspection of training records.
4. Examination of the audits and inspections done on studies using the computer system, including the number of studies reviewed and the responses to any findings.

25.5.3 Education

An additional task that might be performed by the QAU involves the education of all employees regarding the regulations and the impact of the regulations on internal procedures. Both scientists and technicians should be involved in training of this type, as well as the necessary programming staff and their management.

Educational materials that may be generated for each attendee include information on SOPs, the GLP regulations, CVs, and FDA inspection information. Attendance at such a training session should be included as part of each employee's CV.

25.6 Standard Operating Procedures

According to GLP:

...Standard Operating Procedures (SOPs) are detailed operating instructions for personnel involved in the generation and documentation of data derived from nonclinical studies (1).

Traditionally, SOPs have been written for scientific activities as well as the operation, calibration, and maintenance of equipment. SOPs must be extended to those tasks involving computer software and hardware acquisition, development, operation, and maintenance. Laboratories with computer systems must include operations information and use of the software as part of their SOPs.

The FDA has found most notable deficiencies regarding SOPs to be:

1. The absence of SOPs.
2. The failure to amend SOPs when necessary.

25.6.1 General Information

An SOP is an operating manual and is not written to apply to a single study or computer system. Management within each company determines the adequacy of the SOPs. An important reason to have SOPs in the area of software development is to ensure that all programming personnel are familiar with and use the same procedures in the accomplishment of a task. This is a direct corollary to the purpose of an SOP in a laboratory. Ideally the standardization of the job via SOPs, whether in a laboratory or a computer systems development setting, should help prevent the introduction of systematic error in the generation, collection, and reporting of data.

SOPs must be clearly detailed and realistic in scope. They should include routine, repetitive, and unique operations and they must be accessible to those performing the procedure and to those who reference them in their own SOPs. They must be available to the regulatory agency during an inspection.

SOPs should be produced by those responsible for doing the job. They are written according to rules established within each organization regarding format, content and scope. The FDA requires that an historical file of SOPs be maintained and that all revisions be identified according to date.

Keeping all the rules in mind, it is imperative that each organization has a definitive procedure for the generation and maintenance of SOPs including strict archive procedures, ie. an SOP for the writing of SOPs. This document should elaborate on the following general areas:

1. *Preparation:* the forms to be used, numbering scheme, page numbering standards.
2. *Initial review and approval:* responsibilities, handling interdivisional or interdepartmental SOPs.
3. *Issuing and distribution:* by whom, to whom and the tracking of each document.
4. *Revisions:* timeliness, handling major and minor revisions, issuing of update pages or update SOPs, and dating of revisions.
5. *Deletions and consolidations.*
6. *Periodic review.*
7. *Retention:* location and archive of all the current and superceded SOPs.

25.6.2 SOP Building Blocks

A LIMS may be the responsibility of a single or several, departments depending on its implementation. Since the environment may be different at each individual installation, the SOPs required to define the necessary procedures to maintain the system and keep it in control may take different forms. The following are some general categories of information that may be included in SOPs, whether the procedure is handled within the laboratory or within separate systems and operations groups.

If programs are developed or modified, then a systems development life cycle (SDLC) should be defined and followed. The FDA is aware that quality software is produced if an organized development process is followed. Therefore programming standards and development procedures should be in place. These might include the following: concepts applicable to security, accounting, change control, coding standards, and documentation requirements. Additionally, there should be validation standards that define those programs requiring validation, when validation is done, the validation protocol, responsibilities, various validation techniques, supporting documentation, approvals, sign-offs, and periodic revalidation criteria.

Whether the hardware is operated by the laboratory staff or by an operations group outside the department, SOPs for the operation of the computer must exist. Included in the SOP are diagrams and schematics of the hardware or references to other documents where this information can be found, a description of the function of the system, environmental considerations, and description and maintenance information for any alarms. Administration of the system might include system access, operator training, data back-up procedures, security issues, startup and shutdown, maintenance, acceptance testing and diagnostic procedures, and record keeping.

The documentation that might be generated by the various procedures described in the SOPs includes the following:

- *Project Request* defining the objectives of the system (or the requirements specification).
- *Functional Requirements Document.*
- *Data Requirements* describing technical information about the data, including static data, dynamic data, and derived data.
- *Program Specifications* which describe the system requirements, operational environment, and perhaps performance criteria.
- *Program Maintenance Manual* containing information for the maintenance programmer including linkages, system features, file handling, file layouts, restrictions, dependencies, and storage of programs and documentation.
- *Test Documentation* including a test plan, test descriptions, test results, responsibilities, test schedule, approvals, and record keeping requirements.
- *Change Control Documentation* including historical information regarding changes made to the system, requests, approvals, and information necessary to determine the software actually used to generate any particular data.
- *User Documentation* including user guides and operations manuals to support the user in performing the job.

Vendors should have similar records that can be acquired or, minimally, reviewed. Assurance that a vendor does the job in an organized fashion is as important to the purchaser of a system as it is to an inspector who examines the results of the use of the system.

The documentation to support the operations aspect of a computer system consists of the following:

1. *Operations Logs* detailing the status of the system at any one time, including startup and shutdown.
2. *Maintenance Logs* listing the diagnostic operations including reasons for diagnostics and results.
3. *Data backup* including schedules and locations.

In a sense the computer used in laboratory functions is considered in the same way as any other piece of instrumentation. It is calibrated, maintained, and operated according to specific instructions, and records pertaining to regular as well as unexpected situations are maintained.

25.7 Summary

The FDA and other regulatory agencies are developing expertize in computers in order to deal with the growing number of computers involved in the conduct of non-clinical laboratory studies.

The agency's major concern is that computer systems be acquired, developed, maintained, and used in a state of control and that sufficient measures should be in place to maintain that control.

Several procedures have been discussed. The common ground for each is the generation of sufficient, consistent documentation in support of the procedure. Such documentation includes not only SOPs but also the documents mandated by the SOPs. The documentation itself must be shown to be in control, reviewed and approved, and stored in a manner facilitating retrieval.

Unless company policy prohibits it, most documentation discussed in this chapter may be shown to an FDA inspector. The exceptions might include the QAU audit and its inspection reports.

Each company involved in activities regulated by a government must be aware of its responsibilities, and the amount of risk it is willing to take regarding the impact of these procedures on overall productivity. However, it is possible to overdocument only if the documentation is shown to be false. After all, documentation is the only tangible evidence an inspector can review in order to ascertain whether a computer system is doing the job it is expected to do, is in control, and will remain in control.

25.8 References

1. Department of Health, Education, and Welfare. Food and Drug Administration, 'Nonclinical Laboratory Studies, Good Laboratory Practices', *Federal Register*, 43 (1978) No 247 pp 60018-60019

2. Taylor, D.W., 'Inspection of Computer-Supported Toxicological Data Submitted to the FDA', *Drug Information Journal*, 18 (1984) 189-194

3. Ettorre, L.M., 'Retrospective Validation of Existing Computer Systems', *Drug Information Journal*, 21 (1987) *in press*

Chapter 26

Future Trends

Harmon W. Brown

26.1 Introduction

Today there are hundreds of LIMS operating in laboratories throughout the world. Typical users include analytical facilities in the pharmaceutical, environmental and oil industries. The primary benefits that these systems provide are reduced error rates, quicker delivery of results, and improved access to archival information. There are many other benefits, but these vary with the needs of the lab. Features such as audit trailing, cost accounting, and automatic sample scheduling are often considered essential.

In spite of the numerous benefits that LIMS can provide an analytical laboratory, the process of implementing a system is invariably difficult and frequently painful. Quite naturally, people resist change; they want the LIMS to mimic the way the laboratory has always dealt with the flow of data and information. However, after using the LIMS for six months or so, they begin to realize how much more the system can provide. Clearly, LIMS is an emerging technology; not only from the user's perspective, but also from the supplier's.

26.2 The Changing Needs of the Laboratory

Laboratory automation has made considerable progress in the last few years. There is now increasing use of robots, automatic sample changers, and microprocessor controlled instruments. It is not uncommon to load up a sample tray, put it into an automatic injector, and have a chromatograph produce an almost uninterrupted stream of data for 24 hours. Fortunately, this is one area where the data often can be transmitted directly to a LIMS. But frequently that is not the case, many of the spectroscopic instruments such as FTIR, GCMS, and FTNMR can generate enormous quantities of data. Ways of getting the data directly into the LIMS must be found; this is an industry-wide problem that is begging for a solution.

With the low data rate instruments such as balances, pH meters, and colorimeters, the analyst usually is forced to enter the data manually into the LIMS; this is error prone, tedious, and inefficient. Some progress is being made in the use of bar codes to identify samples, operators, instruments and locations. Here a simple optical device can read the bar codes and transmit the data directly into the LIMS.

Another changing need of the analytical laboratory is the growing requirement for raw data storage. Regulatory agencies are dictating this, so that at a later time the data can be re-analysed, if necessary. There is some disagreement as to what constitutes raw data. Some agencies require a very specific reporting format for the data; for instance, a single GC analysis report may consist of five or ten pages. In this age of common place law-suits, it is imperative to have a fast, accurate means of retrieving archival information for companies that deal with chemicals. Therefore, it is no surprise that there is considerable interest in LIMS in order to collect, collate, report, and retrieve analytical data and information.

Many people approach the idea of LIMS as little more than a means for storing and retrieving large amounts of data. A second stage of use involves re-formulating the data and presenting it in different ways, which could be considered information rather than just data. An example might be a warning report on an analysis that lists the tests that were outside acceptable limits, in contrast to a report that simply lists the concentration of the components.

Yet another level of use is laboratory management. LIMS is a wonderful tool for managing and presenting data, but why not extend it into a comprehensive system for laboratory management? A system could create daily work schedules, monitor performance and even recommend the purchase of additional equipment to alleviate a bottleneck. These and many more tasks encountered by the manager of an analytical laboratory can be aided by modern computer systems.

26.3 Analytical Instrumentation

As mentioned above, analytical instruments are becoming more automated each year. Indeed, it is almost the exception when a new instrument does not contain a microprocessor. The primary function of these devices usually is to control the instrument, but they also provide some level of data processing. However, since the data processing needs are ever-changing, the programs should be sufficiently flexible to cope with the change. A better scheme is to pass the raw data to an external general purpose computer that can be programmed to deal with the data in a more flexible fashion. The other side of the coin is that there are disadvantages to having all the instrument control under an external computer, where one is changing programs and perhaps stopping the computer. A preferred scheme is to pass the control parameters from an external computer to an internal microprocessor, and have the best of both worlds.

A consequence of this approach is the existence of a digital connector on the back of the analytical instrument. The two most common types are RS-232C and IEEE-488. This might lead you to think that it should be straightforward to connect these directly to a LIMS. Unfortunately, this is not so. RS-232C and its variants RS-422 and RS-423 have no universal protocols. For example, there are several software and hardware 'hand-shaking' methods; the word length may be 7 or 8 bits; the number of wires in the connecting cables may be 3, 4, 9, 15, or 25.

IEEE-488 is a much more rigorous standard. There are no questions as to how the electrical connections are made or the length of the data word (8 bits). But there is no industry standard communications format for analytical instruments. The electronic test and measurement equipment manufacturers are much more advanced in this regard. It is really quite simple to take an electronic function generator, a digital voltmeter, an oscilloscope, and a desk-top computer, connect them all together with IEEE-488 cables, then write some simple software and have a working electronic test system. It would be wonderful if we could do the equivalent in the analytical chemistry laboratory. Take a sample preparation system, an autosampler, a chromatograph, a data station, and a LIMS (all from different vendors), connect them together, write some simple software, and then have the whole thing play. It could happen. It should happen.

26.4 Computer Hardware

Nearly everyone involved with technology has some appreciation of the rapid rate of change in the world of computer hardware. Let us review some of the highlights. Obviously, computers are getting cheaper. There are perhaps three main factors responsible for this: improved technology, higher volume usage, and fierce competition. Computers are becoming much faster not only because of higher speed devices, but also because of architecture changes such as parallel processing, reduced instruction set (RISC) processors, and increased use of cache memory. Today the low end of the market minicomputers and the high end work stations operate in the range of one to three MIPS (millions of instructions per second). That performance can be expected to double every few years.

Random access memory (RAM) has evolved extremely rapidly. In 1980, 64KB chips were novel; today 256KB chips are a commodity, and 1MB chips are gathering momentum. Chip manufacturers are already testing 4MB devices, and a Japanese company recently announced the development of 16MB memory chips. The pace is breathtaking. But what does this mean to the LIMS user? It comes down to speed. With LIMS large amounts of data are manipulated (sorting, searching, collating, and coalescing), more than would normally fit into the available memory. So the computer is required to make repeated data fetches from the disc, each of which takes 20 or 30 milliseconds. If all the data were in RAM, it might take a fraction of a microsecond to obtain a given piece of data. The net result is that a user sitting in front of the terminal would observe almost instant response to a query instead of having to wait many seconds.

Mass storage in the form of hard (Winchester) disks is improving rapidly also in terms of price to performance. Ten megabytes used to be considered a substantial size for a disk. Now hundreds of megabytes are available at a much reduced cost per byte. Gigabytes are just around the corner. Although RAM memory is must faster than disk memory, disks are still needed for their large capacities and long term storage. As LIMS storage requirements become ever more grandiose, the immense capacity of the hard disks will be welcomed.

Another form of mass storage that is receiving much attention is the optical disk, especially the 'write once' type. This uses a tiny laser beam to burn an indelible spot on the surface of a reflective disk. Since the data cannot be erased it is ideal for the storage of archival data, particularly for those industries that must conform to government regulations. An optical disk can store about a gigabyte of data, and is removable. The access times are somewhat slower than those of a hard disk, but much faster than those for a nine track magnetic tape (which is the current alternative). Very soon these write once optical disks (WORM type) will become standard components of Laboratory Information Management Systems.

Computer networking is another very hot topic. A network is a combination of hardware and software that permits computers to communicate with one another, which in turn gives users the ability to share data and programs. Perhaps the most widely used local area network is Ethernet, developed by a consortium of computer companies. This network supports many computers and peripherals at a speed of ten megabits per second. As with any emerging technology, there are numerous variations such as broadband, baseband, thin Ethernet, DECNET, etc. It is beyond the scope of this chapter to define each of these, but suffice to say that they all work well and there are price/performance reasons for selecting one over another. There are also some networks designed for the PC class of computers that are less expensive, for example the PCnet and the token ring network. The MAP network was developed by General Motors and is being used by many organizations in manufacturing environments. The potential benefits of networking to LIMS are in the areas of distributed processing and data transfer. The future laboratory might be one where the chromatograph and its associated autosampler are under the control of a local microprocessor; as data are produced they would be transferred via a network to a local work station for processing and validation by the analyst; finally, the data would move through the network into the database management system.

26.5 Software

Although software developments do not seem to be as glamorous as hardware developments, there is a burst of activity that will lead to improved products for the chemist. There are new software tools that will aid the applications programmer and new programs that can be used directly by the chemist. Furthermore, much of the new software is portable across a wide range of hardware.

One area of intense activity has been relational database software. Programmers no longer have to worry about creating files with connecting pointers, or sorting algorithms, or user interfaces, or flexible report generators. These tools are all part of modern relational database products such as Oracle, Ingres, Informix, and dBase III which run on a wide range of hardware. A crucial aspect of any database is the way a user can retrieve information. This can be a very complex process. Fortunately, a query language called SQL is evolving into a standard. Normally SQL would be transparent to the user. The applications program would build SQL statements in response to inputs from the user, and these would be used to interrogate the database.

An operating system is a collection of programs that keeps track of files, loads them into memory, and does numerous house-keeping chores. MS-DOS has emerged as the standard for PC computers, but there is no clear winner for mini and mainframe computers. The contender for the universal operating system is UNIX. If this becomes the standard, then programs written to work with UNIX will become portable across many types of computers. This has important consequences for the customer in terms of user training and hardware upgrading. It is also important for the software developer who would then have to write just one version of the software, and it would work on a wide range of hardware. UNIX still has some problems; there are several different versions competing for supremacy, the size and complexity of the operating system are considerable, and there is some difficulty in dealing with real-time applications. Nevertheless, there is reason to be optimistic about UNIX becoming a very widely used operating system.

Artificial intelligence has been a popular topic for some time, but practical applications have been slower to materialize than had been predicted. The most successful applications have been in the use of expert systems technology, particularly in dealing with problems of fault finding and diagnosis. The best known example in the field of chemistry is the Dendral program written at Stanford; this is able to deduce molecular structure information from spectroscopic data. Intellicorp has developed a product that will allow a person to examine a database (Oracle) in very friendly ways. It permits browsing, and also helps you find very specific information. For example, you could ask for the names of all samples from project XYZ that had benzene concentrations greater than ten ppm that were collected during the month of June, 1987. The expert system software creates, from the user inputs, an SQL statement, which may be very complex and convoluted. This SQL statement then finds the desired information. This technology will be used in LIMS before long.

Third party packages for statistical analysis and plotting are getting better and better. Lotus 123 and RS/1 (RS/E) are good examples. It is very important that modern LIMS are able to utilize these powerful software products rather than vendors developing their own.

26.6 Summary

LIMS is in a state of flux. Most analysts recognize the need for a Laboratory Information Management System, and some feel considerable urgency to get going. Nearly everyone has difficulty defining exactly what is needed. And tomorrow's needs almost certainly will be different.

Vendors find themselves trying to hit a moving target. It is rather like buying clothes for a child entering its teens: if it fits today, it will not fit tomorrow. If you anticipate tomorrow, and buy something larger, it is not right today.

Laboratory Information Management Systems are riding the wave of computer technology. We are benefitting from developments in relational database software, faster computers, lower cost mass storage, and networking.

These are the major trends as perceived by the author; perhaps they are obvious to the reader, or perhaps they are incorrect. In any event, it is fun to gaze into the crystal ball and try to guess what will happen in this exciting field of Laboratory Information Management Systems.

Glossary

Alan Braithwaite and R.D. McDowall

ADC Analogue to digital converter. A device for changing a continuously variable signal into a series of numbers that can be used by a computer.
ANSI The American National Standards Institute.
ASCII American Standard Code for Information Interchange. An almost universal code by which letters, digits and symbols are stored in computers. Each character is given a unique number which identifies it. 'A' is 65, 'B' is 66, 'space' is 32,etc.
Access The operation of seeking, reading, or writing data on a storage unit.
Access time The time that elapses between an instruction being given to access some data and that data becoming available for use.
Add-on Any piece of equipment which can be attached to a computer to enhance its capabilities.
Address An identification (number, name, label) for a location in which data is stored.
Algorithm A computational procedure. The method or set of rules to perform a task.
Alphanumeric This is a contraction of Alphabetic-Numeric and refers to the character set consisting of the letters of the alphabet and the ten decimal digits.
Analogue The representation of numerical quantities by means of physical variables, eg. voltage.
Architecture The fundamental design of a computer's hardware.
Array A series of items arranged in a meaningful pattern.
Asynchronous (Of devices) not operating in time coincidence, eg., in asynchronous counters, all bits changing at any transition do not do so simultaneously. (Of transmission) mode in which intervals between transmitted units of information may be of unequal length. Control and time reference is achieved through start and stop elements added to the transmitted data.
Attribute Usually says what condition applies to a file and characters, eg. whether a file is read only.
Audit trail A program that monitors all changes to a LIMS database; successive alterations of the same file are recorded separately and not overwritten as with a journal or transaction log.

BCD Binary Coded Decimal. A means of representing decimal numbers where each individual decimal digit is replaced by its 4 bit binary equivalent.
Backing store A store which extends the computer's memory, such as floppy or hard disks.
Bandwidth The range of frequencies assigned to a channel or cable for undistorted data transmission. Expressed as the difference between the upper and lower limit frequencies, in Hertz.
Bar code reader A device used to read patterns of black and white stripes into a computer system. Each pattern represents a series of encoded characters. Common

coding systems are Code 39 for alphanumeric characters and Interleaved 2 of 5 for numbers only.

BASIC Beginner's All-Purpose Symbolic Instruction Code. A high level, easy to use and easy to learn language for time-sharing or dedicated computer systems.

Batch processing A technique of data processing whereby a program is run in the background and there is no user involvement.

Baud rate The term used to describe the rate at which data is transferred from one place to another, for example, from a computer to a printer. The term is often used for 'effective transfer rate in bits per second'.

Benchmark A test for measuring computer performance. Typically, a standard program or set of programs run on several computers for purposes of comparing speed and efficiency. Benchmarks can be misleading if not interpreted carefully.

Bidirectional Two way communication between devices or computers.

Binary A numbering system using base two where the only digits used are 0 and 1. Counting from 1 to 10 in binary is as follows: 1, 10, 11, 100, 101, 110, 111, 1000, etc.

Binary search A method of searching a sequenced table or file. The procedure involves selecting the upper or lower half based upon an examination of its mid-point value. The selected portion is then similarly halved, until the required item is found.

Bit The smallest possible piece of information. A specification of one out of two possible alternatives. Bits are written as 1 for 'yes' and 0 for 'no'. Word derived from binary digit.

Boolean A variable which has a value of either 'true' or 'false'.

Booting (boot) The term for getting the computer working with the operating system. A program is kept in ROM or on disk and is usually executed automatically when the computer is switched on.

Bootstrap The means by which the system is started. See booting, derived from the phrase 'to pull yourself up by your own bootstraps'.

Bridge A link between two networks of the same type.

Buffer An area of storage which holds data temporarily while it is being received, transmitted, read or written. It is often used to compensate for differences in speed or timing of devices. Buffers are used in terminals, peripheral devices, storage units, and in the CPU, as a buffer can take information from the computer much faster than, say, a printer can print it. A buffer saves the computer from having to wait for the printer.

Bug An error in a program.

Bus This is the main route for data (or control signals) flowing between parts of a computer. New parts can be added to a computer if they can link into its 'bus'; *also* a local area network layout, or topology, where all nodes have access to the same section, or bus, of the network transmission medium.

Byte An eight bit word.

CD-ROM Compact disk, read only memory; an optical disk for storing information. It cannot be written to and is used for storing information such as literature databases which are updated relatively infrequently.

COBOL Common Business Oriented Language. A high level language developed in the early 1960s and used primarily for business applications.

cps Characters per second. This is the measurement used to show the speed at which a printer works.

CPU Central Processing Unit, the computer's 'brain'.
Cache A section of very high speed memory used as a buffer between the CPU and main memory in a large computer. The main effect is to make the programs run faster.
Card A small circuit board which slots into a computer to provide new facilities.
Catalogue A directory of all files available to the computer.
Chain A database organization in which records or other items of data are strung together by means of pointers.
Character A letter, digit, punctuation mark or even a space.
Checksum A summation of digits or bits used primarily for checking purposes mainly in communication.
Clear To remove data and return all circuitry to an initial condition, usually '0'.
Commands Instructions for the computer to do something.
Communications server Provides a network with access to other networks and public services.
Compaction A technique for reducing the number of bits in data without destroying any information content.
Compatible Describes a program or piece of equipment for one computer which will work on another computer.
Compiler A computer program that translates a high level language into machine code.
Concatenate To link together. A concatenated data set is a collection of logically connected sets. A 'concatenated key' is composed of more than one field.
Configuration The particular arrangement of equipment in a computer system, such as the sort of printer, the amount of RAM memory, the size and number of disk drives and so on. Many programs need to be configured when you first buy them. You do this by telling the program what sort of equipment your computer has.
Console An interactive device for system control.
Contention A condition on a bus or communications channel when two or more sources try to transmit simultaneously.
Control character A character whose occurrence in context initiates, modifies or stops a control function (in ASCII, any character in the first two columns of the table).
Control key This key works just like a shift key to produce a range of special characters which have particular meanings to the computer. These 'control characters' might mean instructions like 'start the printer' or 'delete the last character typed in'.
Crash To lose control of the computer so that it stops running a program or stops responding to your commands. Crashes can be caused by problems with the hardware or by mistakes in a program. The only way out is to use a reset key or switch off and start again.
Cursor A movable marker on a screen which shows where the next letter or number will appear.

DAC Digital to Analogue Converter. A device for changing a numeric version of a quantity into an analogue (voltage or current) representation of the quantity.
DMA Direct Memory Access. A method by which data can be transferred at high speed between external devices and internal memory without intervention by the CPU.

Database A structured collection of information. See relational database.
Data dictionary A catalogue of all data types giving their names and structures.
Data element Synonymous with data item or field.
Data item The smallest unit of data that has meaning in describing information; the smallest unit of named data. Synonymous with data element or field.
Data management A general term that collectively describes those functions of the system that provide creation of, and access to, stored data; enforce data storage conventions; and regulate the use of input/output devices.
Data processing Manipulating data to achieve desired results.
Data set A collection of data records with a logical relation to each other.
Debug The process of checking and correcting any program errors either in writing or in actual function.
Density Floppy disks can be single, double or quadruple density, depending on the number of tracks per inch that they can hold.
Diagnostics Programs which can check that the computer is in working order and brings faults to the user's attention.
Digital Numeric representation of data.
Digitizer A device which converts 'continuous' or analogue information into the numbers that a computer can understand. This could be a graphics tablet which lets you store free-hand drawings in a computer, or a device which takes a photograph and converts it into numbers the computer can process.
Digitize To convert from analogue to digital form.
Directory A table giving the relationships between items of data. Sometimes a table (index) giving the addresses of data.
Disk A storage device in which information is stored by selective magnetization of portions of the magnetic surface coating one or both sides of one or more rotating disks. Also called disk drive.
Disk crash Situation in which 'flying' recording heads physically touch a disk surface, resulting in data corruption and damage to both the surface and heads. Caused by dirt, distortion of cartridge, and general mishandling; usually fatal for heads, disk, and data.
Dot matrix As used in a matrix printer, it is a group of needles used to form letters, numbers and graphics from dots, instead of pre-formed characters. It is a fast but not usually high quality method.
Downtime The elapsed time from breakdown to repair of a computer.
Dump To shift blocks of data from one place to another, as when you 'dump' the contents of a disk to a printer to check on the contents of a disk file.

Editor An editor is a program which manipulates text files. It helps the creation, examination, and modification of data.
EDP Electronic Data Processing.
Emulation A technique that permits one computer to run a program on another, or the ability of a terminal or microcomputer to be seen by a computer as a normal terminal.
Enhancement The addition of extra functions to a system, either as a result of expansion or of missing them from the original specification.
EPROM Eraseable Programmable Read-Only Memory.

Error Any discrepancy between a computed, observed, or measured quantity and the true, specified, or theoretically correct value or condition.
Ethernet A type of local area network developed by Xerox and DEC.
Execute To perform a sequence of program steps.
Execution time The time taken to perform a sequence of program steps.
Expert system A software package that uses artificial intelligence techniques to answer questions and make judgements. A knowledge base is built up by the user from which the software makes its deductions.
Extension The last three letters of a file name following a full stop or period.
Extent A contiguous area of data storage.
External schema A user's or programmer's view of the data in a database.

Field Refers to treating a group of computer characters as a single data item (eg. a 'field' containing the Julian date would consist of six numeric characters).
File A set of similarly constructed records or a collection of data.
Filename The name given to a file.
File server An item of equipment attached to a network node which is used by the network as a central file store for programs and data. Normally a microcomputer with a hard disk.
Flag One or more bits used to indicate the status of an item of equipment or data.
Flat file A two-dimensional array of data items.
Floppy disk Usual term for removable disk memories in computer systems. The smaller size (5¼" diameter) are sometimes referred to as diskettes.
Format A predetermined arrangement of characters, fields, lines, punctuation, page numbers, etc. Also, the act of dividing a disk into sectors prior to use.
FORTRAN FORmula TRANslation. A high level language designed mainly for scientific and engineering applications.
Front end processor A communications computer associated with a (larger) host computer, and performing line control, message handling, code conversions, error checking, job queueing and control, and applications functions. It does not include the major logical, arithmetic and processing functions of the host.
Function keys The keys added to the standard typewriter layout on a keyboard to give special commands to the computer. Can normally be programmed by the user.

Gateway A connection between two dissimilar types of network.
Graphics Information displayed as, graphs, pictures and symbols.

Handshake The signals that pass between two nodes or devices to ensure the correct transfer of data.
Hard copy Computer output on paper.
Hard disk A rigid disk on which data and programs are stored. They are fixed, usually permanently, into a disk drive and have a much larger capacity than a floppy disk. Also known as Winchester drives.
Hard wired A unit or device which performs a logical function (or program) by means of specifically designed and constructed hardware, and which can only perform that one function.
Hardware The mechanical, magnetic, and electronic components of a computer.
Heuristic A trial-and-error or empirical method of problem solving.

Hierarchical file A file in which some records are subordinate to others in a tree structure.
Hierarchical Storage units of a database linked together to form a storage storage subsystem where some keys are fast but small and others are large and slow.
High level language A computer language which allows its users to write instructions rather than program in machine code or binary (*See BASIC, FORTRAN).*

I/O port The communication line through which the CPU communicates with peripherals. See input-output.
Icon A graphical representation of the program to be called and run.
IEEE-488 A bus system used to connect instruments and computers together. It has parallel 8 bit data transfer and handshake status signals.
IEEE The Institute of Electrical and Electronic Engineers. The US professional body responsible for many of the internationally recognized standards.
ISO The International Standards Organization, responsible for the OSI seven-layer model (see OSI) on which most manufacturers are basing their network architecture.
Indexed-sequential A file structure in which records are stored in ascending storage sequence by key.
Initialize Set a hardware or software device to a prescribed state before its use (eg., counters, timers, subroutines, and programs may all be initialized).
Input Putting information or programs into the computer.
Input-output Commonly called I/O which refers to equipment or data involved with the information transferred into the computer and information transferred out.
Instruction A statement that specifies an operation and the values or locations of its operands.
Instruction set The total repertoire of operations available for any one computer. Its size has been taken as a measure of excellence of a machine.
Integer A whole number with no fractions or decimal places.
Intelligent terminal A terminal which may undertake some processing independently of the computer to which it is connected.
Interactive A program with which the user participates while the program is running.
Interface General term for any hardware or software used to connect two or more pieces of equipment or software together.
Internal schema The physical structure of the data within a database.
Interpreter A program which turns a high level program, such as a BASIC program, into the machine code a computer can understand. An interpreter looks at each instruction in turn and works out what needs to be done.
Interrupt A signal demanding that program execution should temporarily cease and that control should pass to another routine. Interrupts are usually associated with peripherals requiring attention, eg. I/O devices that are acquiring data on-line. Implies that control may be restored to the interrupted program.
Inverted file A file structure which permits fast spontaneous searching for previously unspecified information. Independent lists or indices are maintained in records keys which are accessible according to the values of specific fields.
Inverted list A list organized by a secondary key not a primary key.

Justify To move a string of symbols left or right in a field.

KB, KByte One thousand bytes (actually 1024KB or 2^{10} derived from the binary not the decimal number).
Key A data item used to identify or locate a record (or other data grouping).
Key compression A technique for reducing the number of bits in keys; used in making indices occupy less space.
Key, primary A key which uniquely identifies a record (or other data grouping) in a database.
Key, secondary A key which does not uniquely identify a record in a DBMS, ie. more than one record can have the same key value. A key which contains the value of an attribute (data item) other than the unique identifier.
Keyboard Used for typing instructions into the computer, it is similar to a typewriter in style although it often has additional keys to give commands such as 'enter' data into the machine.

LAN Local area network.
Light pen A device which can be used to select certain items displayed on a screen simply by pointing to them. Light pens can be used as a fast method of entering information or drawing diagrams.
LIMS Laboratory Information Management System, a database tailored to a laboratory and intended to integrate the sample information with the results produced from analytical techniques and processes, in order to maintain data integrity and speed the reporting of results.
Load To transfer a program from some kind of store (such as disk or tape) into the main memory of the computer, enabling the program to be used.
Local area network A group of computers in fairly close proximity which are connected together to share common resources.
Logical unit (LU) The Logical Unit (LU) number, specifies the peripheral device or direct-access file on which the requested data transfer or command takes place.
Look-up table A method of retrieving values from a table. Usually used for data input where only defined entries are allowed.
Low level A programming language which uses the most primitive language instructions a computer can understand. Machine code and assembler are both low level languages. Each low level instruction does only one thing.

MB,Megabyte One million bytes. 1000KB.
MS-DOS Operating system for IBM-PC microcomputers and compatible machines, developed by the Microsoft Corporation.
Machine code The only computer language a CPU can understand. All instructions are given as binary numbers, and each can only do one thing at a time. Tedious to use.
Machine language *See* machine code.
Mainframe A 'large' computer from a major manufacturer. Frequently, but not invariably, associated with machines orientated towards 'business' applications.
Memory The internal computer storage space for running programs and data.
Memory map Chart showing the memory allocation of a system.
Microcomputer The name usually given to a small computer which uses a microprocessor for its CPU.

Microprocessor A single, large scale integrated circuit (IC) which has capabilities for performing arithmetic and logical functions.
Minicomputer A small computer system most often used for specific applications.
Modem MOdulator/DEModulator used to send and receive serial data over an audio link, eg. allows communication between computers over an ordinary telephone line.
Monitor It looks like a television and displays information so that you can check what you are doing. *See VDU*.
Mouse A device for pointing, held in the hand and rolled on a flat surface which in turn moves a cursor on a screen. It can have one, two or three buttons which act like certain keys eg. <Return>.
Multi-user A computer system where a group of terminals are connected to a single microcomputer so several people can use the computer simultaneously.
Multiplexor A switching network for analogue or digital signals which selects one of several inputs for onward transmission from its single output. Also a device which takes several inputs and combines all of them in a single output without distortion of the original information content.

Nesting A programming technique involving the embedding of routines within others.
Networking Connecting more than one microcomputer together so that they can share the same resources, such as an expensive hard disk unit or printer, or interchange data.
Node Any point in a network where external equipment gains access to the transmission medium.
Non volatile Memory which will retain its data content after the power supply is removed, eg. ROM.
Normalization The decomposition of more complex data structures into flat files (relations). This forms the basis of relational data bases.

OCR Optical character recognition.
OEM Original Equipment Manufacturer. One who incorporates components into systems or equipment.
OSI Open Systems Interconnection. The ISO's scheme to make all computer systems and networks compatible by having an established standard for interconnections.
Off-line Pertaining to equipment or devices not under the control of a computer.
On-line Pertaining to equipment or devices under computer control, or to the user's ability to interact with a computer.
Operand Data used by machine operations.
Operating system A program in the computer which looks after routine tasks such as accepting messages from the keyboard, showing them on the display and organizing the data on disk.
Optical disk *See* CD-ROM and WORM.

PABX Private Automatic Branch Exchange. Used by some manufacturers as the hub of a star network, the typical PABX covers the same area within an organization as an equivalent data network, with the advantage of already having installed (telephone) wiring.
PC Personal computer - now synonymous with IBM PC.
PROM Programmable Read-Only Memory. *See* ROM.

Packet switching A system for efficient transmission of data over networks. Data is split up into 'packets', which are sent to their destination over whatever route may be available at the time, only being reunited on reaching their destination.
Paging In virtual storage systems, the technique of making memory appear larger than it is by transferring blocks (pages) of data or programs into that memory from external storage when they are needed.
Parallel A method of data transfer where several bits of information are sent simultaneously. Often eight bits of data are sent along eight wires which are made into one cable.
Parity The parity bit is often attached to data bytes in serial transmission. It is a simple method of detecting single errors within the byte. When the number of 1s within the byte is even, the parity bit is also set to 1 for even parity, or 0 for odd parity, and *vice versa* if the number of 1s in the byte is odd. If the receiver knows which form or parity is being used, it can perform a comparison check and decide if any errors have occurred during transmission. This is only effective as long as no more than one bit is in error.
Parity check A special form of validity check in which one bit of the binary code is reserved as a parity bit and is set equal to either zero or one to make the total number of one bits in the code consistently even (even parity check) or consistently odd (odd parity check).
Pascal A general purpose high level programming language.
Peripherals All the extra parts that can be connected to the computer, such as monitors, printers, disk drives, modems, etc.
Physical Physical means that it is actually there, as opposed to *logical* or *virtual.*
Pixels The smallest dot on a screen which the computer can control. Each picture cell (pixcell) can be individually set to be 'on' or 'off' or to a particular colour.
Plotter A device for producing graphical hard copy in which lines, curves, and characters are produced by moving a pen according to co-ordinates supplied by a computer. It can work in a 'pen up' mode for moving without drawing and a 'pen down' mode for drawing dots or continuous lines.
Port Facility for connection of a peripheral to a computer; often used to describe the plugs and sockets by which I/O devices, memory modules and so on are connected to the internal bus.
Print server A node within a network configured to receive all data intended for printing. Normally attached to a high-speed printer.
Printer A device which prints output.
Program A program is a set of instructions which, when executed by the machine, performs some useful function. Programs are often referred to as software.
Protocol General term for the set of rules governing a particular data transmission.

RAM Random Access Memory where words may be 'written' (stored) or 'read' (recovered) in any order at random.
ROM Read Only Memory. Permanent data storage in the computer. Its contents cannot be altered by the user. Programmable ROMs (PROM) can be set by the user, unlike ROMs which can only be set at the factory. Eraseable PROMs (EPROM) can be cleared and reprogrammed with special equipment, but not while in use.

RS-232C A common type of interface which connects the computer to devices for serial data transfer, typical examples are printers and plotters.

Read only An attribute that prevents a device being written to.

Real-time A term for which no generally acceptable definition exists. For most purposes, real-time corresponds to time in which a response must be made to a stimulus in order that the progression of a process giving rise to the stimulus is not delayed. Distinction should be made between real-time (ie. process time) and 'strict time' (ie. sidereal time).

Redundancy The use of extra hardware or software as a check against errors or failure.

Relation A flat file. A two-dimensional array of data elements. A file in normalized form.

Relational data A database made up of relations (as defined above).

Relational database Its database management system has the capability to recombine the data elements to form different relations thus giving great flexibility in the usage of data.

Remote job Submission of jobs through an input device which entry (RJE) has access to a computer only through a communications link.

Reset key A 'panic button' which always stops whatever the computer is doing and returns it to its 'just switched on' state so that you can start again. If you get your computer stuck doing the wrong thing, this may be the only way to regain control.

Return key The key, sometimes called <ENTER> or <NEWLINE> which sends signals to the computer telling it that you have finished typing in a line.

Ring A network layout where the transmission medium forms a closed circuit and data circulates in one direction.

Schema A map of the overall logical structure of a database.

Scratch key Memory that has short access time and is used by the system for short term data storage.

Scrolling When the screen of a microcomputer or terminal becomes full of text and the user needs to add more text to it, the text can automatically move upwards so that the top line is lost and space is created at the bottom of the screen.

Search To examine a series of items for any that have a desired property or properties.

Sector The smallest address portion of storage on a disk.

Serial Serial data transmission implies that data is sent, one bit at a time, along a single cable or other transmission medium.

Server A computer on a network which administers controls and allocates resources. See print server, communications server and file server.

Shadowing Writing of data to two independent disks simultaneously, providing two copies of data in the event of failure or disaster.

Software Programs, languages and procedures for a computer system, as contrasted with the physical hardware.

Sort Arrange a file in sequence by a specified key.

Source code The original program, written in assembly or high level language.

Spooling Technique in which output to slow devices is stored in queues in a computer backing store to await transmission when the devices become free. This procedure allows efficient use of system resources, allowing programs using low speed devices to run to completion quickly and to be replaced by others.

Spreadsheet An application program which allows you to enter numbers and values and relate them to each other visually.

SQL Structured Query Language.

Star A type of network where each node is connected directly to a central unit which performs switching and interconnection operations.

Storage A general term for any device capable of retaining information.

Subroutine A distinct group of instructions within a program which do a particular task. They are used by different parts of the program and are brought into use by a command such as 'gosub' in BASIC.

Subschema A map of a programmer's view of the data he uses. It is derived from the global logical view of the data - the schema.

Subsystem A smaller system inside a larger system. Each subsystem can be thought of as a separate system with its own job to do.

Syntax The grammar of a programming language.

TTL Transistor-Transistor Logic.

TTL levels Voltage levels used in TTL systems to indicate logic states, namely less than 0.8V (low) and greater than 2V (high). Usually taken as 0V and 5V.

TTY Teletype.

Terminal A computer terminal is an input (usually a keyboard) or output device (usually a printer or CRT screen) operated by a user.

Time-sharing Refers to the 'simultaneous' use of a computer's resources by several users at terminals performing multivaried tasks.

Token passing Access method used in both ring and bus networks. A node can only transmit if it is in possession of a special data packet, or token, which is passed from node to node in a set pattern.

Topology The shape of a network. The commonest topologies are bus, star and ring.

Transaction An input record applied to an established file. The input record describes some 'event' that will either cause a new file record to be generated, an existing record to be changed, or an existing record to be deleted.

Transaction log Also known as Journaling, this is used as a means of monitoring the changes to the database. As records are created or modified they are copied to the log from which the database could be rebuilt in the event of a disaster. This log is not suitable for Good Laboratory Practice purposes as the log only keeps the latest copies of records, the old copies are overwritten.

Transceiver Transmitter and Receiver. The part of a network node directly connected to the transmission medium.

Tree structure A hierarchy of groups of data within a database, also called a hierarchical structure.

Turnkey system A computer system marked as a complete product, meeting specified needs of a given spectrum of users.

Upwards compatibility Of a range of systems or computers, those in which a program for a particular model can be run on a more powerful machine in the range.

Utilities Programs which do mundane but useful 'housekeeping' tasks such as making copies of disks or changing the numbers of each instruction within a program.

V24 A standard for modem connection used by the computer industry.

VAN Value-Added Network. A network which provides some service as well as interconnection.

VDU Visual Display Unit. A terminal incorporating a cathode ray tube of scanning (television) or storage type for presentation of output. Many operate as teletype replacements. Does not necessarily imply presence of a graphics (line picture) capability, but many provide this.

Validation The production of documented evidence to show a computer system or software package does what it is purported to do, specifically in a regulated environment.

Vapourware The announcement of a computer product or a system before its existence or availability.

Virtual Something that is not actually physically present, but which is almost identical in operation to what it is describing, like a window of the screen which is referred to as a virtual screen.

Virtual memory Memory which can appear to the programs to be larger than it really is because blocks of data or program are rapidly moved to or from secondary storage when needed.

Volatile store Memory device in which the contents are erased when electrical power is removed (eg., semiconductor register). Contrast non-volatile store, usually magnetic in nature.

WAN Wide-Area Network. A network covering large areas, usually connected via radio or telephone links.

WIMP Windows, icons, mouse and pointers, a type of user interface developed by Xerox.

Winchester drive A common type of small hard disk drive, whose disks cannot be removed. It provides more storage and faster access than floppy disks.

WORM Write once read many, a type of optical disk, *see also* CD-ROM. Can be written to whilst connected to the computer; has potential for GLP archiving purposes.

Word processor An application program which provides facilities for the input, processing, storage and output of text-based documents.

Work station A computer on a network which can be used to access the network resources without necessarily having any independent resources of its own.

X25 A standard for packet-switching, which is relevant to LANs where interconnection with other networks is required.

INDEX